Edible Films and Coatings for Food Applications

Milda E. Embuscado • Kerry C. Huber

Editors

Edible Films and Coatings for Food Applications

 Springer

Editors
Milda E. Embuscado
Materials and Process Technology
McCormick and Co. Inc.
Hunt Valley, MD
USA
Milda_Embuscado@mccormick.com

Kerry C. Huber
School of Food Science
College of Agricultural and Life Sciences
University of Idaho
Moscow, ID
USA
huberk@uidaho.edu

ISBN   978-1-4419-2830-6   e-ISBN   978-0-387-92824-1
DOI 10.1007/978-0-387-92824-1
Springer Dordrecht Heidelberg London New York

Printed on acid-free paper

Springer is part of Springer Science+Business Media (www.springer.com)

Preface

The market for edible films and coatings has experienced remarkable growth over the last 5 years according to the U.S. Department of Agriculture. This growth is expected to continue. Accompanying this growth is a vast amount of knowledge on edible films and coatings acquired through research and product development work as well as advances in material science and processing technology. For the past 15 years, there were numerous research articles published and research projects undertaken by the academic and research communities as well as the food and pharmaceutical industry. The impact of sustainability programs, carbon foot printing and heightened interest on the use of renewable resources further propelled and will continue to propel the growth and interests on edible films and coatings. In addition, edible film and coating are adding value to agricultural and food industries by-products. With this backdrop, this book on *Edible Films and Coatings for Food Applications* was organized and created. This book brings together edible film and coating experts from various scientific disciplines from the academic and research institutions and the food industry (protein, carbohydrate/polysaccharide and lipid chemistry, engineering, and manufacturing).

The book starts with a valuable edible films and coatings historical and general overview, followed by four chapters on structure–function relationship of biomaterials used in the preparation of edible films and coatings (proteins, polysaccharide gums, starches and waxes and lipids). These four chapters also include discussion on preparation and properties of edible films and coatings made from these biomaterials. The next six chapters discuss specific applications of edible films and coatings such as protection of fruits and vegetables, meat and poultry and for the delivery of food additives, flavors and active ingredients. The next two chapters deal with mechanical and permeability properties and new advances in analytical techniques for edible films and coatings. Quite unique in this book is the discussion of commercial manufacture of edible films and coatings authored by one of the pioneers in the field. Each author developed his or her chapter in a comprehensive manner such that each chapter can stand on its own. This book was created to help the novice in edible films and coatings as well as those already immersed in the field with the hope that the topics discussed in the book will trigger future novel ideas and processes. Due to the nature of and scope of each chapter, overlapping topics cannot be completely avoided. On the other hand, these overlaps are necessary for each chapter to be able to stand on its own.

A sincere and great appreciation goes to the book chapter authors for their contributions and to all the researchers/scientists/authors who have toiled hard to keep the interest and continued advancement of edible films and coatings alive. A special thanks also goes to Springer Science for their encouragement, patience and their whole-hearted support of this book project. Special thanks to Sauld Embuscado for working on selected figures.

Hunt Valley, MD   Milda E. Embuscado
Moscow, ID   Kerry C. Huber

Contents

Contributors

Gustavo Barbosa-Cánovas, Ph.D.
Center for Nonthermal Processing of Food, Washington State University,
Pullman, WA 99164-6120, USA
Barbosa@wsu.edu

Kirsten Dangaran, Ph.D.
Dairy Processing and Product Research, Eastern Regional Research Center, ARS,
USDA, 600 East Mermaid Lane, Wyndmoor, PA 19038, USA
dangaran.3@osu.edu

Frederic Debeaufort, Ph.D., Eng.
Universitary Institute of Technology, Department of Bioengineering,
Université de Bourgogne, 7 bd Dr Petitjean, BP 17867, F-21078, Dijon, France,
frederic.debeaufort@u-bourgogne.fr

Milda E. Embuscado, Ph.D.
Materials and Process Technology, McCormick and Co. Inc., Hunt Valley,
MD, USA
Milda_Embuscado@mccormick.com

María Alejandra García
Centro de Investigación y Desarrollo en Criotecnología de Alimentos (CIDCA),
CONICET, Universidad Nacional de La Plata (UNLP), 47 y 116 La Plata (1900),
Argentina
magarcia@quimica.unlp.edu.ar

Kerry C. Huber, Ph.D.
School of Food Science & Toxicology, College of Agricultural and
Life Sciences, University of Idaho, Moscow, ID, USA
huberk@uidaho.edu

Michael E. Kramer
Technical Services, Grain Processing Corporation, 1600 Oregon Street,
Muscatine, IA 52761, USA
Mike_kramer@grainprocessing.com

Monique Lacroix
INRS – Institut Armand-Frappier, Research Laboratory in Science Applied to
Food, Canadian Irradiation Centre, 531 des Prairies, Laval, QC, Canada H7V 1B7,
Monique.lacroix@iaf.inrs.ca

Olga Martín-Belloso
Department of Food Technology, University of Lleida, Alcalde Rovira Roure, 191,
Lleida 25198, Spain, omartin@tecal.udl.es

Miriam Martino
Centro de Investigación y Desarrollo en Criotecnología de Alimentos (CIDCA),
CONICET, Universidad Nacional de La Plata (UNLP), 47 y 116 La Plata (1900),
Argentina, mmartino@ing.unlp.edu.ar

Marceliano B. Nieto, Ph.D.
TIC Gums, 4609 Richlynn Drive, Belcamp, MD 21017, USA
mnieto@ticgums.com

Guadalupe Isela Olivas, Ph.D.
Centro de Investigación en Alimentación y Desarrollo (CIAD), Av. Río Conchos
S/N, Cuauhtémoc, Chihuahua 31570, Mexico
golivas@ciad.mx

William Orts, Ph.D.
Western Regional Research Center, ARS, USDA, Albany, CA, USA
orts@pw.usda.gov

Attila Pavlath, Ph.D.
Western Regional Research Center, ARS, USDA, Albany, CA, USA
apavlath@pw.usda.gov

Adriana Pinotti
Centro de Investigación y Desarrollo en Criotecnología de Alimentos (CIDCA),
CONICET, Universidad Nacional de La Plata (UNLP), 47 y 116 La Plata (1900),
Argentina
impronta@arnet.com.ar

Phoebe Qi, Ph.D.
Dairy Processing and Product Research, Eastern Regional Research Center,
ARS, USDA, 600 East Mermaid Lane, Wyndmoor, PA 19038, USA
phoebe.qi@ars.usda.gov

Jesus-Alberto Quezada-Gallo, Ph.D.
Departamento de Ingeniera y Ciencias Qumicas, Universidad Iberoamericana
Ciudad de Mexico, Edificio F segundo piso, Prol. Paseo de la Reforma 880
Lomas de Santa Fe, 01210 Mexico, D.F., Mexico
jesus.quezada@uia.mx

Gary A. Reineccius, Ph.D.
Department of Food Science and Nutrition, 1334 Eckles Avenue, St. Paul,
MN 55108-1038, USA
greinecc@umn.edu

M. Alejandra Rojas-Graü
Department of Food Technology, University of Lleida, Alcalde Rovira Roure, 191,
Lleida 25198, Spain
margrau@tecal.udl.es

James M. Rossman
Rossman Consulting, LLC, 1706 W. Morrison Avenue, Tampa, FL 33606, USA
rossmanj@prodigy.net

Robert Soliva-Fortuny
Department of Food Technology, University of Lleida, Alcalde Rovira Roure,
191, Lleida 25198, Spain
rsoliva@tecal.udl.es

Peggy Tomasula, Ph.D.
Dairy Processing and Product Research, Eastern Regional Research Center,
ARS, USDA, 600 East Mermaid Lane, Wyndmoor, PA 19038, USA
Peggy.Tomasula@ars.usda.gov

Zey Ustunol, Ph.D.
Department of Food Science & Human Nutrition, Michigan State University,
2105 South Anthony Hall, Lansing, MI 48824-1225, USA
Ustunol@anr.msu.edu

Andree Voilley, Ph.D., Eng.
EMMA Department, Université de Bourgogne – ENSBANA,
1 esplanade Erasme, F-21000, Dijon, France
voilley@u-bourgogne.fr

Michael Zapf
Technical Innovation Center, McCormick & Company, Inc., 204 Wight Avenue,
Hunt Valley, MD 21031-1501, USA
Mike_Zapf@mccormick.com

Noemí Zaritzky
Centro de Investigación y Desarrollo en Criotecnología de Alimentos (CIDCA),
CONICET, Universidad Nacional de La Plata (UNLP), 47 y 116 La Plata (1900),
Argentina
zaritzky@ing.unlp.edu.ar

Chapter 1
Edible Films and Coatings: Why, What, and How?

Attila E. Pavlath and William Orts

1.1 History and Background

Edible films and coatings, such as wax on various fruits, have been used for centuries to prevent loss of moisture and to create a shiny fruit surface for aesthetic purposes. These practices were accepted long before their associated chemistries were understood, and are still carried out in the present day. The term, edible film, has been related to food applications only in the past 50 years. One semi-sarcastic tale was that spies' instructions were written on edible films, so that in the off-chance they were captured, they could easily destroy their secrets by eating them. In most cases, the terms film and coating are used interchangeably to indicate that the surface of a food is covered by relatively thin layer of material of certain composition. However, a film is occasionally differentiated from a coating by the notion that it is a stand-alone wrapping material, whereas a coating is applied and formed directly on food surface itself. As recently as 1967, edible films had very little commercial use, and were limited mostly to wax layers on fruits. During intervening years, a significant business grew out of this concept (i.e., in 1986, there were little more than ten companies offering such products, while by 1996, numbers grew to 600 companies). Today, edible film use has expanded rapidly for retaining quality of a wide variety of foods, with total annual revenue exceeding $100 million.[1]

Why do we need edible films? Most food consumed comes directly from nature, where many of them can be eaten immediately as we take them from the tree, vine or ground. However, with increased transportation distribution systems, storage needs, and advent of ever larger supermarkets and warehouse stores, foods are not consumed just in the orchard, on the field, in the farmhouse, or close to processing facilities. It takes considerable time for a food product to reach the table of the

---

[1] http://www.am-fe.ift.org/cms/?pid = 1000355

A.E. Pavlath (✉) and W. Orts
Western Regional Research Center, ARS, USDA, Albany, CA, USA
e-mail: apavlath@pw.usda.gov

M.E. Embuscado and K.C. Huber (eds.), *Edible Films and Coatings for Food Applications*,
DOI 10.1007/978-0-387-92824-1_1, © Springer Science+Business Media, LLC 2009

consumer. During time-consuming steps involved in handling, storage and transportation, products start to dehydrate, deteriorate, and lose appearance, flavor and nutritional value. If no special protection is provided, damage can occur within hours or days, even if this damage is not immediately visible.

As early as twelfth century, citrus fruits from Southern China were preserved for the Emperor's table by placing them in boxes, pouring molten wax over them, and sending them by caravan to the North (Hardenburg 1967). While their quality would not have been acceptable to our modern selective society, the method was quite effective for its time, and was used for centuries for lack of more efficient ones . In Europe, the process was known as "larding" – storing various fruits in wax or fats for later consumption (Contreras-Medellin and Labuza 1981). While such protection prevented water losses, the tight, thick layer interfered with natural gas exchange, and resulted in lower quality products. Larding was a compromise between maintaining moisture content and losing various qualities, including optimal taste and texture. Later in the fifteenth century, an edible film, Yuba, made from skin of boiled soy milk was used in Japan for maintaining food quality and improving appearance (Biquet and Guilbert 1986; Gennadios et al. 1993). In the nineteenth century, a US patent was issued for preservation of various meat products by gelatin (Havard and Harmony 1869). Other early preservation methods included smoking and/or keeping products cool in iceboxes or in underground cellars. Today various modern methods, including combinations of these, such as refrigeration, controlled atmosphere storage, and sterilization by both UV and gamma radiation are used to keep our food safe. Nevertheless, for many kinds of food, coating with edible film continues to be one of the most cost-effective ways to maintain their quality and safety.

1.2 Definition

1.2.1 Edible Films and Coatings

Any type of material used for enrobing (i.e., coating or wrapping) various food to extend shelf life of the product that may be eaten together with food with or without further removal is considered an edible film or coating. Edible films provide replacement and/or fortification of natural layers to prevent moisture losses, while selectively allowing for controlled exchange of important gases, such as oxygen, carbon dioxide, and ethylene, which are involved in respiration processes. A film or coating can also provide surface sterility and prevent loss of other important components. Generally, its thickness is less than 0.3 mm.

1.2.2 Generally Recommended as Safe (GRAS)

Items which are edible or are in contact with food should be generally recognized by qualified experts as being safe under conditions of its intended use, with amount applied in accordance with good manufacturing practices. These food-safe materials must typically have approval of the Food and Drug Administration (FDA). Since it is impractical for FDA to have an all-inclusive list of every potential food ingredient, there are also other opportunities to acquire GRAS status – i.e., manufacturers can petition for approval of an ingredient or food composite provided that this petition is supported by considerable studies. There are three types of GRAS designations (1) Self-affirmed, where the manufacturer has carried out necessary work and is ready to defend GRAS status if challenged, (2) FDA pending, where results of research have been submitted to FDA for approval, and (3) No comment- which is the response of FDA if after review, it has no challenges. More detailed information on procedures to acquire such designations for industrial purposes can be found at the FDA website.[2] However, GRAS status does not guarantee complete product safety, especially for consumers who have food allergies or sensitivities, such as lactose intolerance (milk) and Celiac disease (wheat gluten).

1.2.3   Shelf Life

The time period, whereby a product is not only safe to eat, but still has acceptable taste, texture and appearance after being removed from its natural environment, is defined as shelf life. For practical purposes, a period of at least 2 weeks is required for processed food to remain wholesome, allowing for packaging, transportation, distribution, and display prior to consumption.

1.2.4   Light Processing

'Light Processing' includes a wide variety of processes used to prepare an original product or commodity for consumption, without affecting the original, "fresh-like" quality of the product (Shewfelt 1987). Light processing includes cleaning, washing, paring, coring, and dicing, for fruits and vegetables specifically, and may also include removing waste and undesirable parts from a wide variety of food products (such as de-boning of meat, etc). Although these steps are relatively unobtrusive, nature and rate of respiration changes immediately after most light processing, with the product becoming immediately more perishable. In most cases, these processes cause disruption of cell tissues and breakdown of cell membranes, creating many membrane-related problems (Davies 1987).

1.2.5   Respiration

Any type of food, whether in its natural environment or otherwise, continuously undergoes various biochemical/biological/physiological processes, which use up

---

[2] http://www.cfsan.fda.gov/~dms/grasguid.html

and/or release oxygen and carbon dioxide (colloquially labeled "breathing"). Depending on oxygen level, respiration can be aerobic or anaerobic. Respiration activity of a product is influenced by storage temperature, type of processing, oxygen to carbon dioxide ratio, and absolute value of oxygen concentration itself. If a wax layer is applied, oxygen content of the internal atmosphere will decrease as a function of thickness of the layer, while carbon dioxide content and anaerobic respiration will rise (Eaks and Ludi 1960). As a rule of thumb, when oxygen level drops below 3%, anaerobic respiration will start replacing the Krebs cycle, with the resulting glycolytic pathway releasing unacceptable flavors and causing other problems, such as changes in color and texture. High oxygen levels (>8%) and low carbon dioxide levels (<5%) can prevent or delay senescence in horticultural products, thus maintaining food quality (Kader 1986).

1.2.6 Transpiration

In addition to gases, food products may contain various liquid or solid components; water is most prevalent, but oils, various flavor components, and nutrients are also present. Depending on environment, these components will migrate out of or throughout the product if there is concentration difference acting as a driving force.

1.2.7 Controlled and Modified Atmosphere Storage

In order to control respiration (i.e., transfer of various gases in and out of the product), food can be stored in an environment filled with various gases at appropriate, optimal temperatures. The right gas combination can slow respiratory metabolism, and delay compositional changes in color, flavor and texture. It can also inhibit or delay microbial growth. However, this method can be quite expensive in other than large-scale stationary storage, i.e. controlled atmosphere situations. Modified atmosphere packaging (MAP) is where product is enclosed in a sealed box or bag filled with required atmosphere. Temperature, however, is critical and must be maintained at constant level to avoid in-pack condensation leading to spoilage.

1.3 Effect of Edible Films and Coatings or the Lack Thereof

Today, the most widely used commercial method for long-range protection is interim storage at low temperature (4–8°C), especially for lightly-processed food. Lowering temperature generally decreases undesirable enzyme activities, although temperature decreases down to 0–5°C may actually lead to increases in respiration rate and ethylene production (Eaks 1980). Below 0°C, growth of mold is inhibited, but even this low temperature does not fully eliminate undesirable chemical and physicochemical reactions (Fennema 1993). For example, fruits and vegetables native to tropical climates experience harmful chilling effects such as damage to cell membranes at temperatures of 10–12°C. In addition, some cold-tolerant pathogenic microorganisms are able to grow even under refrigeration.

Accordingly, an increasing amount of research has been conducted over the past 50 years to encase a food product, such that rates of migration of molecules involved in degradative processes are maintained at natural levels and/or minimized.

Edible films are being used for a variety of purposes within a multitude of food systems, even though this fact might not be fully realized by consumers. The shiny surface of an apple in a supermarket is not provided by nature. Some candies (e.g., M&Ms) are coated with shellac to increase product shelf life and provide desired glaze. Medicine pills are often coated to prevent crumbling, to hide any bitter or undesirable taste before swallowing, and to provide controllable timed-release of medications. Even French fries are frequently coated to provide protection during cold storage before frying, control of water losses during frying, and stability against wilting and/or loss of crispness under infrared lamps between frying and serving. Edible films may also be used to limit uptake of oil and fat during frying processes (Feeney et al., 1992; Polanski, 1993). Table 1.1. provides an informative, and non-comprehensive list of examples of commercial edible coatings.

Most fresh fruits and vegetables contain considerable percentage of water, the amount of which is maintained during production in their natural environment. Respiration is maintained naturally at an appropriate equilibrium between oxygen, carbon dioxide and water by skin, which controls transmission to and from surrounding environment. However, as soon as fruits and vegetables are separated from their native production environment (i.e., harvested), the delicate balance is upset. Water activity (a_w) will change, and various physiological reactions, which were previously kept under control in developing fruit or vegetable, will accelerate

Table 1.1 List of commercially used coatings

Name	Main component	Uses
Freshseel™	Sucrose esters	Extending shelf life of melon
Fry Shield™	Calcium pectinate	Reduces fat uptake during frying fish, potatoes, and other vegetables
Nature Seal™	Calcium ascorbate	Apples, avocado, carrot, and other vegetables
Nutrasave™	N,O-Carboxymethyl chitosan	Reduces loss of water in avocado, retains firmness
Opta Glaze™	Wheat gluten	Replaces raw egg based coating to prevent microbial growth
Seal gum, Spray gum™	Calcium acetate	Prevents darkening of potato during frying
Semperfresh™	Sucrose esters	Protect pome fruits from losing water and discoloration
Z*Coat™	Corn protein	Extends shelf-life of nut meats, pecan, and chocolate covered peanut

post-harvest. Respiration will be affected, resulting in altered ratios of oxygen and carbon dioxide. Increased ethylene formation will accelerate ripening, thus changing color, flavor, texture and nutritional characteristics of the fruit or vegetable (Wong et al. 1994a).

Some protection against undesired changes can be obtained using controlled atmospheres and low temperatures during transportation and storage. Controlled atmosphere with high relative humidity can prevent loss of water and, with maintenance of appropriate carbon dioxide and oxygen concentrations, slow down senescence. However, with all attempts to optimize storage through controlled environments, they are still no match for Mother Nature's quality control in the field. Various undesirable changes can still occur once food has been picked and processed, and aging begins. More importantly, cost of maintaining controlled-storage environments can be very prohibitive.

These problems are not limited to just fruits and vegetables. Quality of various other food products, such as meat, pies and confectionery can also suffer before reaching the consumer. Major difficulties are deterioration of vital food components including flavor chemicals, lipids, and vitamins through oxidation. The patent issued in the nineteenth century to coat meat with gelatin to delay microbe formation and loss of water was just the beginning (Havard and Harmony 1869). Numerous research studies were carried out with reasonable success, paving the way for possible commercial applications. Loss of water which affects juiciness of meat products, was reduced by coating meat surfaces with polysaccharides to maintain desired moisture levels (Allen et al. 1963; Shaw et al. 1980). Microbes that contaminate meat can cause serious health safety problems, which can be mediated to some extent by edible films and coatings. Alginate coatings have been shown to prevent microbial growth on beef, pork, lamb and poultry (Lazarus et al. 1976; Williams et al. 1978). When frying battered or breaded meat products, too much oil may be adsorbed if batter does not adhere adequately to meat surface. By coating meat first with cellulose derivatives, oil uptake is considerably lessened (Feeney et al. 1992; Polanski 1993). Infusion of oil into chocolate presents quality problems in confectionaries, i.e., stickiness, moisture absorption, and oxidation (Paulinka 1986; Nelson and Fennema 1991). Similarly, some foods have to be protected against moisture uptake that can otherwise result in loss of crispness. Hydration of certain snack foods can be prevented by coating with sucrose fatty acid esters (Kester et al. 1990), and dried fruit can have longer shelf life when coated with zein (Cosler 1957). In case of chewing gums, an edible film is needed to prevent loss of moisture (Meyers 1994).

Uncontrolled migration of water is generally recognized as the biggest food storage/transportation challenge. Both loss and gain of water are almost always considered undesirable. Even if water loss does not cause immediate visible shrinkage to a food product, it results in economic loss via weight loss. Figure 1.1 shows that even a whole unwaxed apple will show 0.5% of weight loss after 1 day, a loss that will increase more than ten-fold when the fruit is cut in half.

On the other hand, storage in high relative humidity environments may result in water infusion causing variations in the food product between point of preparation

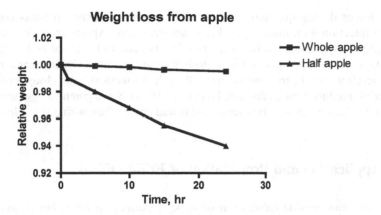

Fig. 1.1 Comparative loss of weight for whole and half apple samples

and sale. This occurrence can be a problem when regulations mandate a minimum solids content. In heterogeneous foods, if various components have different water activities, water transmission during storage can also occur internally between components of the food, causing changes to sensory characteristics. For example, moisture content between the dough and toppings of a pizza are different. If the pizza is not baked immediately after assembly, internal water migration may occur, and the original optimal moisture content of the dough will change. This change will result in different a texture of the baked product. Refrigerated storage is applied to prevent quality and safety changes in toppings (cheese, meat etc.). While refrigeration slows down water migration, water can still migrate even within the frozen product, especially if stored for a long time. Application of a film in some form on surface of the dough, before placing on the toppings, can prevent water migration and retain optimum dough structure. Similar problems exist with pies, where there is need to separate moist filling from crust to prevent sogginess (Labuza 1984), or in chocolate covered nuts, where migration of oil from nuts may retard softening of chocolate covering upon consumption (Murray and Luft 1973; Paulinka 1986; Nelson et al. 1991).

Discoloration or enzymatic browning, caused by polyphenol oxidase, is another frequent food issue. This phenomenon does not affect weight or nutrition, and sometimes not even taste, of food, but it is unsightly and results in loss due to consumer rejection. In addition, changes in gas transition result in various other equally important detrimental characteristics. These include:

- Loss and/or change of flavor and aroma
- Aesthetically unacceptable appearance
- Changes in texture

One obvious solution is to protect food from these changes until they are ready for consumption via application of edible films, which can prevent contamination, microbial growth, and pest infestation. In addition edible films can help alleviate quality issues during quarantines that are frequently invoked on fruits due to possible infection with insect eggs. For example, many exports are held back at ports until costly fumigation is carried out. Fruits need to be stored at 1–2°C for 14–24 days before shipping from Florida to Japan (Ismail 1989). Coatings can reduce oxygen levels within the fruit and kill larva, though this method is not always optimal and may also reduce fruit quality (Hallman et al. 1994). Since potential eggs are laid on natural skin, a better way is to remove skin and enrobe fruit with edible film.

1.4 Application and Preparation of Edible Films

Edible films can provide either clear or milky (opaque) coatings, but consumers generally prefer invisible, clear coatings. Coatings can be obtained in various ways (1) by dipping the product into, or by brushing or spraying it with solution containing film ingredients, so as to deposit the film directly on food surface (Gontard and Guilbert 1994), or (2) by creating stand alone film from solution or through thermoformation for subsequent covering of food surface.

The simplest way to apply a film is directly from solution. Depending on concentration of coating solution, the product will absorb an appropriate amount of coating material necessary to form the desired layer, which when dried, forms a protective layer at the food surface. In most cases, some plasticizer needs to be added to coating solution to keep the developing film from becoming brittle. Possible food grade plasticizers are glycerol, mannitol, sorbitol, and sucrose. If coating cracks, movement of various components will increase by orders of magnitude, resulting in mass flow instead of diffusion. Coatings should have good adhesion to rough surfaces (Hershko et al. 1996). Application of a uniform film or coating layer to cut fruit and vegetable surfaces is generally difficult. Better uniformity can be promoted by adding surfactants to solution to reduce surface tension. This strategy will also reduce the superficial a_w and in turn reduce water loss (Roth and Loncin 1984, 1985). In one standard process, carboxymethylcellulose (CMC) powder was applied to cut fruit surfaces. The CMC adsorbed moisture within pores of the surface, causing the CMC to swell, which not only prevented loss of moisture, but also provided a barrier against oxygen to prevent enzymatic discoloration (DeLong and Shepherd 1972). Coatings derived from non-aqueous media, such as applying an alcohol solution of shellac to candy, result in another level of complexity. For safety reasons the finished coating layer should not contain any solvent residue. Thus, during large scale operation, disposal of exhaust gases may present environmental challenges.

Should a free-standing film be required, it can be prepared from solution by evaporation. It should be pointed out that characteristics of stand-alone films might differ from those of films created on food surfaces (i.e., those produced by dipping in or spraying). Films obtained through evaporation were found to have lower water vapor permeability than those prepared by spraying (Pickard et al. 1972). Varying rate and temperature of evaporation may result in creation of films with differing characteristics. For example, polymer chains may be prematurely immobilized before reaching their optimal structure (by accelerated drying) to affect permeability (Reading and Spring 1984; Greener 1992). When zein films were obtained from solution after drying at 51°C for 10 min, plasticizer was needed to obtain a non-brittle material (Kanig and Goodman 1962). In contrast, drying at 35°C for 24 h yielded flexible zein films without addition of plasticizer (Guilbert 1986, 1988). Films can also be formed by cooling concentrated solutions. However rate of cooling can again result in amorphous, crystalline, or polymorphic films with differing permeabilities. The characteristics of a polymorphic film may be further modified by tempering (Landman et al. 1960; Kester and Fennema 1989a, b). Formation of flexible and stretchable films was also reported from molten acetylated monoglyceride (Feuge et al. 1953). Other possibilities are through precipitation, either by addition of selective miscible solvents that are not a solvent for the film component, or by desalting. In addition, some protein films can form upon heating, causing unfolding of polymeric chains and replacement of intramolecular with intermolecular bonds. This transition affects permeability. In case of proteins, improved films can be obtained by adjusting coating solution pH in relation to the protein isoelectric point, where proteins become least soluble. However, this process does not necessarily improve resistance of films to water transmission (Krochta et al. 1988). The pH can be also adjusted by high pressure liquefied carbon dioxide treatment which although costly, does not leave any salt residue (Tomasul et al. 1997).

When films, comprised of pectin or alginate, are prepared by evaporation from water-soluble components, they are subject to re-dissolution in water or destruction in high humidity conditions. This problem can be avoided by cross-linking polymers at the film surface. Various reactions can be employed to achieve enhanced covalent bonding (e.g., treatment with formaldehyde); however such reactions can create new chemical structures that might necessitate approval by FDA. The most acceptable cross-linking method involves ionic interaction between polymer chains via multivalent ions to form ionomers. While most synthetic films have higher tensile strength than typical edible films, ionomers are exceptions (Pavlath et al. 1999a; Pavlath et al. 1999b). For ionomers, tensile strength of their films is dependent on number of available bonding locations. Ionomer cross-linked films can be used as wrapping materials or, in case of water solubility, as bags that dissolve when immersed in water during food preparation (e.g., soup). In such cases, films do not have to be thin, because they will disappear before tasting and can further act as thickening agents within the food product. This aspect is especially important from an environmental perspective, where disposal is not necessary (becomes consumed as part of the food). Commercial synthetic materials are not generally biodegradable, while edible films typically are. Thus, edible films provide an ideal solution for minimizing packaging waste onboard ships during long voyages, during which maritime regulations forbid throwing of any refuse overboard. While most synthetic commercial packages possess average life times of 200 years in a marine environment, edible films decompose readily providing an environmentally-friendly solution.

Thermoformation is rarely used to create edible films, because most edible components cannot be molded at elevated temperatures without causing irreversible structural changes to the material. Hydroxypropylmethylcellulose and polylactic acid, two biodegradable thermoplastics, are rare exceptions. Any protein-containing cysteine, such as gluten, hair or chicken feather, represents interesting thermoplastic biomaterials, depending on level of cysteine content. When these materials are treated with reducing agents, such as sodium sulfite, disulfide bonds can be cleaved at 90–100°C. This bond cleavage lowers protein molecular weight allowing the material to flow under slight pressure without decomposition. The reduced disulfide bonds can then easily reform resulting in a strong, pliable material (Pallos et al. 2006). Disulfide bonds in some of these proteins are also susceptible to reduction in appropriate solvents with disulfide bridges reformed during solvent evaporation (Beckwith and Wall 1966; Okamoto 1978; Gennadios et al. 1993). As protein molecules undergo unfolding upon heating, new or non-native disulfide bonds may also form to enhance intermolecular cross-linking (Schofield et al. 1983). It was reported that preparation of such films under acidic conditions provided better quality films than under alkaline conditions (Gontard et al. 1992).

Packaging issues are further complicated by strong demand for convenience foods. With fast moving lifestyles of today, consumers desire to spend less time in the kitchen preparing meals. True, an apple can be eaten as a whole fruit but to use it in an apple pie or fruit cocktail, non-edible and/or unappetizing parts of the fruit need to be removed. Consumers do not generally want to waste time with so-called light processing of foods (i.e., skinning and pitting fruits, slicing vegetables, skinning chickens, or just cleaning food surfaces). At the same time, consumers expect appetizing appearance and mouthwatering flavor.

Preparation processes can be tedious and time-consuming. In small-scale commercial operations, such as cafeterias, automatic processing machines can now do much of the light processing work, but cost of such machinery is still too high for typical home kitchens. On the other hand, large-scale commercial processing in a centrally-located factory provides both economical and environmental benefits. In this setting, energy costs are minimized by scale of operation, disposal and/or utilization of waste is carried out efficiently at centralized locations, and volume (cost) benefits are achieved. However, removal of a food from its natural state and environment also accelerates undesirable changes that can lead to deterioration of appearance, texture, and taste. We have all likely had the experience of biting into an apple, setting it down for a few minutes, and then observing how quickly the exposed surface has turned brown. Extensive browning is apparent on apple slices within 30 min, though surface discoloration begins to occur even after just a few minutes (Bolin et al. 1964).

What is the cause of these changes? Natural skin does not hermetically seal fruit from its surroundings. Rather it maintains optimal gas exchange equilibrium to protect against weight loss, discoloration, loss of flavor and texture, among other attributes. While such products are still subject to various aerobic and anaerobic respiratory processes, these processes are maintained in proper balance by natural skin. However, when fruit or vegetable is cut or even just mechanically damaged, cell wall membranes are disrupted, initiating a cascade of various enzymatic processes. Even minimal mechanical damage incurred during handling and transport can stimulate increased formation of ethylene causing physiological disorders and deterioration (i.e., increased cell permeability, loss of compartmentalization, and increased enzyme activities) (Hyodo et al. 1978). Change in the rate of migration of oxygen, carbon dioxide, and ethylene can result in anaerobic fermentation and increased ripening. Even the way in which fruits are cut can make a difference. For example, formation of a white, unappetizing layer on carrot surfaces can be prevented by peeling them with a sharp blade (Bolin and Huxsoll 1991). Water knife cutting can also decrease slicing-related issues (Becker and Gray 1992). Cut surfaces increase chances for growth of microorganisms, causing multiple-fold increase in respiration (Maxcy 1982). Increased respiration rates open up possibilities for cascading biochemical changes, such as degradation of carbohydrates, activation of dormant biological pathways, and facilitation of new, additional enzymatic activities (Uritani and Asahi, 1980) which may induce production of unusual metabolites (Haard and Cody 1978; Griesbach 1987). This circumstance occurs with polyphenol oxidase, an enzyme associated with catalysis of browning in fresh fruits and vegetables, via rapid oxidation of o-quinones and polymerization of oxidized products to melanin (reddish-brown in color).

Nature generally maintains optimum level of water activity (a_w) in most fruit and vegetable products. Conversely, if addition of a protective edible film unduly restricts water migration, resulting increase in water activity may cause undesirable changes. The most frequent problem is growth of microorganisms, such as mold and yeast. Biochemical and enzymatic reactions may also be induced by increasing water activity affecting taste, appearance and crispness. On the other hand, increasing water content can plasticize cellular structure, leading to loss of crispness in some products, as well as increased permeability. Table 1.2, summarizes effect of water activity (a_w) levels as they relates to various undesirable changes associated with fresh fruits and vegetables. For non-enzymatic browning the rate increases until water activity reaches 0.6 and then declines (Rockland and Nishi 1980).

Many food storage issues can be minimized by dehydration or lowering of moisture content of a food product. Unfortunately normal food dehydration procedures may also remove many of the volatile flavor components, which are not restored upon simple rehydration. Interestingly if a food product is coated and then dehydrated osmotically at room temperature (the OSMEM process), only water is removed from the product. This process uses thicker films (1.5–2.0 mm thickness) which, if water soluble will dissolve during rehydration process (Camirand et al. 1968, 1992).

Table 1.2   Quality problems associated with changes in water activity

Water activity	Quality problem	Reference
0.2	Non-enzymatic browning	Labuza (1980)
0.4	Loss of crispness	Katz and Labuza (1981)
0.6	Mold formation	Troller (1980)
0.7	Yeast formation	Troller (1980)
0.8	Bacterial growth	Troller (1980)

1.5 Migration Processes

Several important parameters need to be defined to describe diffusion processes in relation to films. Migration between two adjacent volumes separated by a layer or membrane, occurs in three basic steps. In the first step, the diffusing molecule comes in contact with surface of the layer or membrane, and is adsorbed onto it. In the second step, the molecule then diffuses through the thickness of the layer or membrane. Lastly, once the diffusing molecule reaches the other side of the layer or membrane, it will desorb. Rate of adsorption and desorption is dependent on the affinity between the diffusing molecule and membrane (film) components, especially for water migration. Hydrophilic materials rapidly adsorb water but rate of desorption on the opposite side of the membrane will be controlled by partial pressure of water in that volume.

Despite surface effects noted previously, the most dominant factor in molecular migration is bulk effect – rate of diffusion of molecules while in the membrane or film. In an ideal case, amount of given material (Q) passing through a film can be determined by Fick's law of diffusion (1.1):

$$Q = PA\Delta pt / d, \qquad (1.1)$$

where Q increases in direct proportion to film surface area (A), and decreases with its increasing thickness (d). Increasing partial pressure difference (Δp) of migrating molecule between two sides of the membrane and time (t) also linearly affects total amount of permeate. Permeability coefficient (P) is defined as the product of diffusion coefficient (D) and solubility (S) coefficients ($P = DS$). In an ideal case, P is a constant determined by characteristics and structure of the film.

When a given product is encased in protective film, the outer surface conditions are already given. Ability to maintain low partial pressure difference (for a given component) between the inside of the food product and surrounding external environment is generally limited by economic factors (requires a controlled atmosphere). When a coated product is stored openly in a relatively large area, amount of gas in the surrounding external environment is orders of magnitude higher than amount that is diffused out from the product. If the coated product is placed within an enclosed system (e.g., in a bag or box), equilibrium can be created between the product and surrounding external environment, depending on head-space volume relative to product weight. Film thickness is another important factor in the rate of water and gas transmission, changing not only appearance, but also taste of a food product. While an increasing film thickness will slow down rate of diffusion, there is also a practical limit to consider. If protective coating is edible and is to be ingested together with the product, it must be applied in minimal fashion so that it will not adversely modify original product taste. An exception can be made if a flavor change is considered to be acceptable or product is baked after coating (generally nullifies most additive flavors) or dissolved before being consumed. Therefore, diffusion is mostly dependent on size of the permeate molecule and abundance of "holes" or "channels" in the film molecular structure, through which the permeate can move. Solubility coefficient is influenced by the permeate's ease of condensation and its affinity to components of the film.

In the earlier form of Fick's law (1.1), it was assumed that temperature is constant. Effect of temperature on permeability constant follows Arrhenius law (1.2):

$$P = DS = D_0 S_0 / \text{Exp}((E_{aD} + \Delta H_s) / RT), \qquad (1.2)$$

where D_0 and S_0 are reference values for diffusivity and solubility, E_{aD} is activation energy of diffusion, and ΔH_s is enthalpy of sorption. Since E_{aD} is positive, diffusion increases with increasing temperature. However, since ΔH_s is negative for water, solubility decreases at the same time (Rogers 1985). In hydrophobic films, it was found that water vapor permeability increased with increasing temperature indicating that the controlling factor is diffusion (Hagenmaier and Shaw 1991). Some anomaly was observed with the water vapor transfer rate, which decreased with decreasing temperature over range of 40–20°C, indicating greater absorption. However, water vapor transfer rate remained constant between 20 and 10°C, and permeability increased at 4°C (Kester and Fennema 1989c).

A film or coating must be fairly uniform and free of pinholes, microscopic cracks, and rough surfaces. Therefore preparation conditions (e.g., rate of evaporation and temperature) are important factors in film preparation. Even a small degree of irregularity in a film can exponentially increase rate of diffusion, which is not accounted for by Fick's law. However, rate of diffusion is indirectly proportional to square root of molecular weight of the diffusing molecule; therefore it is a relatively smaller problem for oil, flavor, and other similarly large molecules (relative to water, oxygen, carbon dioxide and ethylene). In the molecular weight range of the most important gases (oxygen, carbon dioxide, water and ethylene), diffusion rate due to film irregularity is by orders of magnitudes faster than that which occurs through the film structure itself. When one pinhole, representing only 0.008% of film surface, was intentionally made, water vapor transmission increased 2.7 times (Kamper and Fennema 1984).

It is evident that Fick's law has additional limitations, especially for cases in which the film includes hydrophilic polymer components. Relative humidity and temperature may alter considerably the diffusion and solubility coefficients of film components (Gontard et al. 1996). At high relative humidity, water uptake can soften film structure through plasticization, making it easier for diffusing molecules to pass through the film. Solubility coefficient of film components also increases (Cairns et al. 1974). Another anomaly, shown by various authors (Landman et al. 1960; Biquet and Labuza 1988; Martin-Polo et al. 1992), concerns influence of thickness of film on water vapor transfer. In very thin films of less than 60 μm, water vapor transfer decreases according to Fick's Law; however, above this thickness level, water vapor transfer rate remains almost constant.

Value for permeability ($P = DS$) is essentially dictated by structure of the film. At the molecular level, diffusing molecules migrate through polymer chains and side chains, which are held in place by hydrogen bonds and van der Waals forces. Side chains can have either beneficial or detrimental influence on permeation coefficient of film. In one sense, a side chain can decrease available intramolecular spacing within the film, which allows migration of permeates, thus making diffusion more difficult. On the other hand, a side chain may decrease degree of crystallinity within a film by introducing more structural irregularities. Crystallinity rigidifies polymer chains, which hinders molecular migration through a film by limiting molecular mobility of film components. Thus, crystalline packing arrangement of polymers and their orientation within a film relative to flow direction especially influence migration of molecules through a film (Fox 1958). These elements regulate "free space" within the film available for migration, and limit molecular movement from cavity to cavity.

In certain cases, protection against migration can be created at surface of a food product without addition of polymeric material. If the surface already has a polymeric structure, these surface polymers can be cross-linked, whereby the product itself provides a new protective layer. For example, this strategy was employed quite successfully for apple slices through creation of an ionomer surface using calcium ascorbate dips. In this scenario, calcium ions interacted with surface pectin molecules, closing the surface, while also incorporating ascorbic acid as an antioxidant (Pavlath et al. 1996; Chen et al. 1999). Calcium formed salt bridges between the carboxylate groups of pectin polymer chains creating cross-links that promoted a more rigid structure (the so-called egg-box model) and decreased permeability at food surface (Grant et al. 1973; Wong et al. 1996). Such treated samples did not brown or lose water over a 2 week period (Chen et al. 1999). Similar protective layers can be formed on potato strips through treatment with calcium acetate and an oxidation inhibitor. Cross-linking at surface of potato strips prevented both oxidation and discoloration, and provided for higher quality French fries upon frying (Mazza and Qi 1991).

There are various synthetic films and packaging materials available for minimizing migration, but with few exceptions, these materials are not edible and cannot be consumed together with food. Thus, they must be removed before consumption. Economics is a primary driver in the choice between use of synthetic and edible films. Edible films are relatively more expensive, with their commercial utilization being limited to convenience food items (e.g., snack foods), where consumers are willing to may pay more. Synthetic, non-edible materials are generally better suited for creating modified packaging environments, creating sealed, box-like enclosures that may be supplemented with special atmosphere. Edible film materials generally decompose more rapidly or lose their quality and integrity faster (oxidation, dissolution, etc.) than those of synthetic origin. On the other hand, synthetic packaging materials have to be removed before eating, leading to challenges in relation to disposal and environment. Many packaging materials end up as litter.

Edible encasing materials must meet standards first, by ensuring that they are labeled GRAS for consumption by the Food and Drug Administration. Secondly, consumers are very selective and finicky in relation to what they choose to eat. Unless different a flavor or texture is intended to enhance consumer acceptance, the film or encasing should not adversely change taste and flavor of the food product, and should also be invisible and undetectable to taste. What materials can then be used to fulfill these difficult requirements?

1.6   Possible Components of Edible Films

The main components of our everyday foods (e.g., proteins, carbohydrates and lipids) can fulfill requirements for preparation of edible films. As a general rule, fats are used to reduce water transmission; polysaccharides are used to control oxygen and other gas transmission, while protein films provide mechanical stability. These materials can be utilized individually or as mixed composite blends to form films provided that they do not adversely alter food flavor. A major objective in preparing films for many foods (e.g., fresh fruit and vegetables) is to ensure that the generated films afford physical and chemical properties necessary to maintain transmission of various gases and liquids at the same rates as they occur within their native systems. Chemical structures of the three major components used to prepare films differ widely, and therefore attributes that each component contributes to overall film properties are different too. The following is a short informative summary of these concepts.

Films from various sources of protein, such as corn, milk, soy, wheat and whey, have been used for years, their major advantage being their physical stability. It should be mentioned, however, that most of these protein sources are in fact mixtures of various proteins comprising a range of molecular weights. If they are used in solution rather than in emulsion, the solution will contain different protein fractions than the emulsion (unless all protein fractions are equally dissolved). Lower molecular weight components are generally more easily solubilized, though they exhibit higher permeabilities than higher molecular weight entities within films. While this limitation can be counteracted with cross-linking, edibility and mouthfeel of a film can be jeopardized by such treatment. When selecting protein for use in an edible film, consideration should extend beyond just protein functionality and GRAS status. It is important to recognize that a given segment of population is allergic to certain proteins, specifically to those of wheat. Consequently, collagen can be extruded to desired shapes such as a casing for sausage links. Collagen replaced traditional casing material (derived from animal intestines) because of its ease of manufacture and scale-up. In general, value of proteins as moisture barriers is low, and they also do not adequately control transfer of oxygen, carbon dioxide and other gases that are important to stability of various foods. Their major advantage is their structural stability, which makes it possible to hold a required form (e.g., sausage casing). Cross-linking can also occur in proteins where the isoelectric point is dependent on interaction of the amino and carboxylic groups of the protein. Thus depending on protein composition, permeability can also be altered. It was reported that depending on the pH of solution from which the film was cast, properties (e.g., color, texture, tensile strength) were markedly different (Gennadios et al. 1993; Gontard et al. 1992.)

Water adsorption occurs readily at surface of polysaccharide films (e.g., those of alginate, carrageenan, cellulose and its derivatives, dextrin, pectin and starch), because of the hydrophilic nature of most polysaccharides. Some polysaccharides such as cellulose derivatives, have lower water transmission than average polysaccharides, though they are still less effective than wax. The primary advantages of polysaccharide films are their structural stability and ability to slow down oxygen transmission. As a general rule films which do not provide protection against water transmission often have desirable properties in preventing oxygen transmission and vice versa (Banker 1966). Resistance to gas transmission can be so effective for polysaccharide films that it can be a challenge to manipulate. For example, permeability for oxygen in high amylose starch films was found to be virtually zero despite addition of plasticizers that were known to increase gas permeability (Mark et al. 1966). Therefore, in spite of their shortcomings with regard to water permeability, polysaccharides can be used to protect food from oxidation. Alginate coating can prevent lipid oxidation and stop rancidity (Mate et al. 1996). Another interesting role of polysaccharide films is to act as "a sacrificing agent" instead of as a barrier. Since most polysaccharides and other hydrophilic materials provide low protection against water transmission (i.e., they are highly hygroscopic), they may be applied as relatively thick films at food surfaces to intentionally absorb water and provide temporary protection against further moisture loss (similar to how a surfer's wet suit takes in water, but provides protection). Carrageenan, a sulfated polysaccharide of D-galactopyranosyl units was found to form a structured gel, which acted as sacrificing agent (Glicksman 1982, 1983). Alginate gels were also reported to possess this function (Shaw et al. 1980). Thus the coated product itself does not lose significant moisture until the sacrificing agent or film itself is dehydrated. If a surfactant is added to the coating, surface water activity can be altered without altering water content inside.

Waxes and fats are the oldest known edible film components. While most waxes are of natural origin, synthetic acetylated monoglycerides have similar characteristics and have been used with the blessing of the FDA in edible films for meat, fish and poultry. Originally, lipid coatings were applied by simply pouring molten paraffin or wax over citrus fruits. This process slowly gave way to adding a thin shiny layer by applying small amount of various wax through dipping or spraying. The hydrophobic fruit surface, which also protects against abrasion during transportation, adds an aesthetic appearance. At the same time, thin wax coatings still allow some breathing to occur. They are excellent barriers to water transmission, while still slowing or altogether preventing other gas migration. Wax will affect oxygen and carbon dioxide transmission, and thus, can result in unwanted physiological processes, such as anaerobic respiration. This process in turn, will diminish quality of the product, resulting in softening of tissue structure, alteration of flavor, delay of ripening, and promotion of microbiological reactions (Eaks et al. 1960). In case of horticultural products with minimal respiration such as root vegetables, thick layers of wax are less harmful and can be used (Hardenburg 1967). It was reported that water vapor transmission through fatty acid monolayers decreased logarithmically as length of the fatty acid hydrocarbon chain increased, though this effect was not indefinite (LaMer et al. 1964). However, there is disagreement about the most efficient chain length. According to one group, the most efficient chain length is C_{12}–C_{14} (Wong et al. 1992; Pavlath et al. 1993; Talbot 1994), while another group found that chain lengths of C_{16}–C_{18} (Hagenmaier and Shaw 1990; Koelsch and Labuza 1992; McHugh and Krochta 1994; Park et al. 1994) gave best results.

However, introduction of double bonds into the chain increases water vapor transmission 80-fold (Roth and Loncin 1985; Hagenmaier and Baker 1997), which is attributed to loss of crystallinity.

Application of edible films is especially difficult when applying lipophilic material to wet surfaces, such as cut fruits and vegetables. Direct application of any lipid to a hydrophilic or wet surface results in weak adhesion at the film-food interface. Dual-coating is one possible solution to this problem, as it provides protection against more than one permeate through use of different laminate layers. For example, the wet cut surface of an apple was first coated with alginate cross-linked via calcium ions. This initial coating provided a more appropriate foundation for subsequent hydrophobic coating with acetylated monoglyceride (Wong et al. 1994a). Unfortunately, two coating processes increase product cost and may also reduce commercial viability. Emulsions represent another approach to apply film, although it is still not clear whether emulsion-cast films are better than dual-coatings. Two conflicting reports, each claiming multiple-fold advantages cite that emulsions are much better than dual coatings (Kamper and Fennema 1984) and vice versa (Martin-Polo et al. 1992, Debeaufort et al. 1993). A more recent study forwards the general belief that multilayer films provide better protection than single layer films from emulsions (Debeaufort et al. 1998).

From an economic point of view, emulsion-cast films have commercial appeal for several reasons.Use of a mixture of fat and carbohydrate components emulsified by protein, allows for direct adhesion of hydrophilic carbohydrate material at food surface and formation of a hydrophobic layer or coating at the external food surface. An aqueous emulsion containing 10% casein, 1% alginic acid and 15% of acetylated monoglyceride by weight resulted in a coating that reduced moisture losses for apple slices by 75% over a 3-day period – relative to uncoated slices. Effectiveness of this coating was unexpected, because alginic acid is hydrophilic and moisture losses should have been less if it were left out from the mixture. However, when the coating mixture did not contain either alginic acid or casein, there was only minimal decrease in water loss. It can be implied that casein provided a bridge between hydrophilic alginic acid and hydrophobic lipid, allowing adhesion to the hydrophilic cut surface (Pavlath et al. 1993; Wong et al. 1994b).

Characteristics of emulsion-derived films however, are sensitive to quality of emulsion – generally requiring droplet sizes of <0.5 µm (Platenius 1939). In fact, droplet size should be in the 10–100 nm range to obtain an optimal thermodynamically stable microemulsion (Das and Kinsellar 1990). However, there are several drawbacks to emulsion-type coatings. These coatings can be wet and difficult to handle and may function more as a sacrificial layer as opposed to a true moisture barrier. Emulsion stability is also sensitive to temperature and its efficiency can be affected by quality of emulsifier used. Such variations represent a handicap to commercial application. One particularly interesting breakthrough that eliminates the need to emulsify polysaccharide-fat coating mixtures is use of amphiphilic compounds, molecules containing both hydrophilic and lipophilic moieties. One such example is sucrose fatty acid ester, which was efficient for retaining crispiness of a snack food (Kester et al. 1990). Casein treated with acyl esters of N-Hydroxysuccinimide yielded fatty acid acylated casein with hydrophobic characteristics (Nippon 1984). Another possibility is chitosan fatty acid salts, where length of the fatty acid chain plays an important role in rate of water vapor transmission. Microstructure of these salts changes dramatically though, when lauric acid is incorporated into the chitosan film instead of shorter or longer fatty acids. Chitosan laureate had the best moisture barrier properties in comparison to both shorter and longer fatty acid chain complexes (Wong et al. 1992). The product is a composite film in which fatty acid molecules are distributed within the chitosan matrix, suggesting the importance of the morphological arrangement of the lipid within the chitosan matrix (Pavlath et al. 1993). This arrangement would have an added advantage, since both chitosan and lauric acid alone are antimicrobial agents (Kabara and Ecklund 1991; Darmadji and Izomimoto 1994). Unfortunately, chitosan is still not fully accepted in the United States for food applications, even though it is approved for use in Canada and Japan.

1.7 Conclusions

Various reviews have been written about properties and potential uses of edible films (Kester and Fennema 1986; Guilbert 1986; Krochta 1992; Krochta et al. 1994; Morillon et al. 2002). An extensive review on antimicrobial films was also published (Cagri et al. 2004), and a special review is available for applications on various types of meat products (Gennadios et al. 1997). Much more research is needed as there is no universal edible film that is applicable for every problem. Obviously, specific barrier requirements and food product specifications will determine the type of layer that is best for a given situation. Products with high moisture contents need fatty layers to prevent loss of water. To prevent possible discoloration, an oxygen barrier component is needed. Unsaturated fats which are easily oxidized require similar protective layers. As stated above, water and oxygen permeability generally are inversely related; thus, many films will need to be composite materials with multiple properties (not unlike Mother Nature's own protection). Ideal edible film should have the following characteristics:

*   Contain no toxic, allergic and non-digestible components
*   Provide structural stability and prevent mechanical damage during transportation, handling, and display
*   Have good adhesion to surface of food to be protected providing uniform coverage
*   Control water migration both in and out of protected food to maintain desired moisture content
*   Provide semi-permeability to maintain internal equilibrium of gases involved in aerobic and anaerobic respiration, thus retarding senescence
*   Prevent loss or uptake of components that stabilize aroma, flavor, nutritional and organoleptic characteristics necessary for consumer acceptance while not adversely altering the taste or appearance
*   Provide biochemical and microbial surface stability while protecting against contamination, pest infestation, microbe proliferation, and other types of decay

- Maintain or enhance aesthetics and sensory attributes (appearance, taste etc.) of product
- Serve as carrier for desirable additives such as flavor, fragrance, coloring, nutrients, and vitamins. Incorporation of antioxidants and antimicrobial agents can be limited to the surface through use of edible films, thus minimizing cost and intrusive taste.
- Last but not least – be easily manufactured and economically viable

References

Allen L, Nelson AI, Steinberg MP, McGill JN (1963) Edible carbohydrate food coatings II. Evaluation of fresh meat products. Food Technol. 17: 1442–1448

Banker GS (1966) Film coating theory and practice. J. Pharm. Sci. 55: 81–89

Becker R, Gray GM (1992) Evaluation of a water jet cutting system for slicing potatoes. J. Food Sci. 57(1): 132–137

Beckwith AC, Wall JS (1966) Reduction and reoxidation of wheat glutenin. Biochim. Biophys. Acta 130: 155–162

Biquet B, Guilbert S (1986) Relative diffusivitives of water in model intermediate moisture foods. Lebensm. Wis. Technol. Food Sci. Technol. 19: 208–214

Biquet B, Labuza TP (1988) Evaluation of moisture permeability characteristics of chocolate films as an edible moisture barrier. J. Food Sci. 53: 989–998

Bolin HR, Huxsoll CC (1991) Control of minimally processed carrot (*Daucus carota*) Surface discoloration caused by abrasion peeling. J. Food Sci. 56: 416–418

Bolin HR, Nury FS, Finkle BJ (1964) An improved process for preservation of fresh peeled apples. Bakers Dig. 38(3): 46–48

Cagri A, Ustonol Z Ryser ET (2004) Review: Antimicrobial edible films and coatings. J. Food Prot. 67(4): 833–848

Cairns JA, Owing CR, Paine FA (1974) Packaging for climate protection. The Institute of Packaging, Newnes-Butterworth, London, England

Camirand WM, Forrey RR, Popper K, Boyle FP, Stanley WL (1968) Dehydration of membrane coated foods by osmosis. J. Sci. Food Agric. 19: 472–474

Camirand WM, Krochta JM, Pavlath AE, Cole ME (1992) Properties of some edible carbohydrate polymer coatings for potential use in osmotic dehydration. Carbohydr polym. 13: 39–49

Chen C, Trezza TA, Wong DWS, Camirand WM, Pavlath AE (1999) Methods of preserving fresh fruits and produce thereof. U.S. Patent. 5, 939, 117

Contreras-Medellin R, Labuza TP (1981) Prediction of moisture protection requirements for foods. Cereal Food World. 26(7): 335–340

Cosler HB (1957) Methods of producing zein coated confectionery. U.S. Patent, 2, 791, 509

Darmadji P, Izumimoto (1994) Effect of chitosan in meat preservation. Meat Sci. 38: 243–254

Das KP, Kinsella JE (1990) Stability of food emulsions: Physicochemical role of protein and non-protein emulsifiers. Adv. Food Nutr. Res. 34: 81–201

Davies E (1987) Plant responses to wounding. In the Biochemistry of Plants Vol. 12, D. D. Davies (ed.), P.K. Stumpf. and E.E. Conn (eds-in-chief) Academic, New York, NY, Chapter 7, pp 243–264

Debeaufort F, Martin-Polo M, Voilley A (1993) Polarity, homogeneity and structure affect water vapor permeability of model edible films. J. Food Sci. 58: 426–429, 434

Debeaufort F, Quezada-Gallo JA, Voilley A (1998) Edible films and coatings: tomorrow's packaging: A review. Crit. Rev. Food Sci. Nutr. 38: 299–313

Delong CF, Shepherd TH (1972) Produce coating. U.S. Patent 3, 669, 691

Eaks IL (1980) Effect of chilling on respirations and volatiles of California lemon fruit. J. Amer. Soc. Hort. Sci. 105: 865–869

Eaks IL, Ludi WA (1960) Effects of temperature, washing and waxing on the composition of internal atmosphere of orange fruits. Proc. Am. Soc. Hortic. Sci. 76: 220–228

Feeney RD, Hararalampu SG, Gross A (1992) Method of collating food with edible oil barrier film and product thereof. U.S. Patent 5, 126, 152

Fennema OR (1993) Frozen foods: challenges of the future. Food Aust. 45(8): 374–380

Feuge RO, Vicknair EJ, Lovegren NV (1953) Modification of vegetable oils XIII. Some additional properties of acetostearin products. JAOCS. 30: 283–290

Fox RC (1958) The relationship of wax crystal structure to water vapor transmission rate of wax films. TAPPI. 41(6): 283–289

Gennadios A Weller CL, Testin RF (1993) Modification of physical and barrier properties of edible wheat gluten-based films. Cereal Chem. 70(4): 426–429

Gennadios A, Hanna MA, Kurth LB (1997) Application of edible coatings on meats, poultry and seafoods: A review. Lebensm. Wiss. Technol. 30: 337–350

Glicksman M (1982) Functional properties. In Food Hydrocolloids, M. Glicksman, (ed.) CRC, Boca Raton, FL Vol. 1 p. 47

Glicksman M (1983) Red seaweed extracts. In Food Hydrocolloids, M. Glicksman, (ed.) CRC, Boca Raton, FL Vol. 2 p. 73

Gontard N, Guilbert S (1994) Biopackaging technology and properties of edible and/or biodegradable material of agricultural origin. In Food packaging and preservations. M. Mathlouthi, (ed.) Blackie Academic and Professional, Glasgow, pp 159–181

Gontard N, Guilbert S, Cuq JL (1992) Edible wheat gluten films: Influence of the main process variables on film properties using response surface methodology, J. Food Sci. 57: 190–199

Gontard N, Thibault R, Cuq B, Guilbert S (1996) Influence of relative humidity and film composition on oxygen and carbon dioxide permeabilities of edible films. J. Agric. Food Chem. 44(4): 1064–1068

Grant GT, Morris ER, Rees DA, Smith PJC, Thom D (1973) Biological interactions between polysaccharides and divalent cations. The eggbox model. FEBS Lett. 32: 195–198

Greener I (1992) Physical Properties of Edible Films and Their Components. Ph.D. dissertation, University of Wisconsin

Griesbach H (1987) New insights into ecological role of secondary plant metabolites. Comments Agric. Food Chem. 1: 27–45

Guilbert S (1986) Technology and application of edible protective films, In Food Packaging and Preservation: Theory and Practice, M. Mathlouthi, (ed.) Elsevier Applied Science Publishing Co., London, UK, pp 371–399

Guilbert S (1988) Use of superficial edible layer to protect intermediate moisture foods: Application to the protection of tropical fruits dehydrated by osmosis. In Food Preservation by Moisture Control, C.C. Seaow, (ed.) Elsevier Applied Science Publishers Ltd. Essex, England, pp 199–219

Haard NF, Cody M (1978) Stress metabolites in postharvest fruits and vegetables role of ethylene. In Postharvest Biology and Biotechnology, H.O. Huttin and M. Milner (eds.) Food and Nutrition Press Inc, Westport, CT.

Hagenmaier RD, Baker RA (1997) Edible coatings from morpholine-free wax microemulsions. J. Agric. Food Chem. 45: 349–352

Hagenmaier RD, Shaw PE (1990) Moisture permeability of edible films made with fatty acid and (hydroxypropyl) methylcellulose. J. Agric. Food Chem. 38: 1799–1803

Hagenmaier RD, Shaw PE (1991) Permeability of coatings made with emulsified polyethylene wax. J. Agric. Food Chem. 39: 1705–1708

Hallman GJ, Nisperos-Carriedo MO, Baldwin EA, Campbell CA (1994) Mortality of Carribean fruit fly (Diptera: Tephritidae) immatures in coated fruits. J. Econ. Entomol. 87: 752–757

Hardenburg RE (1967) Wax related coatings for horticultural products. A bibliography. Agricultural Research Service Bulletin 51–15, United States Department of Agriculture, Washington, DC

Havard C, Harmony MX (1869) Improved process for preserving meat, fowls, fish etc. U.S. Patent 90, 944

Hershko V, Klein E, Nuvissovitch A (1996) Relationships between edible coating and garlic skin J. Food Sci. 61: 769–777

Hyodo H, Kuroda H, Yang SF (1978) Induction of phenylalanine ammonia-lyase and increase in phenolics in lettuce leaves in relation to the development of russet spotting caused by ethylene. Plant Physiol. 62: 31–35

Ismail M (1989) Cold treatment: a successful quarantine method for Florida grapefruit shipped to Japan. Packinghouse Newsletter No. 158: 1–2

Kabara JJ, Ecklund T (1991) Organic acids and esters. In Food Preservatives, N.J. Russell and G.W. Gould (eds.) Blackie and Son, Glasgow, pp 21

Kader AA (1986) Biochemical and physiological basis for effects of controlled and modified atmospheres on fruits and vegetable. Food Technol. 40: 99–104

Kamper SL, Fennema OR (1984) Water vapor permeability of an edible, fatty acid, bilayer film. J. Food Sci. 49: 1482–1485

Kanig JL, Goodman H (1962) Evaluative procedures for film-forming materials used in pharmaceutical applications. J. Pharm. Sci. 51(1): 77–83

Katz EE, Labuza TP (1981) Effect of water activity on the sensory crispness and mechanical deformation of snack food products. J. Food Sci. 46: 403

Kester JJ, Fennema OR (1986) Edible films and coatings: A review. Food Technol. 39: 47

Kester JJ, Fennema OR (1989a) The influence of polymorphic form on oxygen and water vapor transmission through lipid films. JAOCS. 66(8): 1147–1153

Kester JJ, Fennema OR (1989b) Tempering influence on oxygen nd water vapor transmission through a stearyl alcohol film. JAOCS. 66(8): 1154–1157

Kester JJ, Fenema OR (1989c) An edible film of lipids and cellulose ether: barrier properties to moisture transmission and structural evaluation. J Food Sci. 54: 1383–1389

Kester JJ, Bernhardy JJ, Elsen J, Letton J, Fox M (1990) Polyol polyesters as protective moisture barrier for foods. U.S. Patent 4, 960, 600

Koelsch CM, Labuza TP (1992) Functional, physiological and morphological properties of methyl cellulose and fatty acid-based edible barriers. Lebensm Wiss Technol. 25: 404–411

Krochta JM (1992) Control of mass transfer in foods with edible coatings and films. In Advances in Food Engineering, W.E. Speiss and H. Schubert (eds.) Elsevier, Essex, England

Krochta JM, Hudson JS, Camirand WM, Pavlath AE (1988) Edible films for light processed fruits and vegetables, Paper Presented at International Winter Meeting of the American Society of Agricultural Engineers, Chicago, IL

Krochta JM, Baldwin EA, Nisperos-Carriedo MO (1994) Edible Coatings and Films to Improve Food Quality, Technomic Publ. Co., Lancaster, PA, pp 1–379

Labuza TP (1980) The effect of water activity on reaction kinetics of food deterioration. Food Technol. 34(4): 36–40

Labuza TP (1984) Moisture Sorption: Practical Aspects of Isotherm Measurement and Use American Association of Cereal Chemists, St. Paul, MN

LaMer VK, Healy TW, Aylmore LAG (1964) The transport of water through monolayers of long-chain n-paraffinic alcohols. J. Coll. Sci. 19: 673

Landman W, Lovegren NV, Feuge RD (1960) Permeability of some fat products moisture. JAOCS. 37: 1–4

Lazarus CR, West RL, Oblinger JJ, Palmer AZ (1976) Evaluation of calcium alginate coating and protective plastic wrapping for the control of lamb carcass shrinkage. J. Food Sci. 41: 639–642

Mark AM, Roth WB, Mehltretter CL, Rist CE (1966) Oxygen permeability of amylomaize starch films. Food Technol. 20: 75

Mate JI, Saltweil ME, Krochta JM (1996) Peanut and walnut rancidity: effects of oxygen concentration and relative humidity J. Food Sci. 61: 465–472

Martin-Polo M, Mauguin C, Voilley A (1992) Hydrophobic films and their efficiency against moisture transfer. I. Influence of the film preparation technique. J. Agric. Food Chem. 40: 407–412

Maxcy RB (1982) Fate of microbial contaminants in lettuce juice. J. Food Protection. 45: 335–339

Mazza G, Qi H (1991) Control after-cooking darkening in potatoes with edible film-forming products and calcium chloride. J. Agric. Food Chem. 39(10): 2163–2166

McHugh TH, Krochta J (1994) Milk-protein-based edible films and coatings. Food Technol. 48(1): 97–103

Meyers MA (1994) Use of edible films to prolong chewing gum shelf-life. U.S. Patent 5, 286, 502

Morillon V, Debeaufort F, Blond G, Capelle M, Voilley A (2002) Factors affecting the moisture permeability of lipid-based edible films: A review. Critical Reviews in Food Science and Nutrition. 42(1): 67–89

Murray DG, Luft L (1973) Low D.E. Corn starch hydrolysates. Food Technol. 27(3): 32–39

Nelson K, Fennema OR (1991) Methylcellulose films to prevent lipid migration in confectionery products. J. Food Sci. 56(2): 504–508

Nippon Shinyuku Co. Ltd. (1984) Fatty acid acylated casein. Japan Kokai Tokyo Koho, Japan Patent 59, 155, 396

Okamoto S (1978) Factors affecting protein film formation. Cereal Foods World. 23: 256–262

Pallos FM, Robertson GH, Pavlath AE, Orts WJ (2006) Thermoformed wheat gluten biopolymers. J. Agric. Food Chem. 54: 349–352

Park JW, Testin RF, Park HJ, Vergano PJ, Weller CI (1994) Fatty acid concentration effect on tensile strength, elongation and water vapor permeability of laminated edible films. J. Food Sci. 59: 916–919

Paulinka F (1986) Quality consideration in the selection of confectionery fats. The Manufacturing Confectionerie (May) 75–81

Pavlath AE, Wong DSW, Kumosinski TF (1993) New Coatings for cut fruits and vegetables. CHEMTECH. 23(2): 36–40

Pavlath AE, Wong DSW, Hudson JS, Robertson GH (1996) Edible films for the extension of shelf life of lightly processed agricultural products. ACS Symposium, Series No. 647. G. Fuller, T.A. McKeon and D.D. Bills (eds.) Chapter 8, pp 107–110

Pavlath AE, Gossett C, Camirand W, Robertson GH (1999a) Ionomeric films of alginic acid, J. Food Sci. 64(1): 61–63

Pavlath AE, Voisin A, Robertson GH (1999b) Pectin-based biodegradable water insoluble films. Macromol. Symp. 140: 106–113

Pickard JF, Rees JE, Elworthy PH (1972) Water vapour permeability of pored and sprayed polymer films. J. Pharm. Pharmac. 24(Suppl): 39P

Platenius H (1939) Wax emulsions for vegetables. Cornell Univ. Agric. Exp. Sta. Bull. 723: 43

Polanski S (1993) Frying food with reduced frying medium uptake – by cooking food to remove water applying coating of swollen dispersion containing natural edible polymer, drying and frying. U.S. Patent 5, 232, 721

Reading S, Spring M (1984) The effects of binder film characteristics on granule and tablet properties. J. Pharm. Pharmacol. 36: 421–426

Rockland LB, Nishi SK (1980) Influence of water activity on food product quality and stability. Food Technol. 34(4): 42

Rogers CE (1985) Permeation of gases and vapors in polymers. In Polymer Permeability, J. Comyn (ed.) Elsevier Applied Science Publisher, New York, NY, pp 11–73

Roth T, Loncin M (1984) Superficial activity of water. In Engineering and Food. Vol. 1. B.M. McKenna (ed.) Elsevier Applied sciences Publishers, London, England, pp 433

Roth T, Loncin M (1985) Fundamentals of diffusion of water and rate of approach of equilibrium A_w. In Properties of Water in Foods in Relation to Quality and Stability, D. Simatos, JL. Multon (eds.) Martinus Nijhoff Publishing, Dordrecht, The Netherlands, pp. 331

Schofield JD, Bottomley RC, Timms MF, Booth MR (1983) The effect of heat on wheat gluten and the involvement of sulphhydryl-disulphide interchange reactions, J. Cereal Sci. 1: 241–253

Shaw CP, Secrist JL, Tuomy JM (1980) Method of extending the storage life in the frozen state of precooked foods and products thereof. U.S. Patent 4, 196, 219

Shewfelt RL (1987) Quality of minimally processed fruits and vegetables. J. Food Quality. 10: 143–156

Talbot G (1994) Minimisation of moisture migration in food systems, Reprint from F.I.E. Paris

Tomasul PM, Craig JC Jr., Boswell T (1997) A continuous process for casein production using high-pressure carbon dioxide. J. Food Eng. 33(3): 405–419

Troller JA (1980) Influences of water activity on microorganisms in foods. Food Technol. 34 (5): 7

Uritani I, Asahi T (1980) Respiration and related metabolic activity in wounded and infected tissues. In Biochemistry of Plants, Vol. 2. PK Stumpf and EE Conn (eds.) Academic, Orlando, FL, Chapter 11, pp 463–485

Williams SK, Oblinger JL, West RL (1978) Evaluation of calcium alginate film for use on beef cuts. J. Food Sci. 43: 292–301

Wong DWS, Gastineau FA,Gregorski KS, Tillin SJ, Pavlath AE (1992) Chitosan-lipid films: Microstructure and surface energy. J. Agric. Food Chem. 40(4): 540–544

Wong DWS, Camirand WM, Pavlath AE (1994a) Development of edible coatings for minimally processed fruits and vegetables. In Edible Coatings and films to improve Food Quality, Lancaster, PA., Technomic Publ. Co. Chapter 3, 65–88

Wong DWS, Tillin SJ, Hudson JS, Pavlath AE (1994b) Gas exchange in cut apples with bilayer coatings. J. Agric. Food Chem. 42(10): 2278–2285

Wong DWS, Gregorski KS, Hudson JS, Pavlath AE (1996) Calcium alginate films: thermal properties and permeability to sorbate and ascorbate. J. Food Sci. 61(2): 337–341

Chapter 2
Structure and Function of Protein-Based Edible Films and Coatings

Kirsten Dangaran, Peggy M. Tomasula, and Phoebe Qi

2.1 Introduction

Research and development on films and coatings made from various agricultural proteins has been conducted over the past 20 years, but is of heightened interest, due to the demand for environmentally-friendly, renewable replacements for petroleum-based polymeric materials and plastics. To address this demand, films and coatings have been made from renewable resources, such as casein, whey, soy, corn zein, collagen, wheat gluten, keratin and egg albumen. Those made from agricultural proteins create new outlets for agricultural products, byproducts and waste streams, all of which can positively impact the economics of food processes.

Due to casein's ability to form water-resistant films, it was used for hundreds of years in paints and coatings (Gettens and Stout 1984). In the late nineteenth century, casein was converted into a hard plastic material by cross-linking it with formaldehyde. A patent for this technology was issued to Adolf Spitteler in Bavaria (Brother 1940), and it was used for the manufacture of products, such as buttons, umbrella handles, small boxes and pen cases.

Protein-based materials experienced a boom of interest in the early twentieth century. Prior to World War II (WWII), several types of protein-based films, coatings, plastics and textiles were commercially available. Textiles developed from agricultural proteins were categorized as azlons. World War I (WWI) and WWII created a substantial demand for materials to make uniforms, blankets and other supplies for soldiers. Wool was the main textile material, but wool substitutes were also created from such products as casein, soy, corn zein and peanut protein (Brooks 2006). Casein-based protein fibers had silk-like properties, and were commercialized under brand names such as Lanital, Merinova and Arlac developed by the United States Department of Agriculture (USDA). Peanut protein fiber was sold as Ardil, and corn zein was used to make the commercialized textile Vicara (Kiplinger 2003).

K. Dangaran (✉), P.M. Tomasula, and P. Qi
Dairy Processing and Product Research, Eastern Regional Research Center, ARS,
USDA, 600 East Mermaid Lane, Wyndmoor, PA 19038, USA
e-mail: dangaran.3@osu.edu

M.E. Embuscado and K.C. Huber (eds.), *Edible Films and Coatings for Food Applications*,   25
DOI 10.1007/978-0-387-92824-1_2, © Springer Science+Business Media, LLC 2009

The large demand for resources created by WWI and WWII was a catalyst for scientific and technological improvements to protein-based polymers. Additionally, the idea of developing new, non-food uses from agricultural crops received wide support and distinction in the form of the Chemurgy Movement (Finlay 2004). In the 1920s and 1930s, the National Farm Chemurgic Council was formed, promoting the move to generate new products from agricultural commodities that were useful to industry. For a time, this original "green" movement had a growing influence throughout the country. Supporters included important figures like William Hale, a prominent figure within Dow Chemical, and Henry Ford, a leader in the automotive industry, who promoted and used biobased plastics, textiles and fuels in his automobiles. In response to the demand, the USDA established four regional research centers, devoted to creating value-added products, developing renewable biobased materials, and discovering new routes for industrial utilization of agricultural materials (Kelley 1993). However, production and use of agricultural-based polymeric materials declined after WWII due to the development of petroleum-based products, which were cheaper, easier to produce, and generally superior to their agricultural counterparts.

Still today, the four USDA research centers remain mainstays of the Agricultural Research Service (ARS), the research arm of the USDA. While the mission of ARS has expanded to address additional challenges currently facing agriculture, a significant portion of the mission is dedicated to research on value-added foods and bio-based products. Today, there is renewed interest in biobased materials; consumers demand environmentally-friendly products that lower dependence on oil and other non-renewable resources (Comstock et al. 2004). As evidenced by Presidential Initiatives 13101 and 13134, which call for increased use and development of agricultural-based products, biobased product research has become a priority for government, academic, and industrial research institutions.

With their proven track record for function and commercialization, packaging applications are well suited to be made from biobased materials and to incorporate protein-based materials. Food packaging innovation, including films and coatings, is driven by (1) distribution needs, especially as globalization increases the distance, products are shipped; (2) consumer demands for convenient packaging, including easily recyclable packaging; and (3) environmental legislation. During the last 5–10 years, especially in the UK and Japan, government legislation has accelerated the development of environmentally-friendly packaging (Sonneveld 2000; Comstock et al. 2004). Moreover, increasing energy costs also raise the prices of conventional packaging materials. In 2005–2006, food packagers in Europe experienced a 30–80% increase in the cost of packaging materials, mostly due to the escalating cost of petroleum (Anonymous 2006).

The inherent properties of proteins make them excellent starting materials for films and coatings. The distribution of charged, polar and non-polar amino acids (Fig. 2.1a–c) along the protein chain creates chemical potential. Figure 2.2 shows a representation of beta-lactoglobulin, the major protein found in whey. The shading illustrates the domains of polar and non-polar areas along the protein chain. The resulting interactive forces produce a cohesive protein film matrix. Films form and are stabilized through electrostatic interactions, hydrogen bonding, van der Waals forces, covalent bonding, and disulfide bridges (Krochta et al. 1994). Protein

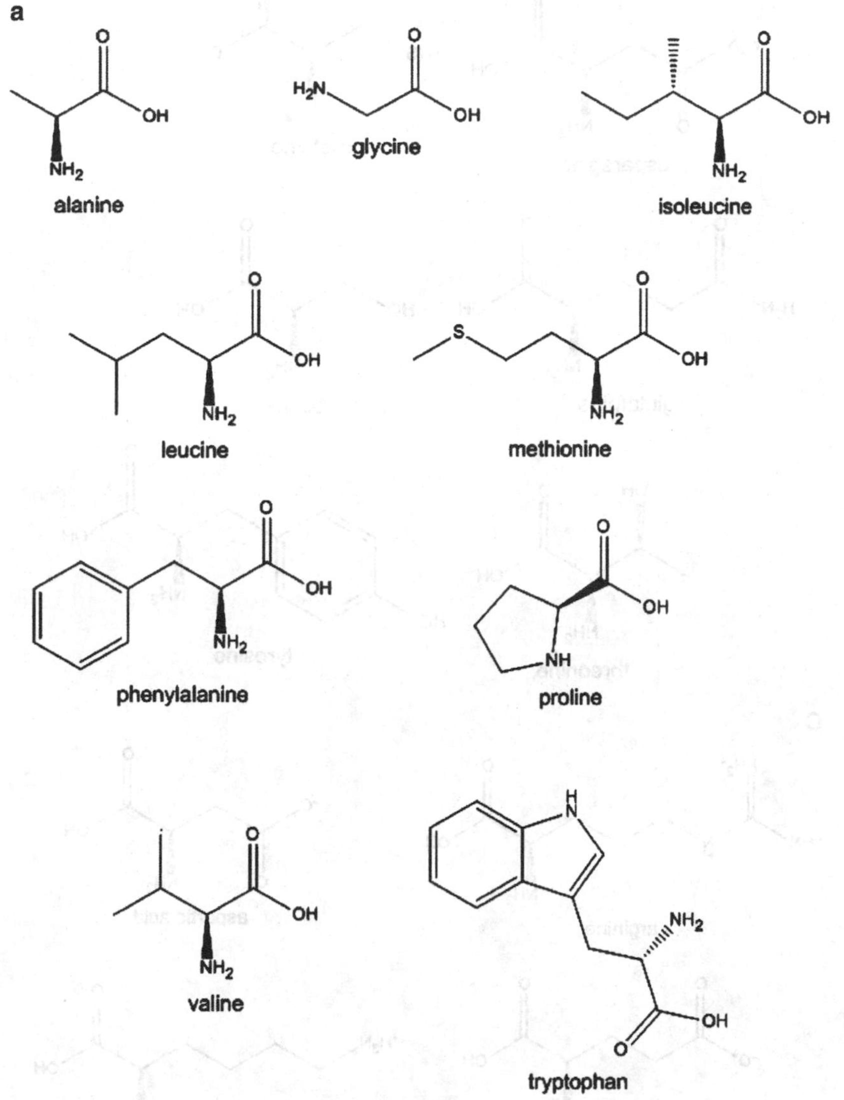

a

Fig. 2.1 (a) Non-polar amino acids. (b) Polar amino acids. (c) Charged amino acids

film-forming capabilities are best demonstrated in emulsified systems in which amphipathic proteins form films at air–water or water–oil interfaces. There are also secondary benefits for using proteins to form films and coatings. Proteins have multiple sites for chemical interaction as a function of their diverse amino acid functional groups, which can allow for property improvement and tailoring. Chemical changes can improve the stability of films and coatings. Cross-linked protein films are often more stable than their polysaccharide-based counterparts and have a longer lifetime (Barone and Schmidt 2006). Figure 2.3 depicts the enzymatic cross-linking of proteins.

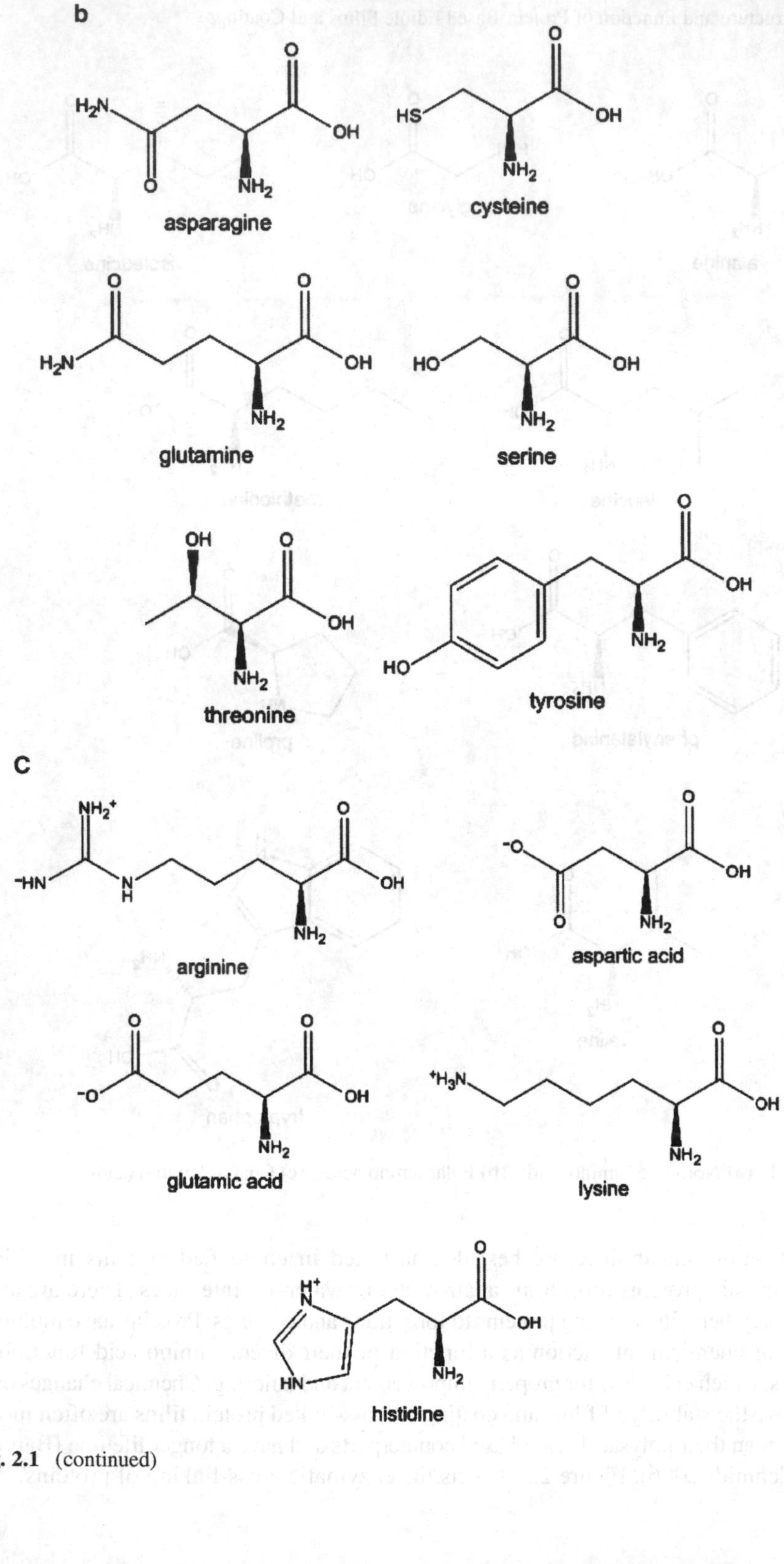

b

asparagine

cysteine

glutamine

serine

threonine

tyrosine

c

arginine

aspartic acid

glutamic acid

lysine

histidine

Fig. 2.1 (continued)

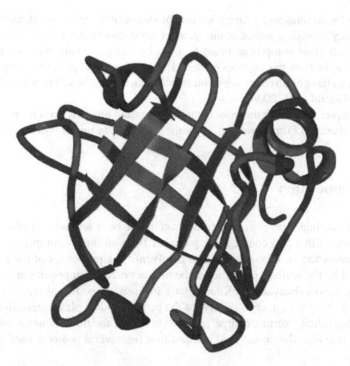

Fig. 2.2 Examples of polar (*light gray*) and non-polar (*dark gray*) domains in proteins as depicted by the beta-lactoglobulin molecule

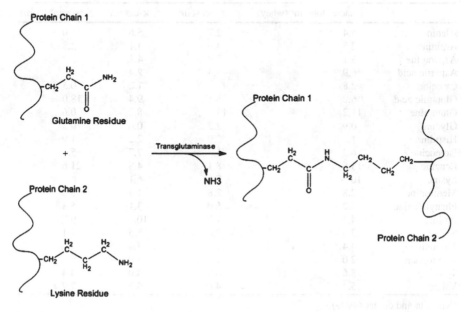

Fig. 2.3 Cross-linking reaction of protein chains catalyzed by the enzyme transglutaminase

Protein-based films and coatings are also biodegradable and compostable. As they degrade, they provide a source of nitrogen, which contributes a fertilizer benefit not available with other non-protein-based films and coatings. Finally, there is emerging evidence that the bioactive peptides produced upon digestion of proteins (dairy sources in particular) have antihypertensive and radical scavenging health benefits (Aimutis 2004; FitzGerald et al. 2004).

This chapter will cover the most commonly studied proteins used for research in films and coatings. Film formation, properties, and future trends will be discussed.

2.2   Composition

Films and coatings may be made from proteins of both animal and plant origin. Protein-based films and coatings are prepared from solutions comprised of three main components: protein, plasticizer and solvent. The properties of the final film are affected by the intrinsic properties of the film or coating components and extrinsic processing factors (Panyam and Kilara 1996). Intrinsic properties of proteins include amino acid composition, crystallinity (of the protein and/or plasticizer), hydrophobicity/hydrophilicity, surface charge, pI, molecular size, and three-dimensional shape. Table 2.1 provides the amino acid composition for several proteins used to make

Table 2.1 Amino acid composition (mol %) of some proteins used to make edible films and coatings

	β-lactoglobulin[a] (whey)	α_{s1}-casein[a]	κ-casein[a]	α-zein[b] (corn)
Alanine	5.4	2.7	5.6	11.0
Arginine	2.5	4.0	4.1	2.8
Asparagine	3.1	3.8	4.2	6.7
Aspartic acid	6.9	3.4	2.4	–
Cysteine	2.8	–	1.0	0.5
Glutamic acid	6.2	8.1	9.4	18.0
Glutamine	11.2	13.1	8.2	0.6
Glycine	0.9	2.2	0.6	0.5
Histidine	1.5	2.9	2.2	1.9
Isoleucine	6.2	5.3	7.7	5.7
Leucine	13.6	8.1	4.8	21.6
Lysine	10.5	7.6	6.1	–
Methionine	2.8	2.8	1.4	3.0
Phenylalanine	3.2	5.0	3.1	5.3
Proline	4.2	7.0	10.2	9.7
Serine	3.3	3.0	5.5	7.1
Threonine	4.4	2.1	7.4	3.2
Tryptophan	2.0	1.6	1.0	–
Tyrosine	3.6	6.9	8.0	4.4
Valine	5.4	4.6	5.7	7.7

[a]Kinsella and others (1989)
[b]Adapted from Larkins and others (1993)

edible films and coatings. The presence of cysteine allows for potential disulfide bridge formation, as noted for beta-lactoglobulin. High concentrations of leucine, alanine and other nonpolar amino acids can create hydrophobic proteins, as seen in α-zein. Extrinsic factors include processing temperature, drying conditions, pH, ionic strength, salt-type, relative humidity during processing and storage, shear and pressure (Damodaran 1996). A brief summary of the properties for some proteins commonly used for making films and coatings is given below.

2.2.1 Animal Proteins

2.2.1.1 Casein

Casein is the major dairy protein group occurring at 24–29 g/l in bovine or goat milk (Audic et al. 2003). There are four main subunits – alpha s1-casein, alpha s2-casein, beta-casein and kappa-casein that make up 38%, 10%, 36% and 13% of casein composition, respectively (Audic et al. 2003). Each of the four protein fractions has unique properties that affect casein's ability to form films. Alpha s1 is a 23.6 kDa protein with a net charge of −21.9 at a pH of 6.6 and a pI of 4.94. It has eight phosphorylated serine residues, 17 proline residues, 25 glutamine residues and no cysteine residues. Alpha s1 has the majority of its charge isolated between the hydrophobic N- and C-terminals, making the protein amphipathic. The phosphorylated serine residues are clustered with glutamine residues, forming calcium-binding sites. Consequently, alpha s1 is Ca^{2+} sensitive, which means that the protein will aggregate and precipitate in low concentrations of the ion. Because it does not contain free cysteine, it cannot participate in disulfide bond formation and cross-linking.

Alpha s2-casein is also Ca^{2+} sensitive. It has a net charge of −13.8, a pI of 5.37, and two regions of high-charge density, making it more hydrophilic than other casein proteins.

Beta-casein is an amphipathic protein with a polar N-terminal, but a large hydrophobic domain. It is also Ca^{2+} sensitive. The solubility of beta-casein increases when the temperature is dropped to 4°C.

Kappa-casein is the other major casein protein; however, it is different from the other three casein fractions, because it is not calcium ion sensitive. Kappa-casein is located at the outer layer of a casein micelle, and is very amphipathic with a polar domain that interacts with polar solvents. It is often described as the "hairy layer" of the casein micelle (Swaisgood 1993, 1996).

2.2.1.2 Whey Protein

Whey is a byproduct of the cheese-making process, and whey proteins are technically defined as those that remain in the milk serum after coagulation of the caseins at a pH of 4.6 and a temperature of 20°C (Morr and Ha 1993). Whey protein is comprised of several individual proteins, with beta-lactoglobulin, alpha-lactalbumin, bovine serum albumin, and immunoglobins being the main proteins (deWit and Klarenbeek 1983; Kinsella 1984). Whey proteins differ from caseins in that their net negative charge is evenly distributed over the protein chain. The hydrophobic, polar and charged amino acids are also uniformly distributed. Consequently, the proteins fold so that most of the hydrophobic groups are buried within the whey protein molecule. Extensive self-association does not occur under neutral conditions, as seen among casein proteins (Swaisgood 1996). The protein interactions that occur between chains determine film network formation and properties.

Beta-lactoglobulin is an 18.3 kDa protein that consists of about 160 amino acids, depending on the genetic variant, and comprises 48–58% of the total whey protein. It does not denature as pH changes, but is thermolabile. Its secondary structure is dominated by an eight-stranded beta-barrel, and also possesses a three-turn alpha-helix. For film formation, there is an important free sulfhydryl group, cysteine 121 (CYS121), and two disulfide bonds between cysteine 66-cysteine 160 and cysteine 106-cysteine 119. In the native form, CYS121 is hidden by the alpha-helix (Sawyer et al. 1999). After thermal denaturation, the thiol group is exposed and available for intermolecular disulfide bond formation. Consequently, edible films made from denatured whey protein are stronger and more cohesive than those made from native protein (Perez-Gago et al. 1999). The denaturation temperature of beta-lactoglobulin is 78°C in a 0.7 M phosphate buffer (pH 6) (deWit and Klarenbeek 1983). The most abundant amino acids in bovine beta-lactoglobulin are leucine, glutamic acid, and lysine. Notably, beta-lactoglobulin is low in tyrosine residues (Etzel 2004), which limits its ability to cross-link through the bi-tyrosine radicals that form when proteins are exposed to irradiation treatment.

Alpha-lactalbumin is the second most abundant whey protein, comprising 13–19% of the total whey protein, with a molecular weight of 14.2 kDa. It contains four internal disulfide bridges. Bovine serum albumin (BSA) is the longest single-chain whey protein, with a molecular weight of 66 kDa. It is prone to precipitation around 60–65°C, due to increased hydrophobic binding between the chains (deWit and Klarenbeek 1983). The immunoglobulins are a mixture of proteins, and are very thermolabile.

2.2.1.3 Meat Proteins

There are three types of meat proteins: sarcoplasmic, stromal and myofibrillar. Enzymes, myoglobulin, and cytoplasmic proteins are examples of sarcoplasmic proteins. Stromal proteins include collagen and elastin, while myofibrillar proteins include myosin, actin, tropomysin, and troponins. Stromal and myofibrillar proteins are used for making edible films and coatings.

Collagen is a fibrous, stromal protein extracted from connective tissue, tendons, skin, bones and vascular system, which are waste products of meat processing. Collagen is made from three parallel alpha-chains, which combine to make a triple-stranded superhelical structure. The amino acid sequence of collagen is dominated by a repeating trio of Gly-X-Y residues (where X and Y are often proline and hydroxyproline, respectively) (Haug et al. 2004). Gelatin is formed when collagen is exposed to a mild heat treatment under acidic or alkaline conditions (Badii and Howell 2006). In this process, collagen is partially denatured, but upon cooling, partially reforms the triple helix structure, though unstructured domains also exist. The resulting protein material is referred to as gelatin. Gelatin contains a large amount of proline, hydroxyproline, lysine, and hydroxylysine, which can react in an aldol-condensation reaction to form intra- and intermolecular cross-links among the protein chains.

Gelatin is used in the food industry to thicken and texturize foods, because of its good gelling properties. It also has excellent foaming properties that translate to it being a good edible film former. Gelatin has been used as a coating by the food and pharmaceutical industry for years. Recently, marine gelatins have attracted increased interest due to the perceived risk and stigma associated with bovine spongiform encephalopathy (BSE), which is a potential risk associated with gelatin obtained from cows. Fish gelatin is also getting more attention as the fish industry tries to find new outlets for their skin and bone by-products. In Alaska alone, over one million pounds of by-product are generated annually (Avena-Bustillos et al. 2006). Fish gelatin from those species that live in cold water environments has amino acid compositions that differ from those of mammalian gelatin. Cold water fish gelatins contain a lower percentage of proline and hydroxyproline; consequently, fish gelatins have a lower melting and gelation temperature (Balian and Bowes 1977).

Myosin and actin make up the major fractions of myofibrillar proteins. Like collagen and gelatin, the secondary structure of myofibrillar proteins possesses the alpha-helix as the dominant secondary structure. The proteins are soluble in salt solutions. There is interest in using myofibrillar proteins for edible films and coatings, because of the need to find economic uses for this by-product-stream originating from the Alaskan fish industry.

2.2.1.4 Egg Albumen

Egg albumen is the second major component of liquid egg white, water being the first, and makes up approximately 10% of the total liquid egg white weight. Within egg albumen, there are five main protein fractions: ovalbumin, ovotransferrin, ovomucoid, ovomucin and lysozyme. Differential scanning calorimetry analysis shows three main endotherms that occur during heating of egg white protein that correspond to the denaturation of ovotransferrin, lysozyme, and ovalbumin. Ovalbumin is the predominant egg white protein fraction, comprising nearly 50% of the overall protein content. Ovalbumin is a 44.5 kDa protein that contains free sulfhydryl groups available for cross-linking (Mine 1995). Ovotransferrin is an iron-binding protein with a weight of 77.7 kDa and a pI of ~6.1. Lysozyme, one of the most studied proteins due to its antimicrobial activity, has been found to be active against gram-negative bacteria.

Egg albumen proteins are thermolabile, and form strong heat-set gels. During heat denaturation, egg white proteins form stable intermolecular beta-sheet structures between ovalbumin, ovotransferrin, and lysozyme. Above 60°C, egg white proteins unfold, exposing their internal sulfhydryl groups that can effect disulfide bond formation, and surface hydrophobicity increases (Mine et al. 1990). Denaturation of egg white proteins can be affected by pH, salt concentration, sucrose, or pre-heat treatment. Over time, as a result of thiol cross-linking, ovalbumin forms s-ovalbumin, which is a more stable form of the protein with a higher denaturation temperature (Ternes 2001).

2.2.1.5 Feather Keratin

Annually, approximately 2.3 billion pounds of feathers are generated as a byproduct-stream of the poultry processing industry. Feather keratin, a fibrous protein, has unique film-forming properties due in part to its disulfide bonds that can be reduced to facilitate intermolecular cross-linking of protein chains. Feather keratin is also semi-crystalline, which adds to its high strength and stiffness (Barone et al. 2005). Chicken feathers are 91% protein, 1% lipid and 8% water (Kock 2006). The most abundant amino acids are cysteine, glycine, proline and serine; there are very few histidine, lysine, or methionine residues (Schmidt 1998).

2.2.2 Plant Proteins

Fruits and vegetables are generally low in protein content, with polysaccharides being the predominant biopolymers obtained from these sources for making edible films and coatings. However, cereal grains, tubers, legumes, and pulses possess higher protein contents than fruits and vegetables. These proteins have been isolated and studied for film formation.

2.2.2.1 Wheat Proteins

Due to the importance of bread products in both food and culture, the breadmaking process has been extensively studied. More specifically, the properties of wheat gluten, which is a major contributor to bread properties, have been investigated in detail. Wheat gluten, which contains a small proportion of charged amino acids (lysine, histidine, arginine) and a high proportion of non-polar amino acids, aggregates easily due to hydrophobic interactions (Haard and Chism 1996). There are four primary wheat protein fractions, classified by their solubility: albumins (water soluble), globulins (soluble in dilute salt solutions), gliadins (soluble in 70–90% ethanol), and glutenin (insoluble under all of the previously mentioned conditions) (Haard and Chism 1996).

Glutenin and gliadin are the major proteins within wheat flour, making up 47% and 34% of the total protein content, respectively (Kinsella 1982). Gliadin is a single chained peptide of approximately 40 kDa, consisting of four distinct fractions: ω_5, $\omega_{1,2}$, α/β, γ, with α/β and γ gliadins containing intramolecular disulfide bonds (Kinsella 1982). These play a role in film formation, strength and elasticity. Glutenin is a mixture of proteins and has a molecular weight distribution between 100 and 1,000 kDa, depending on the number of intermolecular disulfide linkages. Some glutenin molecules are branched. The disulfide bonds present in glutenin and gliadin play an important role in determining the strength of the protein matrix. Oxidizing agents, such as potassium bromate, potassium iodate and ascorbic acid, can be added to affect the strength and elasticity of the matrix. Low molecular weight thiol-containing compounds, like glutathione, also can be present in wheat protein. These small thiol compounds react with the disulfide bonds within and between gluten molecules and cleave them, weakening the wheat protein matrix. Oxidizing agents can selectively react with the small thiol groups, making it easier for protein–protein SS/SH exchange to take place and subsequently, increasing the strength and elasticity of gluten (Kinsella 1982).

The gluten protein matrix is also affected by shear, which causes proteins to unfold, hydrogen bonds to weaken, and a general rearrangement of the proteins to form parallel fibers. These changes increase the elasticity and stretchiness of the resulting matrix. During exposure to shear stress, hydrophobic areas are pulled apart and become exposed to water within the dough matrix, leading to a hydrophobic effect and a re-ordering of water molecules. When the stress is removed, the proteins relax, the hydrophobic areas refold, and the water becomes less structured. Gluten, therefore, can be considered both plastic and elastic.

2.2.2.2   Soy Protein

Soy protein is comprised of a mixture of globular proteins. Approximately 90% of soy proteins separate into either a 2S, 7S, 11S or 15S fraction, based on molecular weight and sedimentation coefficient (Cho and Rhee 2004). The two main globular proteins are beta-conglycinin (7S globulin) and glycinin (11S globulin), which make up 37% and 31% of the soy proteins, respectively. Conglycinin (140–170 kDa) consists of various combinations of three subunits, which are heavily glycosylated (Kunte et al. 1997). Glycinin (340–375 kDa) is made of six AB subunits, which are each comprised of an acidic (A) and a basic (B) polypeptide, linked together via disulfide bonds.

As with other film-forming proteins, glycinin is known to be a gelling agent, emulsifier and foaming agent (Subirade et al. 1998). Heat and alkaline conditions can denature soy proteins, affecting film formation. Similar to beta-lactoglobulin (whey protein), glycinin (11S protein) forms intermolecular disulfide bonds when denatured, which bonds affect the tensile properties of a formed film. Soy protein association and stability is dependent on the pH and ionic strength (Petruccelli and Anon 1994). Beta-conglycinin is less heat stable than glycinin; the proteins have denaturation temperatures of approximately 70°C and 80°C, respectively (Renkema and Vliet 2002).

2.2.2.3   Corn Zein

Zein protein from corn has some unique characteristics compared to most other agricultural proteins used for edible films and coatings. Zein possesses a high percentages of nonpolar amino acids and low proportions of basic and acidic amino acids. The three primary amino acids in corn zein protein are glutamine (21–26%), leucine (20%) and proline (10%) (Shukla and Cheryan 2001). Consequently, corn zein protein is insoluble in water, a characteristic that affects the barrier properties of its films. The two major fractions of zein are alpha-zein and beta-zein. Alpha-zein is soluble in 95% ethanol, and makes up approximately 80% of the total prolamines present in corn, while beta-zein is soluble in 60% ethanol. A helical secondary structure dominates zein proteins (Shukla and Cheryan 2001). When formed into films, zein is glossy, tough and greaseproof, with a low water vapor permeability compared to most other agriculturally-based protein films. Zein has been commercially used as a coating for medical tablets, and has the potential to be used in biodegradable packaging.

2.3   Solvents

Water and ethanol are the solvents of choice for making edible films and coatings, because they are safe for consumption. However, if the application of the film or coating from an agricultural protein is not intended for a food application, other organic solvents may be used. In a study by Yoshino et al. (2002), zein films were made using either aqueous ethanol or aqueous acetone as the solvent system. When ethanol was used as the solvent, resulting films initially had higher tensile strength as compared to those made with acetone; however, they were also more susceptible to moisture and high humidity environments. As the relative humidity of the storage environment increased from 5 to 90%, the tensile strength of the ethanol-solvent zein films dropped over 75% (from 29 to 7 MPa), while the acetone-solvent zein films lost only 15% of their initial strength (from 25 to 21 MPa). Thus, choice of solvent can affect the properties of a formed film.

2.4   Plasticizers

Plasticizers are small molecular-weight compounds that can be added to an edible film or coating solution to improve the flexibility and mechanical properties of the film matrix. Most protein-based films and coatings are very strong, but very brittle when not plasticized (Gennadios et al. 1994); thus, a plasticizer is necessary to improve the application potential of protein-based films. While plasticizers can improve the flexibility and elongation of protein films, they also affect the permeability of the films and coatings (McHugh and Krochta 1994). As a general rule, the addition of a plasticizer increases the permeability of a film or coating.

There are two types of plasticization: internal and external. Internal plasticizers chemically modify a protein chain through addition of substituent groups attached

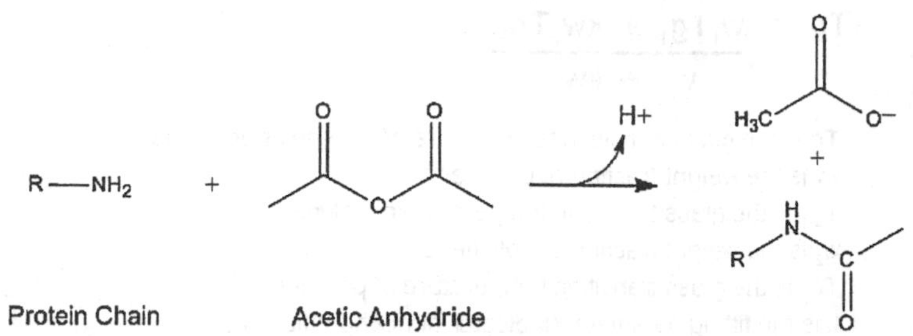

Fig. 2.4 Examples of chemical modification of protein chains

via covalent bonds. Common chemical derivatization schemes are acetylation, succinylation and Maillard reactions with monosaccharides (Fig. 2.4). Internal plasticizers create steric hindrance between the protein chains, leading to increased free volume and improved flexibility. External plasticizers solvate and lubricate the protein chains, lowering the glass transition temperature of the proteins and also increasing the free volume.

Common plasticizers used in edible films and coatings are typically polyols, including glycerol, propylene glycol, polypropylene glycol, sorbitol and sucrose. Fatty acids have also been used as plasticizers in edible films and coatings, though they are not as common. The most effective plasticizer is, of course, water. The effectiveness of a plasticizer is dependent upon three things: size, shape and compatibility with the protein matrix (Sothornvit and Krochta 2001). The state of the plasticizer under normal storage conditions may also affect its permeability and flexibility. Solid plasticizers may have an "antiplasticizing" effect, and decrease matrix flexibility, while improving permeability (Dangaran and Krochta 2007). In traditional polymer-plasticizer systems, antiplasticization occurs when a plasticized system is harder and less flexible than the pure polymer at a temperature below the glass transition or Tg (Chang et al. 2000; Morara et al. 2002).

Several researchers have conducted studies to evaluate the efficiency of different plasticizers in protein-based films, and have developed empirical models to describe the observed phenomena (Sothornvit and Krochta 2000, 2001; Dangaran and Tomasula 2006). Dangaran and Tomasula (2006) used a modified version of the Gordon–Taylor

$$Tg = \frac{w_1Tg_1 + kw_2Tg_2}{w_1 + kw_2}$$

Tg is the glass transition temperature of a blend of polymers
w_1 is the weight fraction of polymer 1
Tg_1 is the glass transition temperature of polymer 1
w_2 is the weight fraction of polymer 2
Tg_2 is the glass transition temperature of polymer 2
k is the fitting parameter (indicator of mixing efficiency)

Fig. 2.5 Gordon–Taylor equation

equation (Fig. 2.5), and applied it to determine the effect of glycerol, propylene glycol, and sorbitol on three types of caseins, each differing in structure and chemical functionality (water solubility). They determined that the effectiveness of a plasticizer was affected by the microstructure and chemical nature of the plasticizer, with glycerol being the most efficient. Sothornvit and Krochta (2001) also found glycerol to be the most efficient plasticizer in a whey protein film matrix. Their empirical model fit the changes in the observed tensile properties and oxygen permeabilities for films with up to 30% plasticizer. They reported an exponential relationship between film properties and plasticizer content. The properties of the protein (crystallinity, hydrophobicity/hydrophilicity) also affect the plasticizer–protein interaction.

2.5 Film and Coating Preparation

There are several ways to form edible films and coatings from agricultural proteins, all of which affect the properties of the final film or coating. Films can be formed via several processes, depending on the starting material. Lipid and wax films can be formed through solidification of the melted material (Fennema and Kester 1991). Biopolymers in solution can form films by changing the conditions of the solution. Applying heat, adding salt or changing the pH may alter conditions in the solution, such that the biopolymers aggregate in a separate film phase. This process is called coacervation (Krochta et al. 1994; Debeaufort et al. 1998).

2.5.1 Solvent Casting

Solvent casting is the preferred method used to form edible protein films for research. Various types of equipment are available for solvent casting of films, from simple casting-plates to more advanced batch and continuous lab coaters (Fig. 2.6a, b). Because it is effective and cost-efficient, the most commonly used method for forming protein film samples for research is by manually spreading dilute film solutions (usually 5–10% solids) of protein and plasticizer into level Petri dishes or

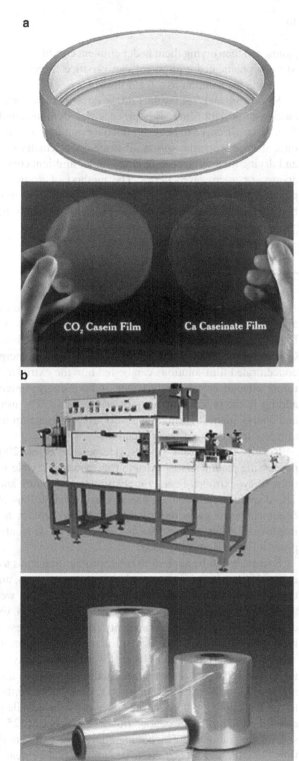

a

CO₂ Casein Film   Ca Caseinate Film

b

Fig. 2.6 (a) Casting plate and casein films. (b) Continuous lab coater and rolls of edible films

plates, and then drying them under ambient conditions or controlled relative humidity. More sophisticated equipment can produce larger protein films by mechanically spreading the solution to a fixed thickness.

Kozempel and Tomasula (2004) developed a continuous process for solvent casting of protein films. This process was demonstrated using casein-based films. Parameters that need to be determined for continuous film production are air temperature, surface properties of the substrate upon which the films are formed, flow rate, and drying time. Films can be dried under ambient conditions, with hot air, infrared energy, or microwave energy. The method of drying can significantly affect the physical properties of the final film, including film morphology, appearance, and barrier and mechanical properties (Perez-Gago and Krochta 2000).

2.5.2   Extrusion of Edible Films

An alternative to solvent casting is extrusion, which uses elevated temperature and shear to soften and melt the polymer, thus allowing a cohesive film matrix to form. Extrusion of proteins into films has certain advantages over solvent casting. Comparatively, extrusion is faster and requires less energy, due to the fact that more concentrated film solutions can be fed into the extruder. For solvent casting, evaporating ethanol or especially water, is both energy intensive and time consuming, adding to the production costs of edible films. The use of extrusion reduces time and energy inputs to bring the cost of biopolymer film formation into a competitive range with synthetic film production.

Results from recent studies (Hernandez et al. 2005, 2006) investigated the processing parameters to extrude transparent, flexible whey protein sheets using a twin-screw extruder. Feed composition, temperature and screw speed were varied. These extruder dimensions and operating conditions allowed for sufficient heat-denaturing and cross-linking of the whey protein to produce sheets that had improved tensile properties, as compared to solvent-cast heat-denatured whey protein films (Hernandez 2007).

Extrusion can be applied to other protein materials as well. In a study by Zhang et al. (2001), sheets were extruded from soy protein, and the effects of plasticizer type, plasticizer amount, and cross-linking agent were measured. The tensile strength and elastic modulus of soy protein sheets were comparable to LDPE, depending on the processing agents (plasticizer, cross-linker) used. Gluten-based films have also been extruded. In a study by Hochstetter et al. (2006), extruded sheets of wheat gluten and glycerol were successfully made. They determined that under the shear and temperature conditions used to make the films, there was an orientation of the protein film matrix. Films tested either parallel or perpendicular to the direction of extruder flow, gave different tensile properties. Films cut in the cross direction of the extruder flow were stronger (25 MPa tensile strength) with shorter elongation (120%), compared to those tested in the direction of the machine flow (17 MPa and 170% for tensile strength and elongation, respectively). However, films were brown in color due to Maillard reactions that occurred during high temperature extrusion (130°C). In a study by Obuz et al. (2001), wheat gluten was extruded with varying proportions of LDPE or metallocene-catalyzed ethylene-butene copolymer (MCEBC). In tests containing up to 25% of the wheat gluten biopolymer, the tensile properties of films were similar to the pure synthetic materials. However, if higher amounts of the biopolymer were incorporated, there was a significant detrimental effect on both elongation and tensile strength. Given the processing advantages of extrusion for producing stand-alone films, more research on extrusion of protein films is warranted.

2.5.3   Spinning

Wet-spinning is a processing technique most commonly used by the textile industry to form fibers (Rampon et al. 1999). During fiber formation, a solution of polymer is passed through a pin hold spinneret under pressure. Frinnault et al. (1997) modified the spinning process to form films from casein, replacing the spinneret with a plate die to form flat films. Protein solutions were extruded into a coagulating bath, and then collected onto a roller. The modified wet-spinning process has also been used to make films from soy protein. Rampon et al. (1999) hypothesized that soy protein chains could be oriented similar to textile fibers using the spinning method. They further theorized that an increased orientation of protein chains within the edible film might facilitate an increase in the elongation values of the films. Though they were successful in forming films, no increase in the orientation of the protein chains was observed. As the modified film formation process used a pressure lower than that commonly used textile processing, the authors suggested that shear could possibly play a role in orientation.

2.5.4   Edible Coating Formation

Edible coatings are formed using the same processes and according to the same mechanisms associated with solvent casting of films. A dilute protein solution is applied to the surface of the food product, and the coating forms upon evaporation of solvent. Typical methods for forming a coating include panning, fluidized-bed processing, spray-coating, and dipping.

Panning is a method used by both pharmaceutical and confectionery industries, and entails putting the product to be coated into a large, rotating bowl, referred to as the pan (Minifie 1989). The coating solution is ladled or sprayed into the rotating pan, and the product is tumbled within the pan to evenly distribute the coating solution over the surface of the food or pharmaceutical material. Forced air, either ambient or of elevated temperature, is used to dry the coating. In a study by Lee et al. (2002a,b), whey protein coatings were applied to chocolate candies to provide a clear, glossy finish to a confectionery product. Edible shellac is currently used to finish chocolates and other candies, such as jelly beans, with a gloss surface. However, shellac is an ethanol-based coating that emits volatile compounds into the air. Whey protein coatings plasticized with sucrose were found to be highly glossy (Dangaran and Krochta 2003), and were preferred to shellac-coated chocolate products in a consumer study (Lee et al. 2002a–c).

Fluidized-bed coating, a method used commonly by the pharmaceutical industry to coat tablets, has been studied for formation of whey protein edible coatings to protect nuts and peanuts. Lin and Krochta (2006) found that alternating the spraying of coating solutions with periods of drying allowed for the gradual formation of the coating. The action of the fluidized-bed during drying of the coating appeared to reduce the formation of clusters of coated product, a problem commonly encountered with formation of coatings via panning.

Spray-coating is used to apply a uniform coating over a food surface, and is a potentially a more controllable method of coating application than pan- or fluidized-bed-coating. However, spray-coating requires that the bottom surface of the product be coated in a separate operation after application of the initial coating and drying. In this scenario, the product must then be turned to expose the bottom for subsequent coating application. Spray-coating is preferred for items possessing a large surface area.

Dipping, the other possible method of forming edible coatings on the surfaces of food, is best-suited for irregularly-shaped food objects. Final formed coatings may be less uniform than coatings applied by other methods, and multiple dipping (with draining and drying steps between dipping operations) may be necessary to ensure full coverage (Krochta et al. 1994).

2.6   Effects of Processing on Properties of Protein-Based Films and Coatings

While there are several methods for forming protein-based films and coatings, solvent casting still predominates. Most of the literature reports have investigated the properties of solvent-casted films. The main properties of interest for films and coatings are tensile properties (tensile strength, elongation at break and elastic modulus), gas permeability, water vapor permeability and appearance. All of these properties can be affected by the extrinsic conditions used to process and produce the films.

2.6.1   Drying Conditions

The effect of drying conditions (e.g., temperature, relative humidity, type of energy source, etc.) on the properties of protein-based films have been investigated for whey protein and casein as well as other protein sources. In all reported studies investigating the films made from either of the dairy proteins, drying effects on film properties were minor, but still measurable. Kozempel et al. (2003) investigated the effects of air temperature, air flow, relative humidity, and initial total solids content on the drying rate and tensile properties of calcium caseinate films. Using convection drying, they found that an air temperature of 34°C and a low relative humidity drastically reduced the period of constant drying from approximately 1,300 to 400 min compared to ambient air temperature, flow and high relative humidity. Also, increasing the initial total solids of solutions from 6 to 15% dropped the constant drying period from 1,300 to 300 min. While drying time was significantly reduced, differences in film tensile properties were not observed. However, tensile strength and elastic modulus did trend upward as the initial total solids were increased. Alcantara et al. (1998) reported similar results for whey protein films dried under different conditions; tensile strength and elastic modulus were marginally increased by an elevated drying air temperature. In a study by Kaya and Kaya (2000), the effect of microwave energy drying was investigated in relation to the properties of whey protein films. They observed no statistically significant increase in film water vapor permeability or tensile properties, compared to films dried under ambient conditions. However, film gloss was significantly increased as a result of microwave drying, and drying time was shortened to 5 min.

2.6.2   Cross-Linking

2.6.2.1   Heat Denaturation

Solvent-cast protein films, for which the solvent is water, are typically formed at room temperature and stabilized through electrostatic interactions, hydrogen bonding, and van der Waals forces among the protein chains. The protein film network may be improved through heat-denaturation, which improves the tensile and barrier properties of solvent casted films by induction of cross-linking between the protein chains. Disulfide bond formation, which occurs with heat-denaturation in protein-based films, is often used to modify film properties (Perez-Gago et al. 1999). Heat-denaturation, and the subsequent polymerization of protein chains, has been studied for two proteins commonly used to make edible films: whey protein and soy protein.

For whey protein, temperature increases exposure of a free thiol group, CYS121 of beta-lactoglobulin, due to alpha-helix shifts. Polymerization occurs through intermolecular disulfide bond exchange, provided that the temperature is held at > 60–65°C (Galani and Apenten 1999). However, polymerization is not the only chemical reaction involved in whey protein film network formation. Noncovalent aggregation also occurs through new hydrophobic, ionic and van der Waals interactions that occur between newly-exposed groups of the heat-denatured whey proteins. These interactions increase as pH is decreased toward the whey protein isoelectric point (Kinsella 1984; Kinsella and Whitehead 1989).

For soy protein films, thermal treatment is also important for the formation of a film matrix. As reported by Renkema and Vliet (2002), soy proteins (with the exception of beta-conglycinin) and purified glycinin participate in heat-induced gel formation, which aggregation is facilitated through formation of disulfide bonds to provide matrix cohesion. In a study by Okamoto (1978), films made from 11S soy protein were stronger than those made from the 7S fraction because of the intermolecular disulfide bonds formed between 11S protein chains.

2.6.2.2 Irradiation

There are other methods for inducing cross-linking of protein chains besides heat denaturation. Irradiation has been successfully used to cross-link casein proteins, as well as soy proteins (Lacroix et al. 2002). Water forms hydroxyl radicals when exposed to gamma-irradiation. Aromatic amino acids, such as phenylalanine and tyrosine, are more likely to react with the hydroxyl radicals than aliphatic amino acids (Sabato et al. 2001). A mechanism hypothesized to explain the radical polymerization process involves the formation of bi-tyrosine linkages between protein chains (Brault et al. 1997). When exposed to 32 kGy of energy, soy protein and soy protein-whey protein blends form stronger films than their non-treated counterparts. Brault et al. (1997) did find that caseinate films cross-linked through irradiation became very brittle and required a plasticizer to improve their properties.

2.6.2.3 Enzymatic and Chemical Cross-linking

Cross-linking of proteins has been induced by both chemical and enzymatic means. Formaldehyde, glutaraldehyde, and lactic acid have been used to cross-link whey proteins through lysine residues (Fig. 2.7). However, the cross-linked products are no longer edible, due to the toxicity of the cross-linking agents (Galietta et al. 1998). More recently, genipin, a small molecule with no cytotoxicity issues, has been investigated as a protein cross-linking agent. In a study by Bigi et al. (2002), gelatin films were cross-linked using genipin. Film elongation decreased drastically and stiffness increased as a result of genipin cross-linking. For 0%, 1%, and 2% genipin addition levels, elongation was 211%, 16% and 13%, respectively, while elastic modulus values were 1 MPa, 5.6 MPa and 6.8 MPa, respectively. Also, genipin cross-linked films did not swell as much as untreated films when exposed to aqueous solutions.

Transglutaminase is a food grade enzyme that uses the acyl-transferase mechanism to link the gamma-carboxyamide (acyl donor) of a glutamine residue to the gamma-amine (acyl acceptor) of lysine residues along protein chains (Mahmoud and Savello 1992). This enzyme is known to improve elasticity in foods. Originally, transglutaminase was extracted from Guinea pig liver, making it expensive and cost-prohibitive for large-scale production (Zhu et al. 1995). Transglutaminase is now posible to be obtained at a lower cost from microbial sources (Chambi and Grosso 2006), and has been used in both homogenous protein systems and mixtures of proteins to affect tensile and permeability properties. Whey protein, casein, soy, egg albumin and wheat gluten have all been investigated for treatment with this enzyme. The molecular

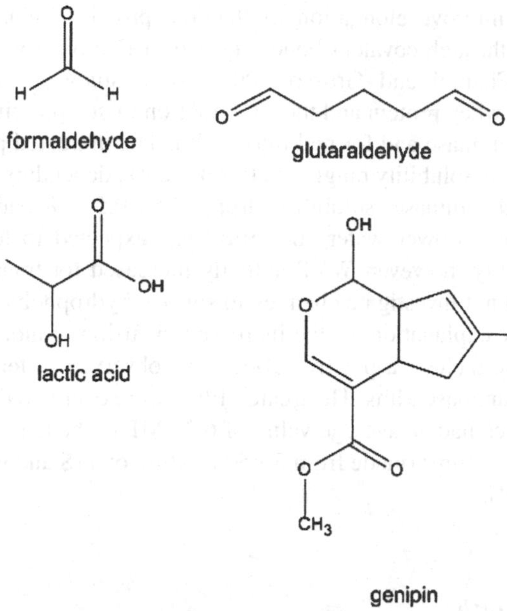

Fig. 2.7 Chemical agents for protein chain cross-linking

weight of alpha-lactalbumin, beta-lactoglobulin and alpha-lactalbumin/beta-lactoglobulin mixtures was shown to increase after transglutaminase treatment, indicative of cross-linking. Moreover, transglutaminase-treated proteins were also more heat stable relative to untreated ones (Truong et al. 2004). The cross-linked protein networks were less soluble, which may improve the water vapor permeability properties of films formed from the cross-linked proteins. Egg white proteins were effectively cross-linked using transglutaminase (Lim et al. 1998), as marked by an increase in protein molecular weight determined by SDS-PAGE. Cross-linking was enhanced as the incubation temperature increased, as more of the proteins were unfolded, exposing more amino acids to the transglutaminase enzyme. Elongation of the transglutaminase-treated egg white films was 50–100%, depending on the glycerol content. Tensile strength reached a maximum of nearly 10 MPa, depending also on the plasticizer content and relative humidity conditions.

Blends of proteins treated with transglutaminase have also been investigated. Chambi and Grosso (2006) made mixtures of gelatin and casein in an attempt to leverage the functionality of both proteins to create an improved film system. According to Howell (1995), three main phenomena can happen when proteins are blended (1) synergistic interaction, (2) precipitation, and (3) phase separation. When treated with transglutaminase, gelatin and casein, in a 1:3 ratio by mass, they had a synergistic effect on film elongation. Elongation increased to nearly 55% in the enzyme-treated mixture, compared to 14–17% elongation for pure individual casein and gelatin enzyme-treated film systems.

Cross-linking can improve elongation in that the protein chains of the film matrix are associated through covalent bonds instead of relatively weaker van der Waals interactions (Chambi and Grosso 2006). In a study by Yildirim and Hettiarachchy (1998), whey protein and the 11S fraction of soy protein were cross-linked using transglutaminase and formed into a film. The untreated protein in the control films had a water-solubility ranging between 325%, depending on pH. After treatment with transglutaminase, solubility dropped to 0.5–6% under the same experimental conditions. Lower water solubility was expected to lead to lower water vapor permeability; however, WVP actually increased for transglutaminase films. The authors did not investigate changes to surface hydrophobicity or microstructure for a possible explanation for the increased diffusion of moisture through the treated-films. Nevertheless, a positive effect was observed in tensile strength values of the transglutaminase films. The treated films were over two times stronger than control films which had an average value of 6.26 MPa. The tensile strength of transglutaminase-treated films (made from 50:50 mixture of 11S and whey protein) increased to 17.86 MPa.

2.6.3 Ionic Strength

The properties of the protein-based films can be affected by the solubility of the protein in a given solvent system. For example, increasing the water solubility of a protein can improve the appearance and increase the elongation of protein films. Similarly, the moisture barrier properties of a film can be improved by decreasing the solubility of the protein in water. Changing the ionic strength of the solution from which a film is cast represents one of the many ways to affect the solubility of the final film product. In a study by Qi et al. (2006), the solubility of various caseins was analyzed as a function of salt (NaCl) concentration. CO_2-precipitated casein (CO_2CN) and calcium caseinate were both studied (Tomasula et al. 1995). The results showed that the solubility of CO_2CN increased significantly as a function of salt concentration; however, the solubility of calcium caseinate remained essentially unchanged. Based on transmission electron microscope (TEM) investigations, CO_2CN protein aggregates were observed to disintegrate and became much smaller and of uniform size at elevated salt conditions, suggesting that electrostatic interactions among caseins brought about by pressurized CO_2 processing might be a dominating force in the formation of these aggregates.

2.6.4 Particle Size

Changes in particle size in relation to either the protein matrix or additives to the film system can affect the properties of the film. Non-soluble particles embedded within a film matrix can affect the permeability of gas through a film. Permeability is affected by the solubility of the diffusing compound in the film and by the diffusion coefficient (Rogers 1985). Changing the path of travel for a molecule of gas diffusing through the film can change the rate of diffusion. Thus, in a film matrix, insoluble particles create a hindrance to diffusing molecules migrating through the film. For a given mass of an insoluble additive, such as fat, a more tortuous diffusion path can be created by making the particles smaller, thereby lengthening the migration time and improving the barrier properties of the film (Pérez-Gago and Krochta 2001).

In a study by Dangaran et al. (2006), the tensile and water vapor properties of casein-based films were affected by decreasing the particle size of insoluble CO_2CN particles. Films made from CO_2CN, which are only partially soluble in water, are opaque and hazy. Dangaran et al. (2006) found that the use of high shear could reduce the particle size of the insoluble portion of CO_2CN. The effect of protein particle size reduction on tensile properties, water vapor permeability, and gloss was studied using the ASTM methodology. As the particle size of the CO_2CN was reduced from 126 to 111 μm, the tensile strength and elastic modulus of the films increased, while water vapor permeability decreased. Comparing films made from proteins of the same particle size reduction, gloss increased, on an average, from 55.3 to 73.0 gloss units, but all films were still hazy. With a particle size less than 86 μm, CO_2CN films were glossy and transparent; however, tensile strength decreased and WVP increased.

2.7 Analytical Tools for Determining Protein Structure

Changes in protein structure, due to processing conditions, can be measured and characterized with appropriate instrumentation. Understanding the fundamental changes in the molecular structures of the proteins can be advantageous to improving the functional properties of the protein-based films for various applications. Modern chromatographic, spectroscopic, and microscopic techniques can be used for analyzing the protein film-making process, from the solutions (often concentrated) used to make the films to the actual films themselves. Some commonly used spectroscopic techniques, such as circular dichroism (CD) and Fourier Transform Infrared (FTIR) spectroscopy (Curley et al. 1998; Farrell et al. 2002), used to study changes at the secondary structure level, can also be applied to study protein conformational changes within films. However, one technical requirement of CD is that the film samples must be transparent. After processing treatments, it is possible for protein films to darken in color, due to Maillard browning, or to become hazy, making CD analysis problematic. FTIR, on the other hand, is highly versatile and can be used for samples of various states, including solutions, powders and films. Changes in the hydrophobicity of protein films can be investigated using fluorescence spectroscopy and water-solubility analysis. The Bradford Protein Assay is a commonly-used specrophotometric method that quantitatively determines the concentration of protein in a solution. Tryptophan residues in protein molecules are frequently used as indicators of their environment (polar or non-polar) by yielding characteristic fluorescence spectra. By following changes in their fluorescence behavior, changes in the local environment of these tryptophan residues can be determined (Liu et al. 2005). Changes in complex protein–protein interactions, which may be attributed to denaturation and aggregation, can be studied using high-resolution microscopic imaging techniques, including atomic force microscopy (AFM), scanning electron microscopy (SEM), TEM, and confocal laser scanning microscopy. These changes will presumably cause differences in the observed film properties.

2.8   Function and Applications of Protein-Based Films

Based on the required tensile, barrier and appearance characteristics and properties, protein-based edible films for specific applications can be designed. Table 2.2 provides an extensive list of various protein-based films, including gas barrier, water vapor barrier and tensile properties, as well as potential applications. At the current state of edible film technology, the best application potential lies in protective coatings for foods. When applied to the surfaces of foods as a coating, protein-based edible films can protect food from chemical or microbial damage, thus lengthening product shelf life and maintaining high product quality.

As evident in Table 2.1, most protein-based films provide excellent barriers to oxygen. This characteristic has been utilized to effectively protect high-fat foods, which are known to form rancid off-flavors, due to oxidation. In two studies, performed on peanuts (Lee and Krochta 2002; Lee et al. 2002a–c), the shelf life of peanuts could be extended to 273 days by coating them with whey protein, compared to only 136 days for uncoated samples. The development of hexanal, an indicator of lipid oxidation, was prolonged for 330 days when the peanuts were coated with a whey protein film that contained vitamin E, an antioxidant.

As barriers to mass transfer of carbon dioxide, protein-based coatings have been found to increase the shelf life of eggs and fresh-cut fruits. Egg quality drops over time as carbon dioxide migrates out of the shell, causing changes to the internal pH. These pH changes can deleteriously affect color and yolk quality. A study by Caner (2005) showed that the shelf life of whey protein-coated Grade A eggs could be extended one week longer than uncoated eggs, when stored under ambient laboratory conditions. In the case of fresh-cut produce, browning and ripening times are major factors determining product shelf life. Protein-based edible coatings that have moderate oxygen, carbon dioxide and water vapor permeability can be applied to the surfaces of fresh-cut produce to extend product shelf life by delaying ripening, inhibiting enzymatic browning, reducing water loss, and minimizing aroma loss (Olivas and Barbosa-Canovas 2005). In a study by Le Tien et al. (2001), whey protein coatings significantly delayed browning in fresh-cut Macintosh apples and Russet potatoes.

Edible films used to protect fresh-cut produce have advantages over modified atmosphere packaging or treating the cut surfaces with ascorbic acid (which inhibits

Table 2.2 Barrier and tensile properties of solvent-casted protein films

Protein	Plasticizer	OP	WVP	Tensile strength (MPa)	Elastic modulus (MPa)	Percent elongation (%)	Comments	References
Native WPI	Glycerol (30%)	–	5.06	3.1	100	7	By crosslinking WPI films through heat denaturation, mechanical properties were improved. Improvements in mechanical properties expand the application potential of edible films	Perez-Gago and Krochta (1999)
Denatured WPI	Glycerol (30%)	–	4.96	6.9	199	41	Application Possibility: Pouches for dry ingredients or coatings that add mechanical strength to brittle products such as corn flakes	Perez-Gago and Krochta (1999)
Denatured WPI	Glycerol (15%)	18.5	2.37	29	1,100	4	Increasing plasticizer content increased elongation but increased permeability. Permeability can be	McHugh and Krochta (1994)
Denatured WPI	Glycerol (30%)	76.1	4.96	18	625	5	improved by incorporating a solid plasticizer sorbitol. Application	McHugh and Krochta (1994)
Denatured WPI	Sorbitol (30%)	4.3	–	14	1,040	3	Possibility: The oxygen permeability is 1–2 orders of magnitude better than PE and rivals Nylon. Protein-based edible films with	McHugh and Krochta (1994)
Calcium Caseinate (CaCN)	Sorbitol (30%)	1.3	–	12.8	356	56	the correct plasticizer can be used to coat high fat foods such as nuts and coffee beans to protect them from oxidative rancidity	Dangaran and Tomasula (2007)
Sodium Caseinate	Sorbitol (30%)	6.1	–	7.6	246	27		Dangaran and Tomasula (2007)

browning). Modified atmosphere packaging can slow oxidation; but if oxygen levels are reduced to a very low level, anaerobic conditions could be created, creating the risk of anaerobic bacterial growth. Effective levels of ascorbic acid may be so high that flavor is negatively affected (Perez-Gago et al. 2003).

While edible coatings that hinder mass transfer of oxygen, carbon dioxide, water vapor, or aroma compounds represent the majority of applications investigated for protein-based edible coatings, they also have the potential to improve appearance by adding gloss (Lee et al. 2002a–c) or by protecting food from microbial spoilage if the film contains an active antimicrobial. Such applications have been investigated for protecting meat products, as well as fruits (Vachon et al. 2003; Cagri et al. 2004; Min et al. 2006a, b).

2.9   Future Trends

Much progress has been accomplished in the research and development of edible films and coatings based on agricultural proteins. Table 2.3 provides a brief list of suppliers of agricultural proteins for edible films research, as well as equipment sources. Based on what is currently known about their properties, protein-based films and coatings are on the brink of broad application in both the food and non-food sectors. Concerning commercialization, the best potential for success is for food coatings that protect against chemical and biological damage, thus lengthening product shelf life. Protein-based solutions intended for film-making can also be used in commercial equipment used for coating. Protein-based coatings are good gas barriers (especially to oxygen) and can act as vehicles for active compounds like antimicrobials or nutrients. Moreover, they could be combined with thermal or non-thermal processing to further increase product shelf life. These concepts have been successfully proven in the laboratory setting (Ouattara et al. 2002).

Both edible and non-edible protein-based films have the potential to function in "green" packaging applications. The oxygen permeability (OP) properties of casein films are superior to those of LDPE. Tests in our laboratory have demonstrated that coating an LDPE with a casein film reduces its oxygen permeability from 1,560 to 86 cc μm/m^2 d kPa. With the growing trend for increased convenience and individualized portion size, packaging per unit has also increased (Marsh and Bugusu 2007). As companies like Wal-Mart and McDonald's begin to seek biobased materials to fulfill their packaging needs, more biobased products with the functionality of petroleum-based packaging are likely to be produced.

Protein-based films can be replacements for synthetic gas barriers, but their mechanical properties still need improvement. Nanotechnology is a viable route to improve both tensile and barrier properties. Nanoparticles and variations at the nanometer scale affect electronic and atomic interactions without changing the chemistry (Lagaron et al. 2005). The water vapor permeability (WVP) of wheat gluten

Table 2.3 Ingredient suppliers and equipment sources

Ingredient/Equipment	Supplier	Address
Whey protein	Davisco	11000 West 78th St Suite 210 Eden Prairie, MN 55344 phone 800-757-7611
Whey protein	Proliant, Inc. (USA)	2425 SE Oak Tree Court Ankeny, IA 50021 USA Phone: 800-369-2672 Fax: 515-289-5110
Whey protein, casein/caseinates	Fonterra (USA)	Sales Office 100 Corporate Center Drive Suite 101 Camp Hill PA 17011-1758 Ph + 1 717 920 4000
Casein	American Casein Company	109 Elbow Lane Burlington, New Jersey 08016-4123 USA Incorporated: 1956 Telephone: 609-387-3130 FAX: 609-387-7204 Sales Email: Sales@109elbow.com
Wheat protein	Tate & Lyle	2200 East Eldorado Street Decatur, IL 62525 United States
Soy protein	The Solae Company	PO Box 88940 St. Louis, MO 63188 Local: (314)659-3000 Phone: (800)325-7108 Tel: 1 717 920 4000
Soy protein	Cargill	Cargill Health & Food Technologies P.O. Box 9300 MS 110 Minneapolis, MN 55440-9300 E-mail us at hft@cargill.com (866) 456-8872
Soy protein	ADM	Soy Proteins 217-451-5141 866-545-8200 4666 Faries Parkway Decatur, IL 62526 Email: Sales@109elbow.com
Laboratory film equipment	Werner Mathis USA Inc	2260 HWY 49 N.E. P.O.Box 1626 CONCORD N.C. 28026 Tel: + 1 7047 866 157 Fax: + 1 7047 866 159 E-Mail: usa@mathisag.com
Permeability testing equipment	Mocon, Inc.	7500 Boone Ave. N. Minneapolis. MN 55428 USA Phone: (763) 493-6370 Fax: (763) 493-6358

This is just a partial list and does not indicate endorsement of these companies by the USDA

films was improved by including montmorillonite clay nanoparticles, while tensile strength and elastic modulus were also increased (Olabarrieta et al. 2006).

Edible films and coatings from proteins (as well as other biobased materials) fit well into the Environmental Protection Agency's (EPA) suggested plan for improved municipal waste management and reduction. As stated in its summary on municipal solid waste in the US (2006), recycling has increased from 10% in 1980 to 32% in 2005. For further and continued improvements of U.S. waste management practices, the EPA suggests reducing the initial amount of packaging, and promotes composting. For reduction, the EPA suggests designing packaging systems that reduce the amount of environmentally toxic materials used in packaging to make it easier to reuse or compost them. They also suggest packaging that reduces the amount of damage or spoilage to food products. Protein-based edible films and coatings fit both criteria. They can reduce the complexity of packaging systems, making them either easier to recycle or compostable. Edible coatings can also protect food from chemical or biological damage, thereby lengthening product shelf life.

References

Aimutis WR (2004) Bioactive properties of milk proteins with particular focus on anticariogenesis. J Nutr 134: 989S–995S

Alcantara CR, Rumsey TR, Krochta JM (1998) Drying rate effect on the properties of whey protein films. J Food Proc Eng 21: 387–405

Anonymous (2006) Bioplastics demand experiencing boom in Europe. Food Navigator.com

Audic JL, Chaufer B, Daufin G (2003) Non-food applications of milk components and dairy co-products: A review. Lair 83: 417–438

Avena-Bustillos RJ, Olsen CW, Olson DA, Chiou B, Yee E, Bechtel PJ, McHugh TH (2006) Water vapor permeability of mammalian and fish gelatin films. J Food Sci 71(4): E202–E207

Badii F, Howell NK (2006) Fish gelatin: structure, gelling properties and interaction with egg albumen proteins. Food Hydrocolloids 20: 630–640

Balian G, Bowes J (1977) The structure and properties of collagen. In: Ward AG, Courts A (eds) The Science and Technology of Gelatin, London, Academic Press, pp 1–31

Barone JR, Schmidt WF (2006) Nonfood applications of proteinaceous renewable materials. J Chem Education 83(7): 1003–1009

Barone JR, Schmidt WF, Liebner CFE (2005) Thermally process keratin films. J Applied Polymer Sci 97: 1644–1651.

Bigi A, Cojazzi G, Panzavolta S, Roveri N, Rubini K (2002) Stabilization of gelatin films by crosslinking with genipin. Biomaterials 23: 4827–4832

Brault D, D''Aprano G, Lacroix M (1997) Formation of free-standing sterilized edible films from irradiated caseinates. J Agric Food Chem 45: 2964–2969

Brooks M (2006) Forgotten fibres: issues in the collecting and conservation of regenerated protein fibres. In: Rogerson C, Garside P (eds) The Future of the Twentieth Century: Collecting, Interpreting and Conserving Modern Materials, London, Archetype, pp 33–40

Brother GH (1940) Casein plastics. Ind Eng Chem January 32(1): 31–34

Cagri A, Ustunol Z, Ryser ET (2004) Antimicrobial edible films and coatings. J Food Protection 67(4): 833–848

Caner C (2005) Whey protein isolate coating and concentration effects on egg shelf life. J Sci Food and Agric 85: 2143–2148

Chambi H, Grosso C (2006) Edible films produced with gelatin and casein cross-linked with transglutaminase. Food Research International 39: 458–466

Chang YP, Cheah PB, Seow CC (2000) Plasticizing-antiplasticizing effects of water on physical properties of tapioca starch films in the glassy state. J Food Sci 65(3): 445–451

Cho SY, Rhee C (2004) Mechanical properties and water vapor permeability of edible films made from fractionated soy proteins with ultrafiltration. Lebenem Wiss u Technol 37: 833–839

Comstock KD, Farrell D, Godwin C, Xi Y (2004) From hydrocarbons to carbohydrates: Food packaging for the future. Journal name volume pages ?Retrieved on January 30, 2006, from the University of Washington Environmental Management Certificate Program website: http://depts.washington.edu/poeweb/gradprograms/envmgt/index.html

Curley DM, Unruh JJ, Farrell HM (1998) Changes in secondary structure of bovine casein by FTIR spectroscopy. J Dairy Sci 81: 3154–3162

Damodaran S (1996) Amino acids, peptides and proteins. In: Fennema O (ed) Food Chemistry. New York, NY, Marcel Dekker, pp 321–430

Dangaran KL, Krochta JM (2003) Aqueous whey protein coatings for panned products. Manufacturing Confectioner 83(1): 61–65

Dangaran KL, Krochta JM (2007) Preventing the loss of tensile, barrier and appearance properties caused by plasticizer crystallization in whey protein films. Internation J Food Sci Technol 42(9): 1094–1100

Dangaran KL , Tomasula PM (2006) Empirical modeling (in preparation)

Dangaran KL, Cooke P, Tomasula PM (2006) The effect of protein particle size reduction on the physical properties of CO_2-precipitated casein films. J Food Sci 71(4): E196–E201

Debeaufort F, Quezada-Gallo JA, Voilley A (1998) Edible films and coatings: Tomorrow's packagings: A review. Crit Reviews Food Sci and Nutr 38(4): 299–313

deWit JN, Klarenbeek G (1983) Effects of various heat treatments on structure and solubility of whey proteins. Journal of Dairy Science 67: 2701–2710

EPA (2006) Municipal Solid Waste in the United States: 2005 Facts and Figures (Executive Summary)

Etzel MR (2004) Manufacture and use of dairy protein fractions. J Nutrition 134: 996S–1002S

Farrell HM, Qi PX, Wickham ED, Unruh JJ (2002) Secondary structural studies of bovine caseins: structure and temperature dependence of beta-casein phosophopeptide (1-25). J Protein Chem 21: 307–321

Fennema O, Kester JJ (1991) Resistance of lipid films to transmission of water vapor and oxygen. In: Levine H, Slade L (eds) Water Relationships in Food., New York, NY, Springer, pp 703–719

Finlay MR (2004) Old efforts at news uses: a brief history of the chemurgy movement and the American search for biobased materials. J Industrial Ecol.ogy 7: 33–46

FitzGerald RJ, Murray BA, Walsh DJ (2004). Hypotensive peptides from milk proteins. The Journal of Nutrition 134: 980S–988S

Frinnault A, Gallant DJ, Bouchet B, Dumont JP (1997) Preparation of casein films by a modified wet spinning process. J Food Sci 62(4): 744–747

Galani D, Apenten RKO (1999) Heat-induced denaturation and aggregation of beta-lactoglobulin: kinetics of formation of hydrophobic and disulphide-linked aggregates. International J Food Sci Technol 34: 467–476.

Galietta G, Gioia LD, Guilbert S, BCuq B (1998) Mechanical and thermomechanical properties of films based on whey proteins as affected by plasticizer and crosslinking agents. J Dairy Sci 81: 3123–3130

Gennadios A, McHugh TH, Weller CL, Krochta JM (1994) Edible coatings and films based on proteins. In: Krochta JM, Baldwin EA, Nisperos-Carriedo MO (eds) Edible Coatings and Films to Improve Food Quality., Lancaster, PA, Technomic, pp 201–278

Gettens RJ, Stout GL (1984) Painting Materials: A Short Encyclopedia. Courier Dover Publications, New York, NY

Haard NF, Chism GW (1996) Characteristics of edible plant tissue. In: Fennema O (ed) Food Chemistry. Marcel Dekker, New York, NY, p 1067

Haug IJ, Draget KI, Smidsrod O (2004) Physical and rheological properties of fish gelatin compared to mamalian gelatin. Food Hydrocolloids 18: 203–213

Hernandez VM (2007) Thermal properties, extrusion and heat-sealing of whey protein edible films. Ph.D. thesisUniversity of California, Davis

Hernandez VM, Reid DS, McHugh TH, Berrios J, Olson D, Krochta JM (2005) Thermal transitions and extrusion of glycerol-plasticized whey protein mixtures. IFT Annual Meeting and Food Expo, New Orleans, LA

Hernandez VM, McHugh TH, Berrios J, Olson D, Pan J, Krochta JM (2006) Glycerol content effect on the tensile properties of whey protein sheets formed by twin-screw extrusion. IFT Annual Meeting and Food Expo, Orlando, FL

Hochstetter A, Talja RA, Helen HJ, Hyvonen L, Jouppila K (2006) Properties of gluten-based sheet produced by twin-screw extruder. Lebenam Wiss u Technol 39: 893–901

Howell NK (1995) Synergism and interactions in mixed protein systems. In: Harding E, Hill SE, Mitchell JR (eds) Biopolymer Mixtures. Nottingham University Press, Nottingham pp 329–347

Kaya S, Kaya A (2000) Microwave drying effects on properties of whey protein isolate edible films. J Food Eng 43: 91–96

Kelley HW (1993) Always something new. USDA Miscellaneous Publication Number 1507

Kinsella JE (1982) Relationships between structure and functional properties of food proteins. In: Fon and Codon ed. Food Proteins, Springer 51–104

Kinsella JE (1984) Milk proteins: Physicochemical and functional properties. Crit Reviews Food Sci and Nutr 21(3): 197–262

Kinsella JE, Whitehead DM (1989) Proteins in whey: chemical, physical, and functional properties. Advances in Food and Nutrition Research 33: 343–427

Kinsella JE, Whitehead DM, Brady J, Bringe NA (1989) Milk Proteins: Possible relationships of structure and funtion. In PF Fon ed. Developments in Dairy Chemistry, New York, Elsevier 55–96 .

Kiplinger J (2003) Meet the Azlons from A to Z: Regenerated & rejuvenated. Fabrics.net January/ February

Kock J (2006) Physical and Mechanical Properties of Chicken Feather Materials. M.S. thesis, Georgia Tech p 114

Kozempel M, Tomasula PM (2004) Development of a continuous process to make casein films. J Agric Food Chem 52(5): 1190–1195

Kozempel M, McAloon AJ, Tomasula PM (2003) Drying kinetics of calcium caseinate. Journal of Agricultural and Food Chemistry 51(3): 773–776

Krochta JM, Baldwin EA, Nisperos-Carriedo MO (1994) Edible Coatings and Films to Improve Food Quality. CRC Press LLC, Boca Raton, FL

Kunte LA, Gennadios A, Cuppett SL, Hanna MA, Weller CL (1997) Cast films from soy protein isolates and fractions. Cereal Chem 74(2): 115–118

Lacroix M, Le TC, Ouattara B, Yu H, Letendre M, Sabato SF, Mateescu MA, Patterson G (2002) Use of gamma-irradiation to produce films from whey, casein, and soya proteins: structure and functionals characteristics. Radiation Physics and Chemistry 63: 827–832

Lagaron JM, Cabedo L, Cava D, Feijoo JL, Gavara R, and Gimenez E (2005) Improving packaged food quality and safety Part 2: Nanocomposites. Food Additives and Contaminants 22(10): 994–998.

Larkins BA, Lending CR, Wallace JC (1993) Modification of maize-seed-protein quality. American Journal of Clinical Nutrition. 58: 264S–269S.

Le Tien C, Vachon C, Mateescu MA, Lacroix M (2001) Milk protein coatings prevent oxidative browning of apples and potatoes. J Food Sci 66(4): 512–516.

Lee SY, Dangaran KL, Krochta JM (2002a) Gloss stability of whey-protein/plasticizer coating formulations on chocolate surface. J Food Sci 67(3): 1121–1125

Lee SY, Dangaran KL, Guinard JX, Krochta JM (2002b) Consumer acceptance of whey-protein-coated as compared with shellac-coated chocolate. J Food Sci 67(7): 2764–2769

Lee SY, Krochta JM (2002) Accelerated shelf-life testing of whey-protein-coated peanuts analyzed by static headspace gas chromatography. J Agric Food Chem 50(7): 2022–2028

Lee SY, Trezza TA, Guinard JX, Krochta JM (2002c) Whey-protein-coated peanuts assessed by sensory evaluation and static headspace gas chromatography. J Food Sci 67(3): 1212

Lin SY, Krochta JM (2006) Whey protein coating efficiency on mechanically roughened hydrophobic peanut surfaces. J Food Science 71(6): E270–E275

Lim LT, Mine Y, Tung MA (1998) Transglutaminase cross-linked egg white protein films: tensile properties and oxygen permeability. J Agric Food Chem 46: 4022–4029.

Liu X, Powers JR, Swanson BG, Hill H, Clark S (2005) Modification of whey protein concentration hydrophobicity by high hydrostatic pressure. Innovative Food Science and Emerging Technology 6: 310–317

Mahmoud R, Savello P (1992) Mechanical properties of and water vapor transferability through whey protein films. J Dairy Science 75: 942–946

Marsh K, Bugusu B (2007) Food packaging – roles, materials, and environmental issues. J Food Science 72(3): R39–R55

McHugh TH, Krochta JM (1994) Plasticized whey protein edible films: water vapor permeability properties. J Agric Food Chem 59(2): 416–419

Min S, Harris LJ, Krochta JM (2006a) Inhibition of Salmonella enterica and Escherichia coli O157:H7 on roasted turkey by edible whey protein coatings incorporating lactoperoxidase system. J of Food Protection 69(4): 784–793

Min S, Rumsey TR, Krochta JM (2006b) Lysozyme diffusion in smoked salmon coated with whey protein films incorporating lysozyme. IFT Annual Meeting and Food Expo, Orlando, FL

Mine Y (1995) Recent advances in the understanding of egg white protein functionality. Trends in Food Sci Technol 6: 225–232

Mine Y, Noutomi T, Haga N (1990) Thermal induced changes in egg white proteins. J Agric Food Chem 38: 2122–2125

Minifie B (1989) Chocolate, Cocoa and Confectionery: Science and Technology. Springer Publishing, New York, NY

Morara CI, Lee TC, Karwe MV, Kokini JL (2002) Plasticizing and antiplasticizing effects of water and polyols on a meat-starch extruded matrix. J Food Science 67(9): 3396–3401

Morr CV, Ha EYW (1993) Whey protein concentrates and isolates: processing and functional properties. Crit Reviews Food Sci and Nutr 33(6): 431–476

Obuz E, Herald TJ, Rausch K (2001) Characterization of extruded plant protein and petroleum-based packaging sheets. Cereal Chemistry 78(1): 97–100

Okamoto S (1978) Factors affecting protein film formation. Cereal Foods World 23: 256–262

Olabarrieta I, Gallstedt M, Ispizua I, Sarasua JR, Hedenqvist MS (2006) Properties of aged montmorrillonite-wheat gluten composite films. J Agric Food Chem 54: 1283–1288

Olivas GI, Barbosa-Canovas GV (2005) Edible coatings for fresh-cut fruits. Crit Reviews Food Sci and Nutr 45: 657–670

Ouattara BS, Sabato SF, Lacroix M (2002) Use of gamma-irradiation technology in combination with edible coating to produce shelf-stable foods. Radiation Physics and Chemistry 63: 305–310

Panyam D, Kilara A (1996) Enhancing the functionality of food proteins by enzymatic modification. Trends in Food Sci Tech 7: 120–125

Perez-Gago MB, Krochta JM (2000) Drying temperature effect on water vapor permeability and mechanical properties of whey protein–lipid emulsion films. J Agric Food Chem 48(7): 2687–2692

Perez-Gago MB, Krochta JM (2001) Lipid particle size effect on water vapor permeability and mechanical properties of whey protein/beeswax emulsion films. J of Agric Food Chem 49(2): 996

Perez-Gago MB, Nadaud P, Krochta JM (1999) Water vapor permeability, solubility, and tensile properties of heat-denatured versus native whey protein films. J Food Sci 64(6): 1034–1037

Perez-Gago MB, Serra M, Alonso M, Mateos M, Del Rio MA (2003) Effect of solid content and lipid content of whey protein isolate-beeswax edible coatings on color change of fresh-cup apples. J Food Sci 68(7): 2186–2191

Petruccelli S, Anon MC (1994) Relationship between the method of obtention and the structural and functional properties of soy protein isolates 1. Structural and hydration properties J Agric Food Chem 42: 2161–2169

Qi PX, Cooke P, Dangaran KL, Tomasula PM (2006) Characterization of casein films made by pressurized carbon dioxide: salt effect on water solubility. 97th AOCS Ann Meeting & Expo

Rampon V, Robert P, Nicolas N, Dufour E (1999) Protein structure and network orientation in edible films prepared by spinning process. J Food Sci 64(2): 313–316

Renkema JM, Vliet TV (2002) Heat-induced gel formation of soy proteins at neutral pH. J Agric Food Chem 50: 1569–1573

Rogers CE (1985) Permeation of gases and vapours in polymers. In: Comyn J (ed) Polymer Permeability. J. Comyn. Kluwer Academic Publishers, London, 11–74

Sabato SF, Ouattara FB, Yu H, D"Aprano G, Le Tien C, Manteescu MA, Lacroix M (2001) Mechanical and barrier properties of cross-linked soy and whey protein based films. J Agric Food Chem 49: 1397–1403

Sawyer L, Kontopidis G, Wu SY (1999) Beta-lactoglobulin – a three-dimensional perspective. Internation J Food Sci Tech 34: 409–418

Schmidt WF (1998) Innovative feather utilization strategies. 1998 National Poultry Waster Management Symposium, Springdale, AR

Shukla R, CHheryan M (2001) Zein: the industrial protein from corn. Industrial Crops and Products 13: 171–192

Sonneveld K (2000) What drives (food) packaging innovation. Packaging Technol Sci 13: 29–35

Sothornvit R, Krochta JM (2000) Plasticizer effect on oxygen permeability of beta-lactoglobulin films. J Agric Food Chem 48: 6298–6302

Sothornvit R, Krochta JM (2001) Plasticizer effect on mechanical properties of beta-lactoglobulin films. J Food Eng 50(3): 149–155

Subirade M, Kelly I, Gueguen J, Pezolet M (1998) Molecular basis of film formation from a soybean protein: Comparison between the conformation of glycinin in aqueous solution and in films. International J Biol Macro 23: 241–249

Swaisgood HE (1993) Review and update of casein chemistry. J Dairy Sci 76: 3054–3061

Swaisgood HE (1996) Characteristics of milk. In: Fennema O (ed) Food chemistry. Marcel Dekker, New York, NY, p 1067

Ternes W (2001) Egg proteins. In: Sikorski Z (ed) Chemical and functional properties of food proteins. CRC Press, Boca Raton, FL, CRC Press, p 415

Tomasula PM, Craig JC, Boswell RT, Cook RD, Kurantz MJ, Maxwell M (1995) Preparation of casein using carbon dioxide. J Dairy Sci 78: 506–514

Truong VD, Clare DA, Catignani GL, Swaisgood HE (2004) Cross-linking and rheological changes of whey proteins treated with microbial transglutaminase. J Agric Food Chem 52: 1170–1176

Vachon C, D''Aprano G, Lacroix M, Letendre M (2003) Effect of edible coating process and irradiation treatment of strawberry fragaria spp. on storage-keeping quality. J Food Sci 68(2): 608–611

Yildirim M, Hettiarachchy NS (1998) Properties of films produced by cross-linking whey proteins and 11S globulin using transglutaminase. J Food Sci 63(2): 248–252

Yoshino T, Isobe S, Maekawa T (2002) Influence of preparation conditions on the physical properties of zein films. J Am Oil Chem Soc 79(4): 345–349

Zhang J, Mugara P, Jane J (2001) Mechanical and thermal properties of extruded soy protein sheets. Polymer 42: 2569–2578

Zhu Y, Rinzema A, Tramper J, Bol J (1995) Microbial transglutaminase–a review of its production and application in food processing. Applied Micro Biotech 44: 277—282

Chapter 3
Structure and Function of Polysaccharide Gum-Based Edible Films and Coatings

Marceliano B. Nieto

3.1 Introduction

Polysaccharide gums are hydrocolloids of considerable molecular weight, and are water-soluble. They dissolve in and form intensive hydrogen bonds with water. Because of the size and configuration of their molecules, these polysaccharides have the ability to thicken and/or gel aqueous solutions as a result of both hydrogen bonding between polymer chains and intermolecular friction when subjected to shear. Gums dissolve in water through the formation of solvent–polymer hydrogen bonds; in solution, polymer molecules may arrange themselves into an ordered structure, called a micelle that is stabilized or fortified by intermolecular hydrogen bonds (Fig. 3.1). The micelle traps and immobilizes water and, depending on the extent of the intermolecular association, the water is either thickened, as measured by a parameter called viscosity, or converted into a gel that possesses both liquid- and solid-like characteristics or viscoelasticity.

When gum solutions are cast on a surface and dried, they leave a film that possesses specific plasticity, tensile strength, clarity, and solubility characteristics (Fig. 3.2). The formation of micelles confers upon gums their ability to form films because these structures are preserved during drying. Attributes of films made from various gums are influenced by the extent of intermolecular hydrogen bonding between polymer chains, arising from differences in gum molecular structures. Structural differences of gums that impact their properties include the presence or absence of branching, electrical charge, substitution (of sugar units), as well as molecular weight. By rule of thumb, linear, high molecular weight, non-ionic polysaccharide gums form strong films that will peel off in one piece at a dried thickness of 5 mils (0.005 in. or 127 µm). Agar and methyl cellulose are good examples of film-forming polysaccharides, whereas highly-branched polymers (e.g., gum arabic, gum ghatti, larch gum), either with or without anionic charge, form weak films that flake rather than peel when removed from the casting surface.

M.B. Nieto
TIC Gums, 4609 Richlynn Drive, Belcamp, MD 21017, USA
e-mail: mnieto@ticgums.com

M.E. Embuscado and K.C. Huber (eds.), *Edible Films and Coatings for Food Applications*, DOI 10.1007/978-0-387-92824-1_3, © Springer Science+Business Media, LLC 2009

Micelle

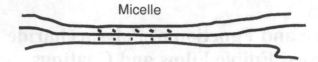

Showing hydrogen bonding with water
and intermolecular hydrogen bonding

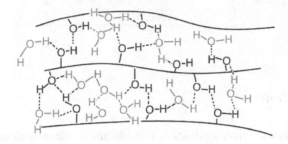

Fig. 3.1 Intermolecular hydrogen bonding formed during hydration of gums in water

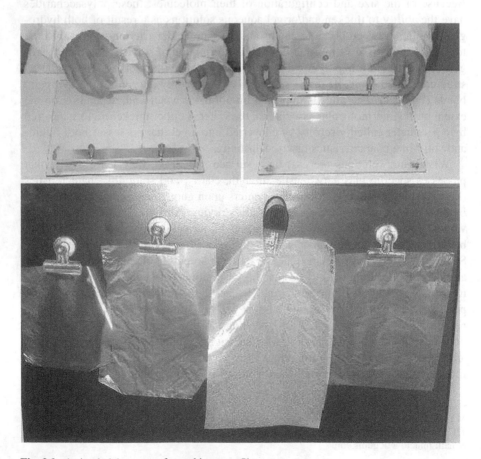

Fig. 3.2 A simple lab process for making gum films

Figure 3.2 shows a simple process of making films from polysaccahide gums. The gum is first dissolved in water at a suitable concentration to make it thick enough to cast, typically between 6,000 and 10,000 cP. Then, using a casting knife, a thin, uniform layer of slurry is drawn on a smooth surface such as a glass plate, a piece of stainless steel, or even a plastic sheet. The film is then dried to the proper moisture or humidity level, and peeled. Whether or not the film peels away from the casting plate depends largely on the type of gum polymer used and also on the thickness of the cast. Interestingly, the solubility of gum-based films takes after the solubility of the original gum polymer from which it is made. Gums that dissolve in cold water like methylcellulose, alginate, carboxymethylcellulose (CMC) and others will produce films that are likewise cold water-soluble. Gums that require heat for dissolution, like agar, will form films that are cold water insoluble. Those gums that are only partially soluble in cold water such as kappa- and iota-carrageenan, gellan and locust bean gum, will produce films with the same similar solubility.

Why and how these polysaccharides produce different types of film structures will be discussed in detail in this chapter. Hydration properties, solubilities and intermolecular chain associations of polysaccahide gums will be discussed in relation to their respective chemical structures to explain variations in film-forming properties.

3.2   Structure and Film Forming Properties of Polysaccharides

The unique means by which gums interact with water is the basis for their more technical name, hydrocolloids. Hydrocolloids are polymers of high average molecular weight, ranging from ~2,000 (e.g., inulin) to >2 million (e.g., xanthan, konjac, guar, β-glucan). Gum polymers are either linear consisting of one sugar monomer or linear consisting of a repeating dimer; linear a sugar side chain on the linear backbone; or branched consisting of a mixture of different sugars. In addition, polysaccharides can exhibit either a neutral charge (e.g., acetate esters, methyl ethers, other neutral sugars), negative charge (e.g., carboxylate, sulfate groups), or positive charge (e.g., amino groups) due to the presence various chemical groups attached to individual monosaccharide units. All of these structural features of polysaccharides contribute to their differences in solubility; synergy or incompatibility with each other or with other ingredients (e.g., proteins, minerals, acids and lipids); thickening, gelling, and emulsifying properties; and, more importantly, their film-forming properties. Table 3.1 summarizes several of these characteristics and properties for some polysaccharide gums.

3.2.1   Linear, Neutral Polysaccharides

Among the various gums, linear and neutral polysaccharides generally exhibit the best film-forming properties, and provide the greatest film strength. Because of

Table 3.1 Chemical and physical properties of gums polysaccharides for use in edible films and coatings

Gum	Viscosity, cP 1% at 25°C	Sugar composition	Other structural features	Ionic charge	Class	Film forming/other properties
Gum Arabic	2–5	Galactose, arabinose, rhamnose, glucuronic acid, Arabino-galactan-protein complex (AGP)	Highly branched	Anionic	Natural	Cold water soluble, dissolves up to 55%, weak non-cohesive film, form flakes
Karaya	400–1,200	Glacturonic aicd, galactose, rhamnose, glucuronic acid	Branched	Anionic	Natural	Cold water soluble, dissolves up to 5–8%, weak film non-cohesive film, form flakes
Ghatti	30–100	Galactose, arabinose, mannose, xylose, glucuronic acid	Branched	Anionic	Natural	Cold water soluble, dissolves up to ~5–8%, weak non-cohesive film, form flakes
Larch	2–4	Arabinose, galactose	Highly branched	Anionic	Natural	Cold water soluble, dissolves up to 60%, weak non-cohesive film, form flakes
ι-Carrageenan	1,000–2,000	L-galactose, 3,6-anhydrogalactose	Linear, sulfate substitution	Anionic	Natural	Thickens cold but requires ~65°C or higher to fully hydrate, dissolves up to 8%, moderately strong film

their linear and non-ionic natures, polymer chains can associate more intimately (through intermolecular hydrogen bonding) to produce stronger films. However, no matter the gum, the first requirement is to ensure that the gum is fully hydrated or dissolved in water, with or without heating. Once hydrated, polymer chains may arrange themselves in an ordered fashion to form intermolecular networks needed for film formation. Several examples of linear, neutral gums possessing appropriate structural features for film formation include agar, curdlan and cereal β-glucan. Other examples include methylcellulose, hydroxypropyl methylcellulose (HPMC), gellan, and konjac glucomannan, all of which have methyl, hydroxypropyl, acyl or acetyl substituents covalently attached to their individual sugar molecules. Though also linear and nonionic, cellulose and microcrystalline cellulose are insoluble in water, and require sodium hydroxide and sodium sulfide treatments at high temperatures and pressures to separate the cellulose fibers and partially hydrate the molecules for film structures to form. Cellulose and microcrystalline cellulose, however, can be used in conjunction with soluble gum polymers to alter the strength, melting characteristics, and solubility of the finished film.

3.2.1.1 Agar [E406; CAS#64-19-7; 21CFR 182.101; FEMA# 2006]

Agar is extracted from two major commercial sources of red seaweed: *Gelidium sp.* and *Gracilaria sp.* It consists of a mixture of agarose (gelling fraction) and agaropectin (non-gelling fraction), which is slightly branched and sulfated. In the commercial manufacture of food grade agar, most of the agaropectin is removed during processing; hence, commercial agars are mainly composed of the agarose fraction. Agarose is a linear polymer with a molecular weight of about 120,000, and is composed of a repeating dimer of D-galactosyl and 3,6-anhydro-L-galactosyl units connected via alternating α-(1→3) and β-(1→4) glycosidic linkages (Labropoulos et al. 2002). Its structural formula is depicted in Fig. 3.3.

Most applications for agar are based on its gelling ability and the fact that it is more stable to low pH and high temperature conditions compared to other gelling systems. Gelation results from formation of a network of agarose double helices, forming junction zones. These double helices are stabilized by the presence of water molecules bound inside the double helical cavity (Labropoulos et al. 2002). Exterior hydroxyl groups allow for aggregation of these double helices via intermolecular hydrogen bonds, upon cooling of agar solution. In this aggregation process suprafibers (consisting of up to 10,000 double helices) are formed, and contribute to the strength of agar films. Agar solution gels at a temperature range of 90–103°C; so, when making an agar film, the solution and casting surface need to be maintained above the agarose gel-setting temperature to avoid premature gelation.

Agar film is clear and strong, but is insoluble in water under ambient conditions, just like the parent agar. The linear, nonionic nature of agar allows hydrated molecules to associate more intimately, forming a network that is stabilized by intensive intermolecular hydrogen bonding during drying of the film. The dried film is strong enough to peel off the belt or casting surface in one piece when a 4% solution of agar

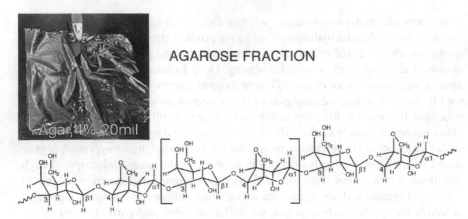

AGAROSE FRACTION

D-Galactose 3,6-anhydro-L-Galactose
DIMER

Fig. 3.3 Molecular structure of agarose and picture of its film, cast from a 4% solution and thickness of 20 mils (0.02 in. or 508 μm)

is cast at a thickness of 20 mils (0.02 in.). Tensile strength, across a 2.5 cm wide strip, and puncture strength values for this film were 2,540 g and 2,150 g of force, respectively. In comparison, films of many other gums would be too weak to peel off in one piece at this concentration and casting thickness, and would be required to be cast at twice the thickness possible for agarose films, or greater. However, it was not possible to produce a solid film with agar by casting its 4% solution at a thickness of 10 mils (0.01 in.), although its film at 20 mils was stronger than that of methylcellulose. In general, agar films are brittle and lack elasticity or stretch.

3.2.1.2 Methylcellulose [E461; CAS#99638-59-2; 21 CFR§182.1480; FEMA 2696]

Cellulose is the most abundant biomass resource on earth. It is the insoluble material found in plant cell walls of all fruits and vegetables, and is present in the leaves, trunks, and barks of higher plants as a structural component. Cellulose is the source of cotton used as clothing material, and also contributes significantly to the insoluble dietary fiber intake of the human diet.

Cellulose derived from pine trees and cotton linters is used as the base material for a number of highly functional water-soluble cellulose derivatives. These derivatives possess wide-ranging properties, which are dependent on the type and degree of derivatization. Chemical modification of cellulose introduces various chemical groups onto the hydroxyl groups of glucose units, posing steric hindrance to native intermolecular interactions of cellulose. Methyl, carboxymethyl, and hydroxypropyl cellulose ethers provide both nonionic and anionic modified derivatives that possess unique functional properties, such as thermal gelation and superior clarity.

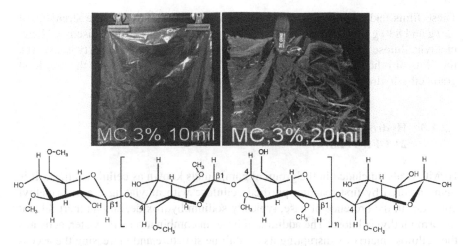

Fig. 3.4 Molecular structure of methylcellulose and picture of its film, cast from a 3% solution and thickness of 10 mils (0.01 in. or 254 μm) and 20 mils (0.02 in. or 508 μm)

Methylcellulose is a chemically-derived cellulose ether with a methyl degree of substitution of 1.6–1.9 on O-6, O-2 or O-3 positions of glucose units (Fig. 3.4). It is both linear and nonionic. The addition of methyl ether groups to the cellulose backbone opens up its structure, and because of steric hindrance, methylcellulose does not re-associate into microfibrils when dried, unlike its insoluble parent cellulose. This chemical substitution is what gives methylcellulose its solubility in cold water, by opening it up for hydration and hydrogen bonding with water. Although substitution reduces the number of free hydroxyl groups on cellulose molecules, methyl ether groups still provide sites for hydrogen bonding with water and intermolecular hydrogen bonding. Because of its linear structure, nonionic nature and high degree of solubility, methylcellulose films are strong and clear.

Methylcellulose is unique in the sense that, although it completely dissolves in cold water and forms a clear solution, it gels upon heating between 48° and 64°C, depending on grade. The gelling phenomenon is a stage in the flocculation process, where methylcellulose molecules are losing their water or hydration, causing molecules to pack or associate more intimately. Depending on the concentration, however, the heated methylcellulose solution can either hold the gel or can completely fall out and settle. For a 2% solution of high viscosity methylcellulose in water (~6,000–8,000 cP), the heated solution will hold an intact gel, but will show some degree of syneresis or weeping. At lower concentrations, or in the presence of competing solutes or interfering insoluble solids, the methylcellulose can completely fall out of solution. As the solution cools, methylcellulose re-dissolves and will thicken again. Methylcellulose is stable over the pH range of 3–11 (Murray 2000).

When a 3% solution of high viscosity methylcellulose is cast at a thickness of 10 mils (0.01 in.), its dried film has enough strength to peel off in one piece, like agar. Pictures of methylcellulose films cast at 10 and 20 mils are shown in Fig. 3.4.

These films had tensile strengths of 760 g and 1,400 g and puncture strengths of 350 g and 830 g, respectively. In contrast to agar and the other films discussed here, methylcellulose films provide some stretch when pulled, which property is also true for films of other derivatized forms of cellulose, such as HPMC and the non-food grade ethylhydroxyethylcellulose (EHEC).

3.2.1.3 Hydroxypropylmethylcellulose [E464; CAS#9004-65-3; 21 CFR 172.874]

HPMC is also included in the group of compounds known as cellulose ethers. It is manufactured by first reacting purified cellulose with alkylating reagent (methyl chloride) in the presence of base, typically sodium hydroxide, and an inert diluent to form methylcellulose. The addition of base, in combination with water, activates the cellulose matrix by disrupting its crystalline structure and increasing the access of polymer hydroxyl groups to the alkylating agent to promote the etherification reaction. This activated matrix is called alkali cellulose (Kirk-Othmer 1993). Once activated, the methylcellulose is then further reacted with the staged addition of an alkylene oxide, which in the case of HPMC, is propylene oxide (Kirk-Othmer 1993; Dow Chemical 2002).

Hydroxypropylmethylcellulose (HPMC) is essentially the same as methylcellulose in structure, except that in addition to methyl substitution, some of the remaining free hydroxyl groups of glucose are substituted with hydroxypropyl groups (Fig. 3.5), which further opens up the cellulose molecule. HPMC is soluble in cold

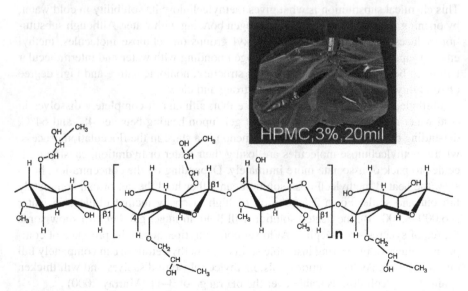

Fig. 3.5 Molecular structure of HPMC and picture of its film, cast from a 3% solution and thickness of 20 mils (0.02 in. or 508 μm)

water like methylcellulose, but its flocculation temperature is higher (~82°C). In contrast to methylcellulose, a 2% HPMC solution in water, when heated to 82°C or higher will cause the hydrated gum to completely fall out of solution. Similar to methylcellulose, however, it re-dissolves when the water cools down to room temperature or lower.

HPMC produces a strong film, but not as strong as that of methylcellulose. Like methylcellulose, its linear structure and nonionic nature are ideal for micelle formation or for film structure development, except that the larger hydroxypropyl groups keep polymer chains farther apart due to steric hindrance. When a 3% solution of HPMC is cast at a thickness of 20 mils (0.02 in.), a clear film is produced. This film is strong enough to peel off the casting surface in one piece (Fig. 3.5), with a tensile strength of ~1,100 g and a puncture strength of ~590 g. These values are relatively low in comparison with those obtained for methycellulose films tested under the same conditions.

3.2.1.4   Konjac Gum [E425; CAS# 37220-17-0]

Konjac gum is a hydrocolloid derived from the root of the plant *Amorphophallus spp.*, and is synonymous with konjac mannan and konjac glucomannan. Konjac is common to traditional Chinese foods that have a history spanning over 2,000 years, and are also popular health foods in Asian markets. Konjac is a heteropolysaccharide consisting of glucose and mannose monosaccharide units in the ratio of 5:8 (Shimahara 1975a), connected by β-(1→4) glycosidic bonds (Fig. 3.6). Physical images, obtained by atomic force spectroscopy and transmission electron

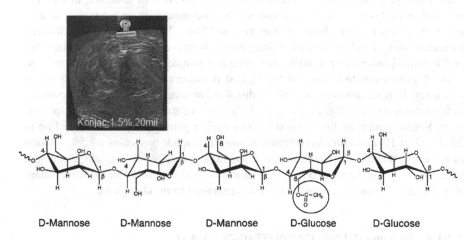

D-Mannose   D-Mannose   D-Mannose   D-Glucose   D-Glucose

Fig. 3.6 Plausible structure of konjac mannan linear chain showing the repeating units (adapted from Maeda et al. 1980). Other proposed repeating structural units are G-G-M-M-M-G-M or G-G-M-G-M-M-M-M (Kato and Matsuda 1969) and G-G-M-M-G-M-M-M-M-M-G-G—M (Maeda et al. 1980; Takahashi et al. 1984). Also shown is a konjac film, cast from a 1.5% solution and thickness of 20 mils (0.02 in. or 508 μm)

microscopy, revealed that the konjac glucomannan molecule is an extended, semi-flexible linear chain without branches. This report is in agreement with konjac glucomannan molecular chain parameters determined using the Mark–Houwink equation by Li et al. (2006). This conclusion, however, is contrary to the molecular structure proposed for konjac mannan in earlier studies by Shimahara (1975b) and Kato (1973), which proposed slight branching, every 50–60 backbone sugar units. In addition, konjac mannan also contains approximately one acetyl ester group per 19 sugar residues (Maekaji 1974). The molecular weight of konjac mannan depends on the species, or even variety, of *Amorphophallus* from which it is extracted and on the extraction method employed. Sugiyama (1972) reported average molecular weight values of 0.67–1.9 million, depending on the *Amorphophallus* variety used, while Li et al. (2006) reported an average molecular weight of 1.04 million Daltons. The proposed molecular structure of konjac is shown in Fig. 3.6.

Konjac solution gels if it is heated after treatment or exposure to alkali. This gelation occurs as a result of the hydrolysis of acetyl groups, which no longer hinder intermolecular hydrogen bonding of chains (Maekaji 1978). A further interesting characteristic of konjac gum lies in its synergy with other hydrocolloids. Takigami (2000) reported synergy between konjac mannan and xanthan, forming an elastic gel, and a significant increase in the gel strength of konjac mannan–carrageenan and konjac mannan-agar mixtures. Furthermore, konjac mannan is unique in that it produces a viscosity as high as 25,000 cP for a 1% solution (Table 3.1), the highest viscosity ever reported for any gum. Solutions of konjac gum are also pH and heat stable, and unlike those of xanthan gum and have smoother texture and flow.

Konjac, like other gums with a linear structure, produces strong films. The acetyl ester groups impart konjac with a slight negative charge and a greater steric hindrance than hydroxyl group, and to a certain degree, this affects film structure. In fact, the removal of acetyl ester groups strengthens the intermolecular association of konjac molecules, such that they form stronger gels and films. This, combined with its high molecular weight and its nonionic nature, makes konjac a good polymer material for edible films. However, a potential disadvantage of konjac mannan is that it is too viscous. A gum needs to be used at the highest possible concentration for processing efficiency. Konjac, however, can only be dissolved in water at relatively lower concentrations compared to other gums, (e.g., 1.5%), before the solution becomes too viscous. Konjac gum is by far the most viscous natural gum, giving a viscosity of up to 25,000 cP at 1% concentration in water. Konjac film, made by casting a 1.5% solution at a casting thickness of 20 mils, has a tensile strength of ~805 g and a puncture strength of ~630 g. These values are remarkable considering how low the konjac mannan gum concentration is compared to films prepared from other gums.

3.2.1.5 Pullulan [E1204; CAS#9257-02-7; GRAS]

Pullulan is a water-soluble, extracellular polysaccharide produced by certain strains of the polymorphic fungus *Aureobasidium pullulans*, formerly known as *Pullularia pullulans*, an organism first described by De Bary in 1866. The basic structure of pullulan was elucidated from the works of several researchers (Bernier 1958; Bender and Wallenfels 1961; Wallenfels et al. 1961, 1965; Bouveng et al. 1962, 1963; Sowa et al. 1963; and Ueda et al. 1963). From the results of these studies, pullulan was characterized as a linear polymer of maltotriose subunits (glucose joined by α-(1→4) linkages) connected via α-(1→6) glycosidic linkages. Catley and coworkers subsequently established the occurrence of a minor percentage of randomly-distributed maltotetraose subunits in pullulan (Catley et al. 1966; Catley 1970; Catley and Whelan 1971; Carolan et al. 1983), and reported that these subunits were also joined by α-(1→6) linkages. Commercial production of pullulan began in 1976 by Hayashibara Company, Ltd., in Okayama, Japan (Tsujisaka and Mitsuhashi 1993), which remains the principal commercial source of pullulan today. This is the same company that petitioned the U.S. FDA for its GRAS status in 2002.

The regular occurrence of α-(1→6) linkages kinks the structure of pullulan and interrupts what would otherwise be a linear starch amylose chain. The unique pattern of α-(1→6) linkages between maltotriose subunits gives the pullulan polymer distinctive physical properties, such as structural flexibility and high water solubility, resulting in distinct film- and fiber-forming characteristics. Leathers (2003) reported that pullulan films and fibers resemble certain synthetic polymers, such as plastics derived from petroleum, and possess the following unique functional characteristics: (1) oxygen impermeability in contrast to other polysaccharide films; (2) being edible and biodegradable, although more expensive than plastics; and (3) highly soluble in water.

Pullulan solutions are of relatively low viscosity compared to most gums. Depending on grade, a 25% solution of pullulan could be ~22,000 cP, which is more viscous than a solution of gum Arabic at the same concentration (~80–120 cP), but is significantly less viscous compared to other gums. Comparing the structures of pullulan and gum Arabic, there is no branching in the structure of pullulan; hence, polymer chains are still able to associate and form a much stronger film than gum Arabic, even if almost every third glycosidic linkage in pullulan is a (1→6) linkage that resembles a substitution.

Pullulan films are formed by drying a pullulan solution (from 5 to 25%, depending on viscosity grade) cast onto an appropriate smooth surface. As shown in Fig. 3.7, a 25% solution of this grade of pullulan and a casting thickness of 20 mils was required to produce a film that peeled off in one piece. The casting solution had a viscosity of ~22,000 cP, and the resulting film exhibited a tensile strength of ~9,000 g and a puncture strength of ~1,800 g. Pullulan films are clear, highly oxygen-impermeable, and have excellent mechanical properties (Yuen 1974). Specialty films may include colors or flavors, and decorative pullulan chips are produced for food uses. Alternatively, pullulan may be applied directly to foods as a protective glaze.

3.2.1.6 Microcrystalline Cellulose (MCC) [E460(i); CAS#99331-82-5; GRAS]

Microcrystalline cellulose (MCC) is produced from cellulose powder, which is subjected to acid hydrolysis to reduce the molecular length of the polymer chains within the cellulose microfibers from 1,000–1,500 DP to 200–300 DP (Iijima and Takeo 2000). The chemical structure of MCC is, therefore, the same as that of

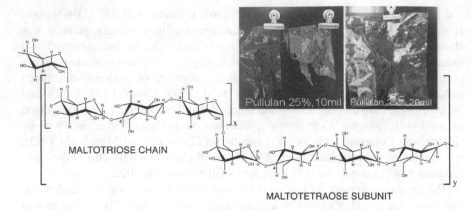

Fig. 3.7 Molecular structure of pullulan and picture of its film, cast from a 25% solution and thickness of 20 mils (0.02 in. or 508 μm)

cellulose, except that MCC chains are a little shorter due to hydrolysis. Any glucose oligosaccharides that are generated during acid hydrolysis are removed by washing. Refined MCC wet cake, which contains only pure crystalline regions of natural cellulose, is mixed with water, neutralized, and dried to produce powdered MCC grades. This grade is insoluble in water and difficult to swell (requires extensive shear to break apart crystalline microfibrils due to intermolecular associations formed on drying). Colloidal grades are also produced by co-processing the wet cake with another hydrocolloid, such as CMC, to coat the crystalline cellulose and minimize re-aggregation of the fibrils during subsequent drying. Hence, this grade can be dispersed uniformly when mixed in water with adequate shear, a property that is used by formulators to mimic the mouthfeel of fat in fat-free and reduced-fat products.

Pure MCC does not form a film (Fig. 3.8). When cast and dried, it leaves a powdery material similar to the starting MCC. This is because the polymer does not dissolve in water, even with heating. Although linear and non-ionic, MCC needs to open first and hydrate to form the structure; however, without a chemical treatment and the harsh pulping process, separation of the individual fibers (or even partial hydration) does not happen. Nevertheless, MCC and cellulose powder can be used in conjunction with other film formers to alter the resulting film's properties.

3.2.1.7 Inulin [CAS#9005-80-5; GRAS]

Inulin is naturally present in many regularly-consumed vegetables and fruits, such as leek, onion, garlic, chicory, artichoke and banana, but is commercially produced from chicory root and Jerusalem artichoke. It consists predominantly of a linear chain of β-(1→2)-linked D-fructofuranose units, attached to a terminal glucose unit in a linkage involving both reducing ends, as in sucrose (Fig. 3.9). Until recently,

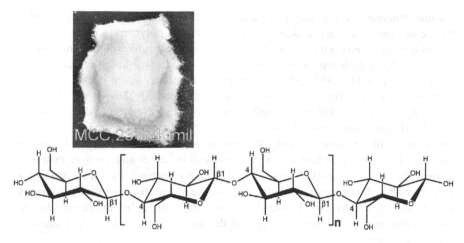

Fig. 3.8 Molecular structure of microcrystalline cellulose and picture showing its lack of film-forming properties

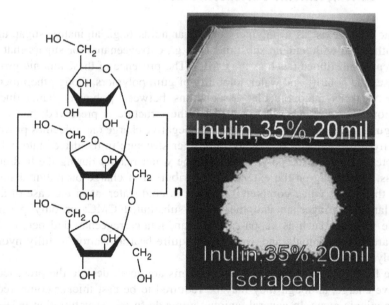

Fig. 3.9 Molecular structure of inulin and picture of its film (scraped and unscraped), cast from a 35% solution and thickness of 20 mils (0.02 in. or 508 µm)

native inulin was considered to be a linear molecule; however, using permethylation analysis, it was shown that native chicory inulin (DP 12) has a very small degree of branching (1–2%). This is also true for inulin from Dahlia (DeLeenheer and Hoebregs 1994). For chicory inulin, the number of fructose units linked to a terminal glucose unit can vary from 2 to 70 (De Leenheer and Hoebregs 1994); Thus, commercial inulin is a mixture of oligomers and polymers.

Inulin is a polymer with a relatively much lower molecular weight compared to other partially-hydrolyzed linear gums, such as low viscosity grades of guar, methylcellulose and CMC. Although it is linear and nonionic, inulin does not form a good film. Its solubility at room temperature is only 20–25% (w/w) in water, depending on grade or DP. In hot water, inulin will dissolve at 50% concentration or higher; but as soon as the solution cools, it starts to crystallize, thicken, and form a consistency of a white sugar icing. When a hot solution of inulin (35%) is cast on a surface, it forms a very uniform, white layer that is harder and stronger than that of sugar; but it also flakes (Fig. 3.9). In confectionery coatings, the traditional sugar shell over chocolate candy, gum balls, or even the softer sugar shell of jelly beans can be strengthened by partial replacement of sugar with inulin. Inulin can also be used in combination with sugar alcohols in the panning process to replace sugar in sugar-free recipes.

3.2.2   Linear, Anionic Polysaccharides

Anionic gums exist as a polymers of: sugar acids (e.g., alginate); sugar units biosynthesized with anionic substituents (e.g., carrageenans); or sugars that are chemically substituted (such as the CMC). The presence of these anionic groups increases the polarity and water-solubility of gum polymers, though the inherent charge weakens intermolecular associations between polymer chains due to repulsion. However, when cast and dried the structure is preserved; hence, all these gum polymers form good films. The negative charge on molecules prevents gums from forming an excessively tight fiber structure in the dried state, which characteristic allows the polymers to imbibe water readily during the hydration process. Depending on the extent and distribution of charge on polymer molecules, the gum is either completely soluble in cold water, as in the case of alginate, lambda-carrageenan and the highly-substituted CMC; or only partially soluble in water, such as kappa-carrageenan, iota-carrageenan and pectin. The latter are all less anionic and, therefore, require heat activation to fully hydrate the polymer molecules.

The film-forming properties of these gums are good, despite the presence of negative charge, although they may be required to be cast thicker compared to most nonionic gums. In general, anionic gums do form a structure that is strong enough to hold the film together during peeling from the casting surface. It appears that the distribution of charge over the length of a linear gum molecule has less of an effect on film structure and strength than the presence of single or multiple neutral sugar side chains substituted along a linear backbone chain. As a general rule, the larger the substitution or the side chain, the greater the steric hindrance and interference to intermolecular association, resulting in a weaker film structure.

3.2.2.1 Sodium Alginate [E401; CAS#9005-38-3; 21CFR184.1724; FEMA 2015]

Alginate is extracted from brown seaweed of the family *Phaeophyceae*. Commercial sources include *Laminaria sp.*, *Macroscystis pyrifera, Ascophyllum nodosum, Eclonia sp., Lessonia nigrescens, Durvillae antarctica* and *Sargassum spp.* (Draget 2000). Alginate is present in seaweed as a salt of sodium, calcium, magnesium, strontium and barium in gelled form; hence, the first step in the extraction process is to apply an acid treatment to convert alginate into alginic acid, followed by an alkali treatment (Na_2CO_3 or NaOH) to produce water-soluble sodium alginate.

Alginates are linear, unbranched polymers containing β-(1→4)-linked D-mannuronic acid (M) and α-(1→4)-linked L-guluronic acid (G) units, and are therefore highly anionic polymers. Alginates are not strictly random copolymers, but are instead block copolymers. They consist of blocks of both similar and alternating sugar units (i.e., MMMMMM, GGGGGG and GMGMGMGM) (Fig. 3.10), with each of these blocks having different conformational structures . For example, the M/G ratio of alginate from *Macrocystis pyrifera* is about 1.6:1, whereas that from *Laminaria hyperborea* is about 0.45:1. Alginates may also be prepared with a wide range of average chain lengths (50–100,000 DP) to suit the application. Commercial grades that are high in guluronic acid are usually labeled HG.

Gelation of alginate, with calcium or a bivalent ion, is instantaneous. The G-block responds to calcium cross-linking faster than the M-blocks, because of its three-dimensional "egg box" molecular conformation, which structure accommodates Ca^{2+} ions to form salt bridges, corresponding to junction zones between adjacent polymer chains (i.e., zipping together of guluronate chains). Nevertheless, mannuronate segments (M-blocks) may also associate through Ca^{2+} salt bridges,

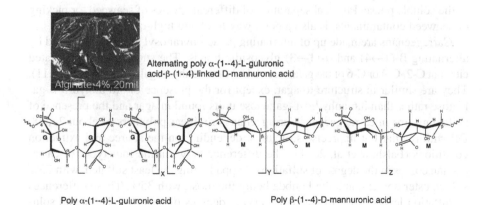

Alternating poly α-(1--4)-L-guluronic
acid-β-(1--4)-linked D-mannuronic acid

Poly α-(1--4)-L-guluronic acid   Poly β-(1--4)-D-mannuronic acid

Fig. 3.10 Structure of alginate showing polyguluronic, polymannuronic and alternating guluronic-mannuronic fractions, and picture of its film, cast from a 4% solution and thickness of 20 mils (0.02 in. or 508 μm)

forming more rigid gels with good heat stability (Donati et al. 2005). When gelation is required, a high guluronic alginate is used, whereas a high mannuronic alginate is used when thickening is the desired attribute. Different viscosity grades of either high G or high M alginates are commercially produced, and some alginate grades incorporate a sequestering agent to reduce calcium sensitivity.

Sodium alginate forms a decent strong film, despite the negative charge on the molecule (Fig. 3.10). The carboxylate groups make alginate very soluble in water, but these negative charges do not have as much impact on alginate's film-forming property as the steric hindrance of a substitution. Comparison of negatively-charged gums with different sized substituent groups seems to indicate that charge repulsion has less effect on film structure than steric hindrance as exemplified by sodium alginate and sodium-CMC, the latter forming a weaker film. When cast at 4% concentration and thickness of 20 mils, the alginate film produced a tensile strength of ~1,500 g and a puncture strength of ~830 g. However, removing the negative charge by cross-linking the alginate with calcium increased the tensile strength of the resulting film, indicating that film structure forms better without the charge repulsion.

3.2.2.2   Carrageenan [E407; CAS# 9000-07-1; 21CFR 172.620; FEMA #2596]

Carrageenan is a collective term for polysaccharides extracted from certain species of red seaweed of the family, *Rhodophycae*. Commercial sources are *Eucheuma spinosum, Eucheuma cottonii, Gigartina spp.* and *Chondrus crispus*. In terms of chemical make-up and structure, there are 4 types of carrageenan extracts: kappa I, kappa II, iota and lambda types. Different seaweeds produce different carrageenans, with one type being more predominant in any species of seaweed. A purer kappa and iota-carrageenan can be produced following potassium chloride precipitation, where kappa and iota fractions are made insoluble, and lambda is removed in the soluble phase. Physical separation of different species of seaweed, or picking of seaweed contaminants, is also a good way to obtain high-quality fractions.

Carrageenans are made up of alternating galactopyranosyl units dimer linked by alternating β-(1→4) and α-(1→3) glycosidic bonds. The sugar units are sulfated either at C-2, C-3 or C-6 of the galactose or C-2 of the anhydrogalactose unit (Fig. 3.11). They are similar in structure to agar, except for the presence of 3,6-anhydro-D-galactose rather than 3,6-anhydro-L-galactose units found in agar and the presence of sulfate groups in carrageenan. Carrageenans are linear polymers of about 25,000 DP with regular but imprecise structures, depending on the source and extraction conditions (Falshaw et al. 2001). The difference between the four types of carrageenan comes in the degree of sulfation – kappa I being the least sulfated, with 25% sulfate ester content, and the lambda being the most, with 35%. These differences in sulfation levels give carrageenans varying degrees of negative charge and solubility in water. Lambda carrageenan is completely cold-water soluble, while kappa-carrageenan is only partially cold-water soluble and requires heating to 82°C for full activation. The iota-carrageenan has solubility in between that of lambda- and kappa-types.

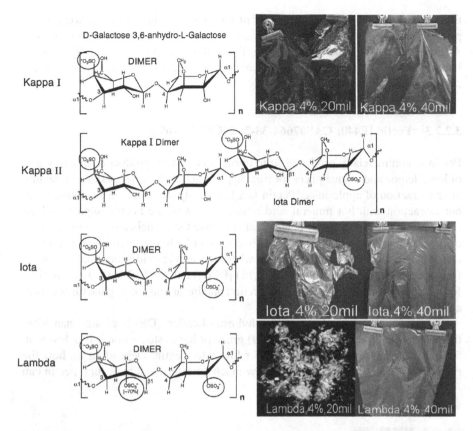

Fig. 3.11 Structural formulas of various carrageenan fractions and pictures of kappa-iota- and lambda-carrageenan films, cast from 4% solutions and thicknesses of 20 mils (0.02 in. or 508 μm) and 40 mils (0.04 in. or 1,016 μm)

Refined carrageenans produce clear solutions and, therefore, clear films. Kappa 1, being the least negatively-charged and least soluble in cold water, requires heating to ~80–82°C to fully hydrate; but hydrated molecules are more free to associate because less repulsion allows for more intermolecular hydrogen bonding, better gelling and slightly stronger film formation upon drying. The dried film is also only partially soluble in cold water, similar to the parent carrageenan, which requires heating to ~50°C to dissolve. Lambda, on the other hand, being the most soluble in cold water and most negatively-charged, produces a slightly weaker film than both kappa and iota carrageenan. As might be expected, lambda film dissolves in cold water.

Carrageenan films of the kappa-, iota-, and lambda-types are shown in Fig. 3.11. When cast at 4% concentration and a thickness of 40 mils (0.04 in.), the tensile strength of the kappa-carrageenan film was the highest at 1,940 g, compared to 944 g for the iota and 659 g for the lambda films. Thus, kappa, due in part to its lowest relative degree of sulfate substitution and negative charge, produced the strongest films. However, the puncture strengths of the three carrageenan types were unexpectedly not significantly different from each other, ranging from 955 to 1,040 g. As expected, solubilities of the resulting films differed markedly, especially between lambda-carrageenan and kappa-carrageenan films, the latter being only sparingly soluble in the mouth.

3.2.2.3   Pectin [E440; CAS#7664-38-2; 21CFR184-69-5]

Pectin is commercially produced from citrus peel as a by-product from extraction of lime, lemon and orange juices; or from apple pomace, the dried residue remaining after extraction of apple juice. Pectin is a heteropolysaccharide in its native state, but extraction with hot mineral acid removes most of the neutral sugars such as rhamnose, galactose, arabinose, etc., that comprise the branched or "hairy" regions of the polymer. Thus, commercial pectin consists of a homopolymeric linear chain of α-(1→4)-D-galacturonic acid units, where the uronic acid group may be either free or esterified with methanol (Fig. 3.12) (Pérez et al. 2000, 2003). By definition, for use in food and pharmaceuticals, pectin contains at least 65% galacturonic acid and methyl galacturonate.

By convention, if the degree of methyl esterification (DE) is greater than 50%, the pectin is called high methoxyl (HM) grade or high ester, while if it is less than 50%, it is called low methoxyl (LM) or low ester pectin. Depending on how the extraction process is controlled and how much de-esterification occurs, pectin can

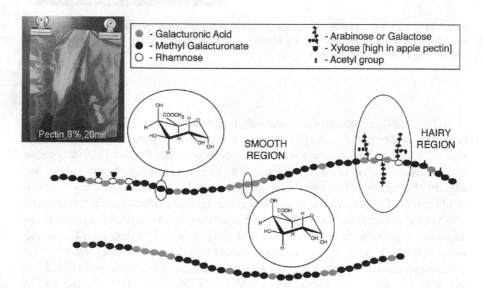

Fig. 3.12 Plausible structure of pectin and picture of its film, cast from an 8% solution and thickness of 20 mils (0.02 in. or 508. μm)

have a degree of esterification as high as 77% or as low as 20%. Low ester pectin with amidation is also produced commercially.

Pectin is used in food mainly for its gelling property, which is influenced by its degree of esterification (DE). Gelation properties also translate into better interpolymer chain associations and film formation; therefore, pectins will generally form stronger films when gelation conditions (pH and/or presence of calcium) are met. With regard to DE, HM pectin that has a DE greater than 69% is a rapid-set pectin, whereas pectin with a DE of 60–61% is a slow-set pectin. When conditions for gelling are met, rapid-set pectin will gel at a higher temperature and, depending on the cooling rate, will set the product faster than a slow-set pectin. In other words, gelation depends not only on pectin DE, but also on both time and temperature. May (2000) reported that HM rapid-set pectin sets at a temperature as high as ~85°C at pH 3.0 or at ~70°C at pH 3.2; while HM slow-set pectin sets at ~58°C at a pH of 2.8 and at ~50°C at pH 3.0. Special grades of HM pectin, with DE between 50 and 60%, are also commercially produced and are referred to as extra slow-set.

To form a gel, HM pectins require a pH below 3.5, with optimum levels around 2.8–3.0, and the presence of sugar solids in the range of 60–65%; heating is also required to fully hydrate and activate pectin molecules. Some adjustment of pH is necessary to cause protonation of unesterified galacturonic acid units to reduce the negative charge and intermolecular repulsion of pectin molecules. The combination of a high sugar concentration and the protonation or neutralization of negative charges on galacturonic acid groups promotes intermolecular associations of polymer chains through intermolecular hydrogen bonding, resulting in gel formation.

Gelation of conventional LM pectins requires addition of calcium, sugar ranging from 20 to 55% and pH levels of 3.0–5.0. Low methoxylpectins (<50% esterified) gel by calcium divalention bridging between adjacent two-fold helical chains, forming the so-called 'egg-box' junction zone structures; a minimum of 14–20 galacturonic acid residues is required at junction zones (Ralet et al. 2001). Gel strength increases with increasing Ca^{2+} concentration, but decreases with increasing temperature and acidity, or reduction of pH lower than 3.0 (Lootens et al. 2003). Gelation of amidated LM pectin, on the other hand, only requires a calcium source and pH adjustment. It works in both sugar-free formulations without sugar solids and in high Brix jelly recipes containing as high as 80% sugar. Gels produced using LM pectin are thermoreversible, meaning that they can be re-melted and reformed on cooling.

Pectin films are generally not as strong as alginate films. Methyl substitution (a larger group than –COOH) and the presence of some remaining "hairy regions" or branching on pectin molecules seem to pose more steric hindrance to film formation than the smaller uronic acid groups of alginate. Pectin films cast at 8% concentration and a thickness of 20 mils (0.02 in.) had a tensile strength of 3,850 g and a puncture strength of 1,120 g. However, compared to alginate at 4% concentration and a thickness of 20 mils, pectin casted at 8% and 10 mils, gave tensile strength of 965 g and puncture strength of 28 g demonstrating a significantly weaker film.

3.2.2.4 Gellan Gum [E418; CAS#7101052-1; 21 CFR 172.665]

Gellan gum is a bacterial exopolysaccharide, like xanthan gum, that is produced from aerobic fermentation of a carbohydrate substrate by the bacterium, *Sphyngomonas elodea*, previously called *Pseudomonas elodea*. It is a linear, anionic polymer of about 50,000 DP, consisting of a tetrasaccharide repeat of →4)-α-L-rhamnopyranosyl-(1→3)-β-D-glucopyranosyl-(1→4)-D-glucuronopyranosyl-(1→4)-β-D-glucopyranosyl-(1→. The (1→3)-linked glucose unit possesses a L-glyceryl attached through C-2 and acetyl at C-6 (Chandrasekaran and Radha 1995) (Fig. 3.13). Gellan gum may or may not be de-esterified by alkali treatment to produce the low acyl and high acyl grades, respectively.

The functionality of gellan gum depends on its degree of acylation. High acyl gellan forms soft, very elastic, transparent, flexible gels, while low acyl gellan forms hard, non-elastic, brittle gels (Sworn 2000). High acyl gellan gel is more flexible than xanthan–LBG gel and firmer and more brittle than agar gel. Both grades produce thermoreversible gels.

Though insoluble in cold water, gellan, like carrageenan, swells and thickens as the solution is heated, but loses viscosity as it fully dissolves. Sworn (2000) summarizes other differences between low acyl and high acyl gellans. These include (1) low acyl gellan is fully hydrated between 80°–95°C; high acyl gellan is fully hydrated at 70°C or higher; (2) low acyl gellan sets between 10 and 60°C, depending on the concentration and presence of ions; high acyl gellan sets at 70°–80°C, both grades showing no hysteresis; (3) low acyl gellan shows improved gelling when acids and mono- and divalent ions are added; high acyl gellan is not affected

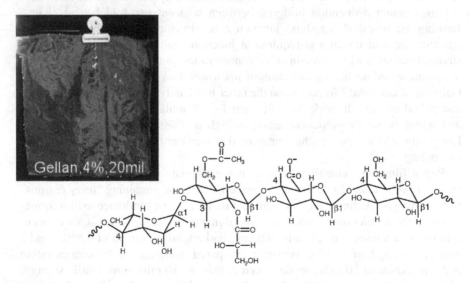

Fig. 3.13 Molecular structure of gellan gum and picture of its film, cast from a 4% solution and thickness of 50 mils (0.05 in. or 1,270 μm)

by these ions; (4) low acyl gellan experiences reduced firmness and modified texture when sugar is added; high acyl gellan experiences increased gel strength under the same conditions.

Gellan gum films (Fig. 3.13) are not as strong as those of agar. At 4% concentration and a casting thickness of 20 mils, gellan gives a tensile strength of ~700 g and a puncture strength of ~250 g, values that are significantly less than those observed for agar film (see Sect. 3.2.1.1). Like agar film, however, gellan film is clear and, like the parent material, is insoluble in cold water. Intuitively, a low acyl gellan gum that produces stronger gel will also produce a stronger film. If gellan is de-esterified, or all of its acyl groups are removed, the linear gellan polymer could become completely insoluble in cold water like agar, and would require higher temperature or boiling to dissolve; but de-esterification will result in stronger films.

3.2.2.5   Carboxymethylcellulose or Cellulose Gum (CMC) [E466; CAS#:9004-32-4; 21CFR 182.1745]

Carboxymethylcellulose (CMC) is a cellulose derivative that is mainly used in many food applications for its viscosity, water binding properties, and solution clarity. There are several available viscosity grades of CMC, ranging from ~50 (2% concentration) to 13,000 cP (1% concentration) in water. For general thickening applications, high viscosity grades are chosen for economic reasons, as the price of CMC is dictated more by the uniformity of esterification rather than the viscosity each grade provides. However, in other applications like flavored syrups, the rheology of CMC and absence of graininess in solution, due to unsubstituted and insoluble regions in the molecule, are sufficient reasons to use the lower viscosity grades, which are less prone to this problem. In film applications, low viscosity CMC is preferred over viscous grades, because higher gum concentrations can be achieved in casting solutions.

The CMC structure involves carboxymethyl substitution of the native cellulose polymer at C-2, C-3 or C-6 positions of anhydroglucose units (Fig. 3.14). The degree of substitution (DS) is generally in the range of 0.6–0.95, but the legal limit is a DS of 1.0. The higher the DS, the more soluble and stable the CMC solution. However, the uniformity of substitution along the cellulose backbone also influences the solubility and smoothness of CMC solutions.

CMC is anionic due to the terminal carboxylate group that ionizes to COO⁻ when CMC is dissolved in water at neutral pH. It is commercially produced as the sodium salt. At acidic pH levels of 3.0 or less, CMC becomes insoluble and loses water binding properties. Different viscosity grades for CMC vary in degree of polymerization (DP) and molecular weight. Smaller molecules, or low DP products, produce lower viscosity solutions than larger molecules, or high DP molecules, due to effects of chain length. To a certain extent, molecular weight affects film strength; but, as long as there is sufficient length in the polymer chain, film structure will form. For CMC, a DP of 50 is capable of forming a cohesive film at 8% concentration.

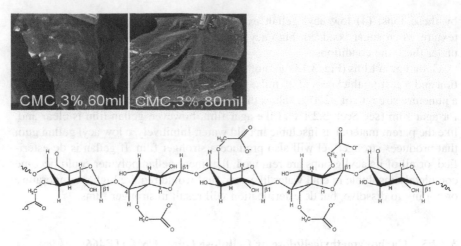

Fig. 3.14 Structure of carboxymethylcellulose and picture of its film, cast from a 3% solution with thicknesses of 60 mils (0.06 in. or 1,524 μm) and 80 mils (0.08 in. or 2,032 μm)

Figure 3.14 shows CMC films based on a 3% solution and casting thicknesses of 60 (0.06 in.) and 80 mils (0.08 in.). The 60-mil film had a tensile strength of 1,920 g and a puncture strength of 1,170 g. In addition to steric hindrance due to carboxymethyl substitution, negative repulsion between chains further weakens intermolecular associations and the structure of CMC films. This scenario explains why a 3% CMC solution had to be cast at 80 mils to form a cohesive film – compared to alginate and methylcellulose (nonionic), which form cohesive films at 27 and 10 mil film thicknesses, respectively, using identical concentrations of gum.

3.2.2.6   Propylene Glycol Alginate [E405; CAS#9005-38-3; 21CFR172.858; FEMA#2941]

Sodium alginate has poor acid stability and is highly sensitive to calcium ions. These properties are drawbacks to many applications because of issues regarding processing and timing with premature gelation. These limitations are removed through chemical modification of alginate to propylene glycol alginate (PGA). In fact, this chemical modification gives entirely new properties to the native gum.

Propylene glycol alginate is chemically derived by treatment of alginate with propylene oxide under acidic conditions. This treatment introduces a propylene glycol ester group bonded to the carboxyl groups of guluronic and mannuronic units (Fig. 3.15). Conversion of alginate, or alginic acid, to PGA reduces the molecule charge to a minimum, and changes the behavior of the polymer in many ways compared to standard alginate. Commercial PGAs are esterified at ~90% or higher. Esterification makes PGA less calcium sensitive; it thickens to some extent in the presence of calcium ions, but does not gel. PGA also tolerates low

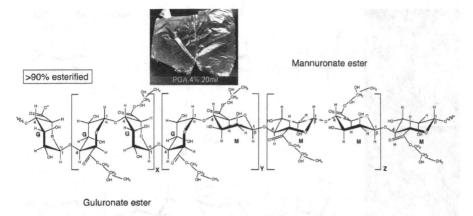

Fig. 3.15 Molecular structure of propylene glycol alginate and picture of its film, cast from a 4% solution and thickness of 20 mils (0.02 in. or 0,508 μm)

pH conditions, and has superior emulsifying and foaming properties relative to native alginate.

Reduction of negative charge, however, does not confer additional tensile strength to the film, because the propylene glycol ester group is bulkier (relative to the carboxyl group of native alginate) and impedes inter-polymer chain association. Film formed with PGA is quite strong and peels off in one piece when cast at a 4% concentration and thickness of 20 mils (0.02 in.) (Fig. 3.15). Tensile strength of this film was 990 g and puncture strength was 445 g, which values are lower or weaker compared to those of the film made with sodium alginate using the same experimental parameters.

3.2.3 Linear, Cationic Polysaccharides

3.2.3.1 Chitosan [CAS#9012-76-4]

Chitosan is the principal derivative of chitin, the material comprising the exoskeletons of crustaceans and mollusks, and is produced by alkaline deacetylation of chitin. Chitosan is poly-β-(1→4)-2-amino-deoxy-D-glucopyranose; its idealized structure is shown in Fig. 3.16. Typical commercial chitosan is about 85% deacetylated. In solution, chitosan forms micelle-like aggregates from fully acetylated segments of polysaccharide chains, interconnected by blocks of almost fully deacetylated polysaccharide, stretched by electrostatic repulsion (Pedroni et al. 2003). Chitosan in the free amine form is insoluble in water at neutral pH. However, it is soluble in glacial acetic acid and dilute HCl, but insoluble in dilute sulfuric acid at room temperature. Chitosan carries a large number of amino groups along its chain and is, thus, capable of forming multiple complexes. At acid pH, the

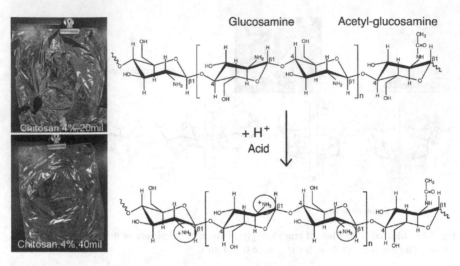

Fig. 3.16 Structure of chitosan and picture of its film, cast from a 4% solution and thicknesses of 20 mils (0.02 in. or 508 μm) and 40 mils (0.04 in. or 1,016 μm)

protonation of $-NH_2$ groups converts them to $-NH_3^+$, which can associate with polyanions to form complexes and bind anionic sites at bacterial and fungal cell wall surfaces. At higher pH levels (>4), chitosan can form complexes with colorants and heavy metals. These appealing features make chitosan widely applicable in wound healing, production of artificial skin, food preservation, cosmetics, and wastewater treatment (Risbud et al. 2000; Juang and Shao 2002).

Chitosan forms strong films. Figure 3.16 shows films made with chitosan at a concentration of 4% and casting thicknesses of 20 (0.02 in.) and 40 mils (0.04 in.). Film tensile strengths were 1,780 g and 2,970 g, and puncture strengths were 1,500 g and 2,035 g, respectively. When a plastic wrap made from chitosan was used for storage of fresh mangoes, the shelf life of the produce was extended up to 18 days, without any microbial growth and off-flavor (Srinivasa et al. 2002).

3.2.4 Linear, Substituted, Neutral Polysaccharides

Linear, substituted gums are polysaccharides with linear backbones and sugar substituents that occur mainly on C-6, as in the case of seed gums like fenugreek, guar, tara, locust bean and cassia gum, or C-3 as in the case of xanthan. The sugar substituent in seed gums is galactose, a monosaccharide, which is present on 16–50% of backbone sugar units, and in xanthan, the substituent is a trisaccharide, a larger side chain. Presence of these sugar substituent groups, their frequency or number, and size ultimately impact these gums' solubility, viscosity, and film-forming properties.

3.2.4.1 Fenugreek [CAS#977155-29-5; FEMA 2484; GRAS as Botanical Substance to Supplement Diet]

Fenugreek,[1] *Trigonella foenum-graecum*, is a legume that used to be grown only in the Mediterranean region, but today it is grown all over the world. It grows up to 60 cm in height, branching off with trefoil leaves and little white flowers. It produces sickle-shaped pods, containing 10–20 brownish diamond-shaped seeds with a diameter of 3–4 mm and a deep groove in the middle. Traditionally, fenugreek has been used in folk medicine for tonic, nutrition, as an appetizer and an antifebrile. It has also been shown to lower cholesterol better than other seed gums, such as guar and locust bean (Evans et al. 1992).

Fenugreek gum is simply the milled endosperm of fenugreek seeds, and possesses a similar sugar composition as guar, tara, locust bean, and cassia gums. Fenugreek is a galactomannan, consisting of a linear chain of mannose linked by β-(1→4) glycosidic bonds, with galactose substitution at the C-6 position (Fig. 3.17). The ratio of mannose to galactose is ~1:1. A linear mannan polymer is completely insoluble in water, just like native cellulose. However, galactose substitution in fenugreek prevents polymer chains from associating intimately during drying, making it completely soluble in cold water. In fact, fenugreek is the most soluble of the seed gums.

However, galactose substitution on every mannose unit in the linear backbone weakens intermolecular hydrogen bonding between polymer chains; thus, it does not form a cohesive film at the same concentration and thickness as methycellulose

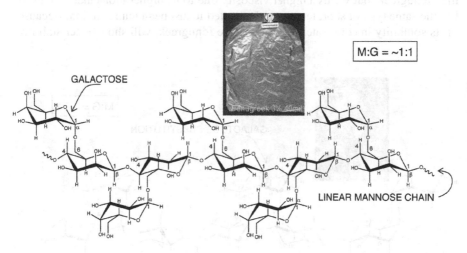

M:G = ~1:1

GALACTOSE

LINEAR MANNOSE CHAIN

Fig. 3.17 Structure of fenugreek and picture of its film, cast from a 3% solution and thickness of 40 mils (0.04 in. or 1,016 μm)

---

[1] http://www.airgreen.co.jp/fenugreek/index_e.html

or any of the other gums described earlier (Fig. 3.17). Furthermore, the absence of electrostatic charge on the fenugreek molecule makes this polymer hydrate slower; nevertheless, fenugreek, like guar gum, achieves full hydration in cold water with time. The finer the mesh, the faster the fenugreek gum hydrates.

The fenugreek film shown in Fig. 3.17 was prepared using a 3% solution and a casting thickness of 40 mils (0.4 in.). This film had a tensile strength of 2,330 g and a puncture of strength of 230 g. Fenugreek possesses the weakest film structure of all seed gums, due to its relatively denser galactose substitution, that poses more steric hindrance to inter-polymer chain association.

3.2.4.2 Guar Gum [E412, CAS#9000-30-0; 21CFR184.1339; FEMA#2537]

Guar gum is the ground endosperm of guar seed from the leguminous shrub *Cyamopsis tetragonoloba*, and is a galactomannan similar to fenugreek, tara and locust bean gum. It consists of a linear mannose chain linked by ß-(1→4) glycosidic bonds, with galactose substitution at the C-6 position of approximately every other mannose unit (Fig. 3.18). Guar gum is a nonionic polymer, consists of longer chain molecules (DP 10,000) than those found in locust bean gum. Higher galactose substitution of guar gum compared to locust bean gum (Petkowicz et al. 1998) increases the solubility of this galactomannan; hence, like fenugreek, guar also achieves full hydration in cold water; however, guar dissolves slower.

Standard grade of guar gum has a viscosity of ~4,500 cP at 1% in water, making it by far the most economical thickener for stabilizing foods. It hydrates slower than fenugreek, but yields a higher viscosity due to its higher molecular weight. It has the same type of short texture as the heated locust bean gum solution. Because of its solubility in cold water, guar gum, like fenugreek, will show better stability

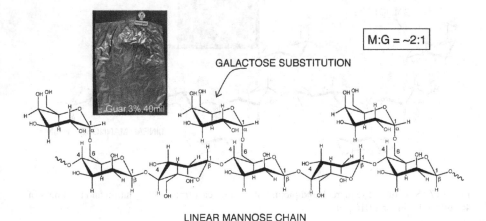

Fig. 3.18 Structure of guar gum and a guar gum film, cast from a 3% solution and thickness of 40 mils (0.04 in. or 1,016 μm)

to freeze-thaw cycles than locust bean gum. On thawing, guar gum will regain its water binding property, whereas locust bean gum will sparingly rehydrate. As a general rule, gums that are cold water soluble are also freeze-thaw stable.

Galactose substitution in guar also poses steric hindrance, reducing intermolecular hydrogen bonding, but not to the same extent as in fenugreek. Figure 3.18 shows guar film, cast at 3% concentration and with a casting thickness of 40 mils (0.04 in.). It gave a tensile strength of 4,580 g and a puncture strength of 655 g, values that are significantly higher than those of fenugreek film produced under the same conditions.

3.2.4.3   Tara Gum [E417; CAS#39300-88-4; GRAS]

Tara gum is the ground endosperm of seeds from the tara tree (*Cesalpinia spinosa* Lin). The plant is native to South America, where it grows as a tree or a bush. Today, tara is also cultivated in Morocco and East Africa. Like fenugreek, guar gum and locust bean gum, tara gum is a galactomannan, comprised of approximately 25% of galactose and 75% mannose. Galactose substitution in tara gum occurs on approximately every third mannose unit (Fig. 3.19).

Tara gum is partially soluble in cold water, and its solution is more viscous than those of either guar gum or carob bean gum at the same concentration. Like carob bean gum, tara gum forms a composite gel, together with xanthan. It also exhibits gelling synergy with both agar and carrageenan.

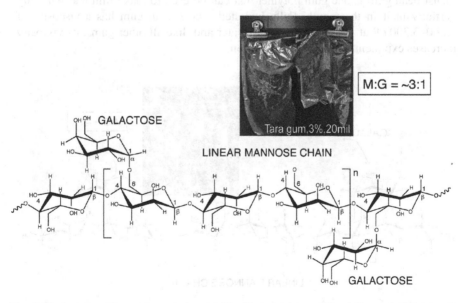

Fig. 3.19 Structure of tara gum and picture of its film, cast from a 3% solution and thickness of 20 mils (0.02 in. and 508 μm)

Based on steric hindrance, tara gum, which possesses less galactose substitution than fenugreek and guar, will, therefore, produce a stronger film. A film prepared using a 3% tara gum and a casting thickness of 20 mils (0.02 in.) as shown in Fig. 3.19 peeled off the surface in one cohesive piece, which was not possible with the other two seed gums prepared at the same casting thickness. The tensile strength of this film was 1,365 g, and the puncture strength was 390 g.

3.2.4.4 Locust Bean Gum (LBG) [E410; CAS#9000-40-2; 21CFR184.1343; FEMA#2648]

Locust bean gum, or carob bean gum, is derived from the endosperm of carob seeds from the tree *Ceratonia siliqua*. Chemically, it is a nonionic polysaccharide that consists of a linear backbone of mannose linked by β-(1→4)-glycosidic bonds, with the galactose substitution occurring at the C-6 position of every fourth mannose unit on average (Fig. 3.20). There are regions of the molecule that are densely substituted and regions where galactose substitution occurs on only every 10–11 mannose units (Wielinga 2000). The galactose content of locust bean gum is reported to be between 17 and 26% (Wielinga 2000). The linear mannose chain is quite similar to cellulose; hence, without galactose substitution, this gum would be insoluble in water. The degree and sporadic nature of galactose substitution make it sparingly soluble in cold water. Heating to 82°C is required to fully activate locust bean gum; thus, when it is frozen or the solution is re-dried, locust bean gum reverses to being almost insoluble in cold water, like the original material. Therefore, locust bean gum is one gum polymer that can be used to make film that will only partially melt in the mouth. Fully hydrated locust bean gum has a viscosity of 2,500–3,300 cP at 1% concentration in water and, like all other gums, its viscosity increases exponentially with concentration.

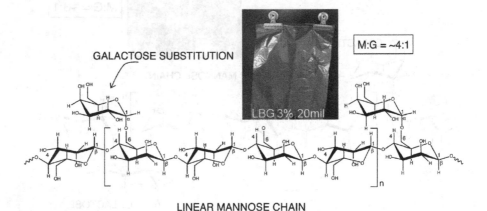

Fig. 3.20 Structure of locust bean gum and picture of its film, cast from a 3% solution and thickness of 20 mils (0.02 in. and 508 μm)

Locust bean gum exhibits strong synergy with xanthan gum. Neither gum forms a gel when used alone, but together, their water binding properties are dramatically improved. A 60/40 to 50/50 blend of LBG/xanthan yields the most synergistic ratio (Sworn 2000). At concentrations of up to 0.4% of the total gum blend, its heated solution shows a pseudoplastic, gel-like behavior. Between 0.4 and 1.0% total gum concentrations, it is a soft gel; at 1% total gum concentration or higher, it forms an elastic gel that is not prone to syneresis.

Locust bean gum is also synergistic with kappa-carrageenan, with the optimum ratio being around 20/80 LBG/carrageenan. LBG makes the carrageenan gel more elastic, as demonstrated by the shift in the fracture point of the gel during compression testing. This synergy also gives stability to the carrageenan gel, allowing its use in acidic applications, between pH 3.0 and 5.0. Similar synergy is exhibited by LBG and agar.

There is yet another seed gum with the same sugar components that deserves to be mentioned in this section – cassia gum. It has a lower degree of galactose substitution than locust bean gum. It is obtained from *Cassia tora* (also known as *Cassia obtusifolia*), and has a mannose to galactose ratio between five and seven. It has a fairly regular substitution pattern (Daas et al. 2000) and even lower solubility in cold water than locust bean gum. It is not currently allowed for general food use in the U.S., but it is allowed in pet foods.

Periodically, supply of locust bean suffers a cycle of shortage, dramatically increasing its cost. Attempts to engineer guar gum, a low-cost seed gum in the same chemical family, as locust bean gum, by specifically removing some of the pendant galactose groups using certain α-galactosidases, is a promising approach for creating a viable locust bean gum substitute in times of shortage.

Expectedly, locust bean gum film, like tara gum, film peeled off easily when cast at a 3% concentration and thickness of 20 mils (Fig. 3.20). With galactose substitution that occurs at about every fourth mannose, vs. every third in tara gum, inter-polymer chain association is less restricted with locust bean gum, resulting in a stronger film structure. Tensile strength of its film was 2,000 g and puncture strength was 460 g, values that are higher than other seed gum films prepared under the same conditions.

3.2.5 Linear, Substituted, Anionic Polysaccharides

3.2.5.1 Xanthan Gum [E415; CAS#11138-66-2; 21CFR172.695;]

Xanthan gum is a polysaccharide that is produced by the bacterium, *Xanthomonas campestris*, during a fermentation process involving a carbohydrate substrate and other growth-supporting nutrients. This gum is actually an excretion to protect bacterial cells when the pH of the fermentation medium becomes too low and unfavorable for growth. It is likely because of this native protective function that xanthan gum demonstrates superior stability to acid over other gums. It is stable over

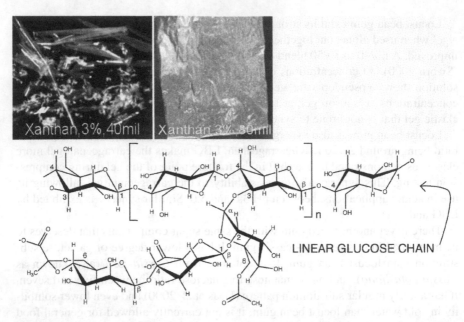

LINEAR GLUCOSE CHAIN

Fig. 3.21 Structure of xanthan gum as adapted from (Jansson et al. 1975 and Melton et al. 1976) and a picture of its film, cast from a 3% solution and thicknesses of 40 mils (0.04 in. or 1,016 μm) and 80 mils (0.08 in. or 2,032 μm)

a wide range of pH, from 1 to 13; very resistant to enzyme hydrolysis; tolerant of high salt, high sugar and high alcohol systems; stable at boiling temperatures; and tolerates retort processing better than most other gums.

Xanthan gum is an anionic polymer with a molecular structure that consists of a linear glucose chain linked by β-(1→4) glycosidic bonds; it also possesses a trisaccharide side chain attached through O-3 of alternate glucose units of the main chain (Jansson et al. 1975; Melton et al. 1976). The trisaccharide side chain is composed of an inner mannose unit, partially acetylated at the C-6 position; a glucuronic acid unit; and a terminal mannose unit, partially substituted with pyruvic acid in the form of a 4, 6-cyclic acetal (Fig. 3.21). Glucuronic acid and pyruvate groups confer a negative charge to the xanthan molecule. The repeating unit of xanthan, therefore, is a pentamer. Each molecule consists of about 7,000 pentamers.

Xanthan gum is soluble in cold water and, like most powders that thicken, it forms lumps if mixed in water with insufficient agitation. It forms a moderately viscous solution with water and, at 1% concentration, gives a viscosity between 1,200 and 1,600 cP. Xanthan solution is highly pseudoplastic and shows a drastic decrease in viscosity with increasing shear rate. It can have a soft, gel-like consistency, depending on concentration and shear applied, typical of a pseudoplastic fluid. The superior suspending property of xanthan gum is attributed to its weak-gel, shear-thinning properties. It hydrates rapidly in cold water, even in the presence of high levels of sugar, salt or alcohol. The consistent water-holding ability of xanthan may be used for control of syneresis and to retard ice recrystallization in freeze-thaw situations. Xanthan gum shows superior freeze-thaw stability (Giannouli and Morris 2000). Its viscosity is relatively unaffected by ionic strength, pH level (1–13), shear or temperature.

Xanthan gum is capable of synergistic interactions with glucomannan like konjac gum, and with galactomannans such as guar, tara and locust bean gums (previously discussed in Sect. 3.2.4.4). With guar, a boost in water binding capacity and viscosity is achieved at all ratios, with the optimum ratio being between 75:25 and 80:20 – guar to xanthan. With locust bean gum, optimal synergy and gelation is achieved at 50:50 LBG to xanthan (Goycoolea et al. 2001). Tara gum exhibits synergy with xanthan intermediate of that achieved with guar and locust bean gums. An extreme synergy between xanthan and konjac mannan takes place at an optimal ratio of 80:20 konjac to xanthan, respectively, with viscosity build-up as high as 161,000 cP at a 1% total gum concentration (Takigami 2000). This noted synergy allows for very low usage levels of this gum combination in food applications.

Figure 3.21 illustrates a cohesive xanthan film formed at 3% concentration and casting thickness of 80 mils (0.08 in.); no film was formed at 40 mils (0.04 in.). By comparison, guar gum, with its smaller galactose side chain compared to the trisaccharide side chain of xanthan, produced a cohesive film at 40 mils thickness under similar parameters (0.04 in.).

3.2.6   Branched Polysaccharides

Film-forming properties of gum polymers are greatly affected by branching. Although most branched gums have very large molecular weights, their molecules assume a globular conformation. This shape does not allow the formation of micelles consisting of polymer chains intimately associated with each other. Although some intermolecular bonding does still take place, it occurs only on a limited basis. Consequently, their films do not have the structural strength, except if cast at exceedingly high gum concentrations and film thicknesses. All branched gums – such as those from gum Arabic, ghatti, karaya and larch gum – will not peel off in one piece when cast and dried, but will flake off instead when cast at the same or a significantly higher concentrations and thickness than the linear polysaccharides.

3.2.6.1   Gum Arabic [E414; CAS#9000-01-5; 21 CFR 184.1330; FEMA 2001]

Gum Arabic is an exudate from the stems and branches of sub-Saharan *Acacia senegal* and *Acacia seyal* trees that either naturally exude sap or are tapped to form large nodules of gum to seal wounds in the bark of the tree. The chemical structure of gum Arabic has been reviewed by Verbeken et al. (2003). Gum Arabic is a complex and variable mixture of arabinogalactan oligosaccharides, polysaccharides, and glycoproteins. Depending on the source, the glycan components contain a greater proportion of L-arabinose relative to D-galactose (*A. seyal*) or D-galactose relative to L-arabinose (*A. senegal*). The gum extracted from *A. seyal* also contains significantly more 4-*O*-methyl-D-glucuronic acid, but less L-rhamnose and unsubstituted D-glucuronic acid, than that from *A. senegal* (Williams and Phillips 2000).

Goodrum et al. (2000), as cited by Chaplin (2008), reported that gum Arabic is a highly-branched polymer with a main chain of (1→3)-linked β-D-galactopyranosyl units, with side chains of (1→3) β-D-galactopyranosyl units joined to it by (1→6) links and with a molecular weight of ~250,000. The side chains are 2–5 units in length. Both the main chain and the side chains have attached units of α-L-arabinofuranosyl, α-L-rhamnopyranosyl, β-D-glucuronopyranosyl and 4-*O*-methyl-β-D-glucuronopyranosyl units. In addition, gum Arabic has a minor fraction of an arabinogalactan-protein (AGP) complex with an MW of ~2.5 × 10^6 Da. The covalently-linked protein, ~2% of the total gum, is rich in hydroxyproline, containing a repetitive and almost symmetrical 19-residue consensus motif – ser-hyp[a]-hyp[a]-hyp[a]-thr-leu-ser-hyp[b]-ser-hyp[b]-thr-hyp-thr-hyp[a]-hyp[a]-hyp[a]-gly-pro-his – with contiguous hydroxyprolines (a) attached to oligo-α-1→3-L-arabinofurans and non-contiguous hydroxyprolines (b) attached to galactose residues of oligo-arabinogalactans combining a β-1→3-galactopyran core with rhamnoglucuronoarabinogalactose pentasaccharide side chains.

Figure 3.22 shows the plausible structure of gum Arabic. The high degree of branching makes gum Arabic assume a more globular shape despite its high molecular weight, a configuration that prevents the formation of micelles. In solution, gum Arabic molecules have limited sites for intermolecular hydrogen bonding, due to the extensive branching that deters such interactions from developing. Although the gum dissolves and forms hydrogen bonds with water, it does not effectively immobilize water. The result is a poor thickening property, compared to most linear gums and the lack of film strength when its solution is cast and dried.

Comparing *A. Senegal* and *A. seyal* gum sources, Naouli (2006) determined that the latter is more branched using a Mark–Houwink plot of log intrinsic viscosity vs. log molecular weight. Corollary to this finding, when cast at the same thickness and concentration, the film of *A. seyal* is weak and cracks on drying, whereas the film obtained from *A. senegal*, though it also does not peel from the casting surface in one piece, does not crack on drying. Rather, it remains uniform, coating the surface like a laminate. The greater branching of the *A. seyal* is likely to be responsible for its weaker, non-cohesive film. Solubility of both films is very good. The low viscosity of gum Arabic and the lack of film strength increases its dissolution in the mouth because it disintegrates faster in the saliva.

3.2.6.2   Gum Ghatti [E419; CAS#9000-28-6; 21 CFR 184.1333; FEMA#2519]

Gum ghatti, also called Indian gum, is an exudate gum from the tree *Anoggeissus latifolia*, found in India and Sri Lanka. Its applications are similar to those of gum Arabic in

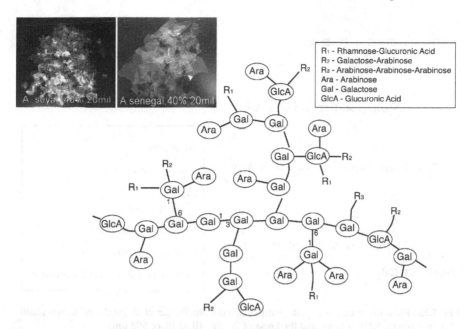

Fig. 3.22 Plausible structure of the gum Arabic molecule, showing the main polysaccharide component (AGP complex not shown) and picture of *A. seyal* and *A. senegal* films, cast from a 35% solution and thickness of 20 mils (0.02 in. or 508 μm)

the food and pharmaceutical industries, where it is used as an emulsifier except that this gum is much more viscous than gum Arabic and is difficult to spray dry for commercial production. Commercial ghatti, therefore, is mainly milled.

The structure of gum ghatti has been extensively studied (Aspinall et al. 1955, 1958a; Aspinall et al. 1958b, 1965; Aspinall and Christensen 1965). It was found that ghatti consists of L-arabinose, D-galactose, D-mannose, D-xylose and D-glucuronic acid, in a molar ratio of 48:29:10:5:10, respectively; it also has traces of rhamnose. Tischer et al. (2002) showed that gum ghatti is made up of arabinose, galactose, manose, glucuronic acid and rhamnose, in a molar ratio of 61:39:6:10:6, but xylose was not detected. The polysaccharide has a backbone chain of (1→6)-linked ß-D-galactopyranosyl units, with some (1→4)-D-glucopyranosyluronic acid units, some joining (1→2)-D-mannopyranosyl units and some L-arabinofuranose units (Fig. 3.23). Hanna and Shaw (1941) reported a mean molecular weight of the soluble portion of gum ghatti to be ~12,000 Da, as determined by osmotic pressure measurements.

The gum ghatti film does not crack when dried, and provides a uniform laminate on the casting surface prior to scraping. However, like gum Arabic and other branched gum films, it is not strong enough to peel off in one piece when cast at a concentration of 5.5% and casting thickness of 20 mils (Fig. 3.23). Because it provides much higher viscosity than gum Arabic, dissolution of ghatti film is not as fast.

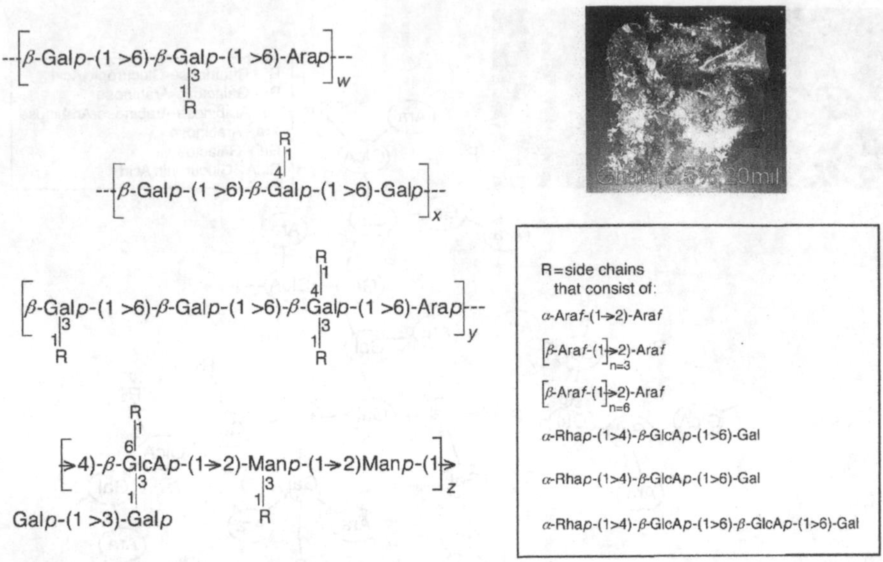

$$--[\beta\text{-Gal}p\text{-(1>6)-}\beta\text{-Gal}p\text{-(1>6)-Ara}p]_w---$$
$$\overset{\overset{\text{R}}{|}}{\underset{1}{3}}$$

$$--[\beta\text{-Gal}p\text{-(1>6)-}\beta\text{-Gal}p\text{-(1>6)-Gal}p]_x---$$
$$\overset{\overset{\text{R}}{|}}{\underset{1}{4}}$$

$$[\beta\text{-Gal}p\text{-(1>6)-}\beta\text{-Gal}p\text{-(1>6)-}\beta\text{-Gal}p\text{-(1>6)-Ara}p]_y---$$

$$[>4)\text{-}\beta\text{-GlcA}p\text{-(1>2)-Man}p\text{-(1>2)Man}p\text{-(1>}]_z$$

Gal$p$-(1 >3)-Gal$p$

R = side chains
that consist of:

α-Ara$f$-(1→2)-Ara$f$

$[\beta\text{-Ara}f\text{-(1→2)-Ara}f]_{n=3}$

$[\beta\text{-Ara}f\text{-(1→2)-Ara}f]_{n=6}$

α-Rha$p$-(1>4)-β-GlcAp-(1>6)-Gal

α-Rha$p$-(1>4)-β-GlcAp-(1>6)-Gal

α-Rha$p$-(1>4)-β-GlcAp-(1>6)-β-GlcAp-(1>6)-Gal

Fig. 3.23 Plausible structure of gum ghatti (adapted from Tischer et al. 2002) and a gum ghatti film, cast from a 5.5% solution and thickness of 20 mils (0.02 in. or 508 μm)

3.2.6.3 Karaya [E416; CAS#9000-36-6; 21 CFR 184.1349; FEMA#2605]

Gum karaya is exuded from *Sterculia urens*, a bushy tree found in dry regions of North India. Native gum karaya is a branched, acetylated polysaccharide with a reported molecular weight of 9–16 million (Le Cerf et al. 1990). Structurally, karaya is heavily acetylated, containing α-D-galacturonic acid and α-L-rhamnose units as the main chains, with O-4 of the acid and O-2 of rhamnose linkages, and the acid is linked by 1, 2 linkage of β-D-galactose or by 1, 3 linkage of β-D-glucuronic acid on side chains, where one half of the rhamnose units carry at O-4 by 1,4 linkage of β-D-galactose units (Stephen and Churms 1995) as shown in Fig. 3.24. Because of its lesser degree of branching and much larger molecular weight compared to gum Arabic, karaya solutions exhibit high viscosity, between 400 and 1,200 cP for a 1% solution by weight.

Karaya film possesses the same limitation as the ghatti film; it lacks adequate tensile strength to peel off from the casting surface in one piece. However, it does coat surfaces uniformly without cracking when the film dries. It is also tacky and adheres well to the casting surface, a characteristic of most tree exudates. Karaya film is soluble in cold water, but will dissolve slower in the mouth, due to its high viscosity, much like ghatti film.

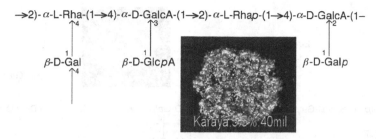

Fig. 3.24 Plausible structure of karaya gum and picture of its film, cast from a 5.5% solution and thickness of 40 mils (0.04 in. or 1,016 μm)

3.2.6.4   Larch Arabinogalactan [CAS#9036-66-2; 21 CFR 172.610; FEMA#3254]

Larch gum, also known as arabinogalactan, is the polysaccharide derived from wood of the Western Larch, or *Larix occidentalis*. Arabinogalactans occur in other types of larch, but the grade that is marketed for supplement use comes from Western Larch. Larch gum, like other tree exudates, exists as a polydisperse, densely-branched polysaccharide with a molecular weight ranging from 10,000 to 120,000 Da.

Using compositional and methylation analyses in conjunction with NMR spectroscopy, Ebringerova et al. (2005) reported that larch gum has the following structural features: (a) most main-chain Gal*p* residues carry a side chain on O-6; (b) about half of the side chains are β-(1→6)-linked Gal*p* dimers; (c) about a quarter of the side chains are single Gal*p* residues; (d) the rest of the side chains contain three or more Gal*p* residues and include most of the arabinose present; and (e) arabinose occurs as single α-Ara*f* units and dimeric side chains of β-L-Ara*p*-(1→3)-α-L-Ara*f*-(1→6). The presumed structure of larch gum is shown in Fig. 3.25.

Larch gum, like gum Arabic, has very high solubility in water, due to its high degree of branching. A high concentration of larch gum, e.g. as high as 55–60%, can be dissolved in water with ease. It is tacky, like gum Arabic and the other tree exudates, and shows good adhesion and casting properties. The resulting film, however, is very weak as expected and cannot support its own structure, shown as flakes in Fig. 3.25. It is very soluble in cold water; and, therefore, its rate of dissolution in the mouth is also fast, producing a rapid melting sensation contributed partly by its low viscosity.

3.3   Composite Films and Flakes

Gum polymers can hold insoluble particulates, soluble solutes, drugs, flavors and other active ingredients in the film matrix. Other polymers, such as proteins and lipids, can be used in conjunction with the gums to alter solubility and water vapor

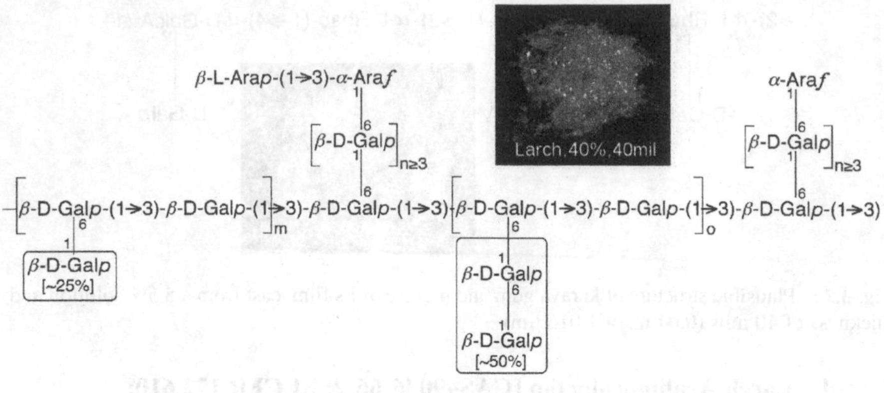

β-L-Ara$p$-(1→3)-α-Ara$f$ α-Ara$f$

Larch, 40%, 40mil

$\left[\beta\text{-D-Gal}p\right]_{n\geq3}$ $\left[\beta\text{-D-Gal}p\right]_{n\geq3}$

$-\left[\beta\text{-D-Gal}p\text{-(1→3)-}\beta\text{-D-Gal}p\text{-(1→3)-}\beta\text{-D-Gal}p\text{-(1→3)}\right]_{m}\left[\beta\text{-D-Gal}p\text{-(1→3)-}\beta\text{-D-Gal}p\text{-(1→3)-}\beta\text{-D-Gal}p\text{-(1→3)}\right]_{o}$

β-D-Gal$p$ [~25%] β-D-Gal$p$

β-D-Gal$p$ [~50%]

Fig. 3.25 Plausible molecular structure of larch gum and a larch gum film, cast from a 40% solution and thickness of 40 mils (0.04 in. or 1,016 μm)

and oxygen permeabilities of a film. Plasticizers – such as glycerol, propylene glycol and sugar alcohols – can also be used to increase plasticity of a film. There is a wide choice of gum polymers, including their various combinations that can be used in film applications – such as breath film; sore throat and cold-flu film strips; fresh produce coatings; and confection coatings and glazes. Whether one gum polymer or a combination of gums is used, the property of the final film is dictated by both the gum polymers used and by composite effects of all ingredients used to make the film.

Below is a guideline for using polysaccharide gums in film applications:

1. Choose the gum or gum combination that will achieve the desired film property. Many film applications – such as breath film, sore throat film, and cold and flu films – require that the film dissolve rapidly in the mouth and that the gum polymer is strong enough to provide adequate structural support to the film, so that it can be peeled off in one piece from the production belt. In other applications, such as cake glitters, films do not have to be as strong, just strong enough that, when scraped at the end of the production belt after drying, they come off as flakes of desired size. Another requirement might be that a film needs to be shiny; while most gums give sheen in the cast state, some are shinier than others. For instance, gum Arabic gives a shiny film that readily flakes off the belt.

2. Know the viscosity of the gum polymer being used. Depending on the gum or gum combination, the formulator needs to vary gum concentration, based on the available water in the recipe, to provide proper casting viscosity. It is important to control viscosity of the film slurry for uniform casting and final film thickness. There is a period of about 5–10 s before the cast slurry enters the drying tunnel; during this period, the cast slurry could spread if it is too thin, resulting in uneven film thickness. Viscosity is especially critical for drug delivery films that have strict specifications on the dose of active per sheet. For machinability, a casting

viscosity between 6,000 and 10,000 cP is ideal and practical for ease of batch preparation and casting uniformity. However, a much thicker slurry – as high as 83,000 cP for a 3% guar and 118,000 cP for 1.5% konjac – can be successfully cast, dried and peeled. In these situations, adjustment of casting speed is needed, i.e., slowing it to allow the slurry to coat the belt or the surface evenly.

3. Know how to hydrate or dissolve the gum polymer. A film slurry requires a much higher concentration of gums than most other food applications, typically 4–8%, or even higher for non-viscous gums. From an economic standpoint, film recipes also require the highest dose or concentration of the other ingredients, so that the slurry has the minimum amount of water needed during casting. Drying water off the slurry is the most expensive part of the process; the less water that is present, the cheaper the process becomes.

All gums that hydrate in cold water – such as CMC, xanthan, guar, alginate – tend to form lumps if added to the batch without sufficient mixing. It is desirable to dry blend gums with other ingredients for ease of dispersion and dissolution, or gums could be slurried in glycerol, alcohol, oil or any suitable solvent that is called for in the film recipe, to minimize lumping. Alternatively, readily-dispersible or agglomerated versions of traditional gum powders may also be used.

It is also important to know whether the gum polymer requires heating or boiling to fully activate the molecule and to know the best way to put the particular gum in solution. If the gum is not hydrated, film structure will not form. For example, traditional agar requires boiling for about 3–5 min to fully dissolve; however, there is a grade of quick-soluble agar that will dissolve between 65 and 82°C that can be used instead. Kappa-carrageenan and locust bean gum require heating to 82°C to fully activate; iota-carrageenan and tara gum require heating to ~65°C to achieve full activation. Methylcellulose and HPMC, which are cold water-soluble gums, are best hydrated by dispersing the gum powder in half of the water that is between 55–60°C for methylcellulose and 82°C for HPMC, and then adding the other half of the water ice-cold. These gums will hydrate with minimal lumping and foaming if dissolved this way.

Table 3.2 summarizes results of a study done by Nieto and Grazaitis (2006), comparing the dissolution times of various film composites (without flavor, color or active ingredients). The films were prepared using sodium alginate and CMC in combination with other gums and a plasticizer. Twelve composite film recipes with different tensile strengths and solubilities are shown. Solubility of the alginate-CMC film is greatly improved by addition of gums that do not form cohesive films (weak gums), such as gum Arabic, inulin, resistant maltodextrin, and polydextrose, as well as insoluble MCC. This phenomenon is mainly due to the weakening of the film matrix structure, leading to faster disintegration of the film. To the contrary, glycerol, a plasticizer, strengthens and increases dissolution time of the resulting film.

Active ingredients can be incorporated directly into these recipes before the film is cast. These active ingredients, which can comprise up to 30% of a film by weight, become locked into the film matrix, and must remain stable until product consumption.

Table 3.2 Tensile and puncture strengths and dissolution times of edible film composites based on the work of Nieto and Grazaitis (2006)

Composite film recipe		Tensile strength (g)	Puncture strength (g)	Dissolution time (s)
Water	91.1%	1,168.7	468.9	40.25
Na Alginate	4.0%			
LV CMC	4.0%			
Glycerin	0.9%			
Water	89.1%	1,360.4	192.9	24.5
Na Alginate	4.0%			
LV CMC 15F	4.0%			
MCC	2.0%			
Glycerin	0.9%			
Water	89.1%	1,701.7	356.9	34.75
Na Alginate	4.0%			
LV CMC	4.0%			
Maltodextrin	2.0%			
Glycerin	0.9%			
Water	95.0%	735.9	189.5	28
Na Alginate	2.5%			
LV CMC	1.25%			
K-Carrageenan	1.25%			
Water	87.6%	1,255.1	244.9	83
LV CMC	8.0%			
K-Carrageenan	3.5%			
Glycerin	0.9%			
Water	91.1%	1,171.3	1,175.5	48.5
Pectin	4.0%			
Na Alginate	4.0%			
Glycerin	0.9%			
Water	95.0%	834.1	341.2	31
Na Alginate	2.5%			
LV CMC	2.5%			
Water	95.0%	713.7	290.2	20
Na Alginate	2.5%			
LV CMC	1.25%			
Gum Arabic	1.25%			
Water	95.0%	456.1	102.0	16
Na Alginate	2.5%			
LV CMC	1.25%			
Insulin	1.25%			
Water	95.0%	804.3	107.5	12.6
Na Alginate	2.5%			
LV CMC	1.25%			
Polydextrose	1.25%			
Water	95.0%	761.9	122.1	14.2
Na Alginate	2.5%			
LV CMC	1.25%			
Maltodextrin	1.25%			
Water	95.0%	355.7	86.88	15.4
Na Alginate	2.5%			
LV CMC	1.25%			
MCC	1.25%			

Tensile strength across a 2.5-cm wide film strip and puncture strength measured using the TA XT Plus Texture Analyzer probes TA-96B and TA-1085-5, respectively. Dissolution times measured using a sensory panel on two ~0.05-mm thick by, 1 cm film samples.

Examples of actives used in film strips include ingredients for oral hygiene, such as mint; alertness, such as caffeine; nutrients and botanicals. Color and flavor may also be added.

3.3.1 Rapid Melt Film Composite

In general, a weak film dissolves and undergoes rapid disintegration in the mouth. With a combination of the right gum and other ingredients, a rapid-melt film can be produced. In applications such as breath film, this fast dissolution attribute is desirable. However, gum polymer or polymers need to provide adequate tensile strength to the film so that adequate machinability is maintained during processing and drying. Any of the recipes from H to L in Table 3.2 can be used or further manipulated to produce films that dissolve faster in the mouth. Addition of gum Arabic, inulin, polydextrose, or resistant maltodextrin can improve the rapid-melt characteristic of films, with only minimal reductions in tensile strength, leaving room for further adjustment to achieve desired solubility in the mouth.

There are many options for making rapid-melt films using various polymers as matrix materials. For instance, pectin, lambda-carrageenan, and methylcellulose can be used together to provide the primary film structure, while other ingredients can be added to weaken the film for rapid melting.

3.3.2 Slow Melt Film Composite

Slow-melt films are easily produced by increasing the solids with gums that enhance film structure, as well as by casting films thicker to improve their strength. In the study by Nieto and Grazaitis (2006), doubling of casting thickness for recipes H to L (Table 3.2) increased dissolution time from 1.4 to 3.3 times.

Use of gums that produce strong, but cold water-soluble films – such as methyl-cellulose, HPMC and konjac – will also increase the tensile strength of the resulting films, allowing them to dissolve more slowly. There is also a viscosity effect that contributes to the slow-melting sensation; for films comprised of gums that form viscous solutions, film melting is significantly hindered, especially with the limited liquid available in the mouth during ingestion.

3.3.3 Insoluble Films

As a general rule, the initial water-solubility of a particular gum determines the solubility of its resulting film. Therefore, agar and gellan are the best choices for making insoluble films, because they require high heat – 100°C and 95°C, respectively – to fully hydrate and solubilize. Though both are relatively expensive, kappa-carrageenan and locust bean gum represent less expensive alternatives for producing films with acceptable insolubility in cold water or in the mouth. Kappa-carrageenan and glycerol produce a film that is only sparingly soluble in cold water. Locust bean gum, because it is practically insoluble in cold water, will also yield a film that is sparingly soluble in the saliva when placed on the tongue. Cross-linking alginate with calcium also yields a strong, clear film that is insoluble in water; this combination has been used to microencapsulate probiotics, such as *Lactobacillus acidophilus*, to increase its survival under gastric conditions. Based on a study by Chandramouli et al. (2004), the viability of bacterial cells within calcium-alginate microcapsules increased as both capsule size and alginate concentration increased. Agar and gellan may also be used in this same application.

3.3.4 Method of Manufacture

In a commercial setting, production of composite films and flakes requires an elaborate setup of machines for automated processing on an industrial scale (see Chap. 14). This section focuses on the required handling and processing of polysaccahide gums to maximize their functionality in film applications.

1. *Preparation of the film slurry*. For proper hydration of gums, a mixing tank with allowance for heating, such as a steam jacket, and high speed mixing are needed since gums and other solid components are normally used at high concentrations to reduce the amount of water to be later removed during drying. Casting viscosities of 6,000–10,000 cP are commonly used, and gum usage is typically as high as 8%, or even higher. Entrainment of air is a problem when mixing a thick slurry; consequently, the mixing kettle is either fitted with a vacuum system, or an anti-foaming agent may be used. Flavors, colors, and active ingredients are generally used at very low levels that do not increase solids in the slurry appreciably. However, composite films may include sweeteners and bulking agents, such as maltodextrin, that can boost solids to make the process energy efficient. In making gum Arabic flakes, a higher concentration (40–50% gum in water) is needed, because its solution at these higher concentrations is just viscous enough to cast.

2. *Casting*. After mixing the gum and other ingredients, slurry is pumped into a reservoir to be fed onto a belt through a casting knife. The casting knife sits on top of the stainless steel belt, which is moved by two drum rollers on each end of the line. The casting knife adjusts the clearance between the knife and the belt and, therefore, thickness of the cast. It is important that the gum is fully hydrated and lump-free at this time. Any lumping will cause uneven cast or slitting and tearing of the film when peeled at the other end of the belt.

3. *Drying*. The cast slurry goes through a temperature- and humidity-controlled tunnel dryer. There is an ideal moisture content or humidity level for the gum

films to peel away from the casting belt (~24–29%). If too dry, gum films become brittle and can easily tear, a shortcoming that is overcome by addition of a plasticizer – glycerol, propylene glycol or sorbitol – if needed. When making flakes, the product is dried on the belt to its final moisture, scraped off and collected for packing. Otherwise, the peeled film is rolled and stored under controlled humidity until further processed.

4. *Slitting and Packing.* The last step in the process involves cutting the film into desired shapes and dimensions for packing, which is often the most elaborate step in the process. It is critical for the film to have the appropriate moisture content at this point. Uniformity of the film structure affects the yield of edible filmstrips; the more uniform the filmstrips, the lower the amount of waste.

3.4   Edible Coatings

3.4.1   Fruits and Vegetable Coatings

The demand to increase product shelf life and enhance microbial safety of fresh fruits and vegetables has prompted the development of edible coating technology based on polysaccharides, lipids and/or proteins. These materials can function as barriers to water vapor, gases, and other solutes, and also as carriers of functional ingredients, including antimicrobial agents and antioxidants. Fruits and vegetables are living tissue up until the time they are consumed or cooked/processed for consumption. Controlling respiration within these living tissues improves storability and extends the shelf life of fresh produce. Lin and Zhao (2007) published a very comprehensive review of innovations in edible coatings for fresh and minimally-processed fruits and vegetables.

Many polysaccharides have been evaluated or used for forming films and coatings, including starch, dextrins, cellulose derivatives, alginates, carrageenan, xanthan, chitosan, pectin and others. The variety and versatility of gum polymers provide a broad choice of materials for producing films for fresh fruits and vegetables that have unique functions, including adhesion and cling, oxygen permeability, plasticity, oil barrier properties, etc.

Cellulose derivatives – such as methylcellulose, CMC, HPC, and HPMC – have been studied as edible coating materials for fruits and vegetables to provide barriers to oxygen, oil, and/or moisture (Sanderson 1981; Morgan 1971; Sacharow 1972; Krumel and Lindsay 1976; Maftoonazad and Ramaswamy 2005). They retain the firmness and crispness of apples, berries, peaches, celery, lettuce, and carrots when used in a dry-coating process (Mason 1969), and are also used to preserve important flavor components of fresh fruits and vegetables (Nisperos-Carriedo and Baldwin 1990). In addition, they reduce oxygen uptake without increasing internal carbon dioxide levels within coated apples and pears by simulating a controlled atmosphere and environment (Lowings and Cutts 1982; Banks 1985; Meheriuk and Lau 1988; Santerre et al. 1989).

Very promising results have been reported regarding oil barrier properties of alginate-based films providing cling, reducing weight loss by acting as a sacrificing moisture coat and sealing volatiles (Fisher and Wong 1972; Cottrell and Kovacks 1980; King 1983; Kester and Fennema 1986; Conca and Yang 1993; Amanatidou et al. 2000; Hershko and Nussinovitch 1998). In addition, other gums, including exudate gums like gum Arabic, are used to coat pecan nuts, to seal in the oil (Arnold 1968), and potatoes, to prevent darkening (Mazza and Qi 1991). Xanthan gum provides a uniform coating with good cling and improved adhesion in wet coating batters (Chen and Nussinovitch 2000). Carrageenan-based coatings have been applied to fresh fruits and vegetables, such as apples, to reduce moisture loss, oxidation, and/or disintegration (Bryan 1972; Lee et al. 2003). Chitosan has been extensively studied as an edible coating for fresh produce, because of its excellent film-forming behavior, oxygen and CO_2 barrier properties, broad antimicrobial activity and compatibility with other substances, such as vitamins, minerals, and antimicrobial agents (Krochta 1997; Shahidi et al. 1999; Park and Zhao 2004; Han et al. 2004a, b; Durango et al. 2006; Vargas et al. 2006; Chien et al. 2007; Ribeiro et al. 2007). A chitosan coating on sliced mango fruit slowed down the rate of water loss and drop in sensory quality; increased the soluble solids content, titratable acidity and ascorbic acid content; and inhibited growth of microorganisms (Chien et al. 2007).

In addition to providing barrier properties, films made from gums have also been studied as carriers of actives, such as antimicrobial agents and nutrients. Kappa-carrageenan films can effectively carry food-grade antimicrobials – such as lysozyme, nisin, grape seed extract, and EDTA – for a wide range of applications in food packaging materials (Choi et al. 2001). These films can also deliver anti-browning agents, such as ascorbic acid, producing positive sensory results and a reduction of microbial levels in minimally-processed apple slices (Lee et al. 2003). Xanthan gum coating was used to carry 5% Gluconal Cal, a mixture of calcium lactate and gluconate, and 0.2% α-tocopherol acetate (vitamin E) (Mei et al. 2002). On its own, chitosan coating has been used for its antifungal properties (Zhang and Quantick 1998; Park et al. 2005). Iverson and Ager (2003) invented a chitosan-based antifungal coating, mixed with an edible wax emulsion and/or a preservative such as sodium benzoate and/or an adhesion additive, such as zinc acetate, and/or a wetting agent to form a composition having a solid content of 15% or higher.

3.4.2 Confection Coatings

Confection coating is applied to candy centers or discreet product pieces in a process called panning. This process is done in a coating pan, a bowl that is motor driven to provide a tumbling action. The product moves continuously during the coating process, and coating material is applied to the product as the pan rotates. Coating material, such as sugar syrup, needs to adhere to the product surface and dry quickly, so that repeated charges of coating material, up to ~60 times, can be applied within a reasonable period of time.

There are three types of panning: soft panning, hard panning and chocolate panning. With soft panning, a sugar shell is built around surfaces of candy centers by wetting them with a sugar syrup, and dusting it with confectioner's sugar or some type of dusting powder. This process produces a soft sugar shell, like that used for jellybeans. Hard panning, on the other hand, produces a harder sugar shell through application of repeated charges of the engrossing and smoothing syrups without dusting. Each individual coat of syrup is dried prior to adding the next, and the result is a denser, more compact sugar shell encasing the candy center (e.g., the sugar shell of gumballs, coated peanuts and chocolate candies). The third type, chocolate panning, consists of building a layer of chocolate around candy centers – such as nuts, maltballs and raisins – by melting the chocolate and spraying it on the centers while tumbling. Coated product is either polished, glazed and packed for sale with a chocolate finish, or hard-panned to add a layer of sugar shell for strength. This shell could also be food-colored and polished for added appeal (Fig. 3.26).

When candy centers contain oil, they must be pre-sealed, or oil will migrate through the chocolate layer to the product surface and will spoil the appearance of the finished confection. Pre-sealing is achieved by coating candy centers with several charges of a 40% concentration of gum Arabic syrup to wet and form a

Fig. 3.26 Confections with hard sugar shell and chocolate finish

film around the centers. Other low-viscosity polysaccharides can also be used, such as larch gum and inulin, separately or in combination with the gum Arabic. High viscosity gums are not suitable for this application, even though they form stronger films or seals, because most of them are already too viscous at 1–2% concentration. Further, wetted centers take too long to dry or can become soggy. In addition, gum Arabic syrup is used for polishing coated product to impart sheen to the surface.

Other than for functions discussed above, gums have limited use in standard confection coatings due to cost. However, for sugar-free recipes, polysaccharides are inevitable replacements for sugar or augmenting ingredients, and are used in combination with sugar alcohols, like maltitol, erythritol, and sorbitol. Again, gum Arabic, larch gum, inulin, and resistant maltodextrin are the most suitable choices due to their low solution viscosities. A gum Arabic and maltitol combination is used for applying a sugar-free shell layer around chewing gums and candies with soft centers. Below is a recipe for sugar-free coated peanuts.

Sugar-free, Chocolate-coated Peanuts

Procedure

Ingredients	lb.
Peanut centers [Batch Size]	100.0
Sealing syrup-40% Gum Arabic	~3.0
Sealing powder (Cocoa/Inulin 50/50 blend)	~5.0
Unsweetened bakers Chocolate for Engrossing	100.0
Sugar-free hard panning:	
Engrossing syrup and smoothing syrup	100.0
Polishing syrup (40% gum Arabic)	~2.0
Shellac glaze	~0.5

1. *Pre-sealing*. Count 50 pieces of candy centers; weigh and record their weight. Place 100 lb. of candy centers in a coating pan, and start rotation of the pan at 25 rpm. Scoop ~500 mL of 40% gum Arabic syrup into the coating pan, and allow the batch to run until all candy center surfaces are wet (about 1 min). Using a scoop, add enough sealing powder to dry the surface. Run until all powder adheres to the candy center surfaces. Repeat the process for the second and third applications. Spread batches onto trays to harden overnight.

2. *Engrossing*. Take 50 pieces of sealed candy centers; weigh and record their weight. Melt the chocolate and keep melted (between 43 and 49°C). If using a clean pan, smear some chocolate inside the pan before engrossing. Start the pan and begin coating with chocolate (by spraying or using a ladle). Use cold air as required to set chocolate and dry product surfaces. Continue coating chocolate to twice the weight of the center. Separate doubles and clusters of candy centers by hand. If needed, run warm air to partially melt chocolate and smooth the surface. Weigh batch and record weights. Spread onto trays to harden in a cool, dry room.

3. *Sugar-Free Hard Panning* (This step is not required for chocolate finish).
   a) Prepare the gum solution, diluting the 40% gum Arabic solution to make a 20% syrup. Prepare the sugar-free dusting powder by mixing maltitol

and inulin at 80:20 ratio. Start the pan and add a charge of gum Arabic syrup to candy centers. Mix for about one minute. Record the amount of syrup added. Add a small amount of dusting powder to the pan in front of the candy centers to get the centers rolling. As candy centers become sticky, add more dusting powder to prevent them from sticking together (begin with just a small amount, and add only enough to keep the centers separate). Dry centers thoroughly with air while centers are tumbling. Add two more charges of gum Arabic syrup, increasing the amount of syrup as needed, and dust in the same way as previously described. Record the volume of syrup used in each charge.

b) Prepare 70% engrossing syrup by dissolving 60 parts maltitol, 3 parts gum Arabic powder, and 7 parts inulin in 30 parts water (by weight). Bring syrup to a boil, and cool it to ~29°C before use. Weigh 50 pieces, and record their weight. If manually ladling syrup, add enough of a charge of engrossing syrup to completely wet the surfaces of the candy centers, distributing syrup throughout the bed while mixing. Record charge size and time on batch sheet. Mix for about 1 min. Apply air to dry the centers thoroughly. If using an automatic spray system dry until humidity in the pan is about 12–16%. Repeat process until half of the desired coating weight has been applied. This is determined by weighing 50 pieces of centers periodically.

c) Prepare 65% smoothing syrup by diluting the 70% engrossing syrup to ~65°Brix. Add a charge of smoothing syrup (1/3 less than the engrossing syrup), and distribute it throughout the bed by mixing. Record the volume added. When candy centers begin to roll freely, apply air to dry them thoroughly as in step 3b. Repeat the process until desired coating weight is achieved. Weigh 50 centers periodically to aid this determination.

d) Add coloring by mixing 20% of color lake dispersion with 80% of smoothing syrup. Apply four charges to candy centers; for each charge, mix until candy centers begin to roll freely, and dry with air thoroughly as in step 3b. Add a fifth charge as follows: divide into two half-charges, apply the first half-charge, and dry without air. When dry, add the second half-charge, but allow candy centers to tumble only until drying starts. Stop the pan and dry centers by jogging the pan every 10 min. Do not leave pan running to prevent scuffing of the coat.

4. *Polishing with gum Arabic.* Use this step only if the candy has a chocolate finish. Place the coated batch in a clean, ribbed pan. Ensure the product is cold and hard. Start the pan and add one wetting of polishing solution of 40% gum Arabic. Stop the pan as soon as the product surface is no longer sticky. Use as little air as needed. Repeat the polishing step one or two more times. Dry thoroughly after the last application. (For sugar-coated product, skip polishing and proceed directly to step 5).

5. *Glazing with Shellac.* Use a clean ribbed pan. Without air, start the pan, and wet batches with a shellac glaze. After wetting, use air to dry. Stop the pan and jog it every minute to prevent pieces from sticking together.

3.4.3 Bakery Coatings-Icings and Glazes

Many bakery products owe their appeal to icings and glazes applied to their surfaces. Toppings on these products are usually what consumers notice first (Fig. 3.27). They not only enhance overall appearance, but also add sweetness and flavor to the finished products, as well as improve texture and shelf life. While consumers enjoy these topical applications, there are many challenges that need to be overcome in ensuring that a commercial icing or glaze, when applied, will remain appealing until the product is consumed. Will glaze melt and disappear in the doughnut on a hot summer day? Will glaze stick to packaging during transport of the product? Will sugar crystallize out of the icing and harden before the product is consumed?

Sugar, the main ingredient in icings and glazes, can be a very problematic ingredient, because it absorbs water from the environment and melts with high humidity and temperature. Therefore, a glaze can easily disappear and sip into a donut as a result of moisture uptake; or even without humidity, it can just melt. A gum like agar is used to minimize melting of the sugar coating, and to give the glaze a set. Either traditional agar or quick-soluble agar can be used in a glaze. However, because agar is quite expensive, kappa-carrageenan can be used as a cheaper alternative in some cases. A gum blend consisting of sodium alginate and gum Arabic, where calcium from gum Arabic forms cross-links between alginate polymers to form a gel and set the glaze.

Agar and alginate-gum Arabic systems are also used in icing to prevent sugar from crystallizing and icing from hardening prematurely. Agar is used in icings at

Fig. 3.27 Gum application in bakery icing and glaze

~0.2% and in doughnut glazes typically at 0.4% or higher to prevent sugar crystallization and reduce tendency of the glaze to chip, crack or "weep". The alginate-gum Arabic system, on the other hand, is used in icing at 1.0%, and may also be used in glazes at 1.0–1.5%. See recipes below.

A. Bakery glaze

Ingredients	%
Water	38.90
Glucose syrup	5.00
Sucrose	55.00
Agar	0.40
Citric acid	0.20
Potassium sorbate	0.10
Flavor/color	q.s. (as much as enough)
	100.00

Procedure

1. Add water and glucose syrup to the mixing tank and start heating.
2. Turn on the mixer, and add potassium sorbate and agar. Bring mixture to a boil, and mix ~3 min until agar is fully dissolved.
3. Add all of the sugar.
4. Continue cooking until Brix is ~60.
5. Lower the temperature to 74°C and add citric acid.
6. Apply glaze warm. (Note: Quick-soluble agar that dissolves at 65–82°C may also be used.)

B. Icing recipe

Ingredients	%
Confectioners sugar (10×)	69.00
Water	28.60
Shortening	2.00
Alginate-gum Arabic blend	1.00
Citric Acid	0.20
Potassium sorbate	0.10
Vanilla	0.10
	100.00

Procedure

1. Add water to batch tank.
2. Add potassium sorbate.
3. Dry blend gums with 10 parts of sugar and add blend to water in the batch tank while mixing.
4. Add vanilla and the rest of the sugar.
5. Add shortening to the mix, and cream.
6. Mix well, and heat to 120°F.

In addition to icings and glazes, another coating applied to bakery products that deserves to be mentioned under this heading is the egg wash. Breads and pastries will look much more appealing with an egg wash applied on the surface before they are baked. A polysaccharide, such as gum Arabic, can actually replace egg wash in this application. Gum Arabic provides sheen that adds shelf appeal; in addition, it is much less prone to burning.

3.5 Regulatory Status

Most gum polymers discussed in this chapter are approved for use in foods and/or are GRAS, except for chitosan and cassia gum, which are only allowed in supplements and pet food applications, respectively. This is good news because regulation of edible films will take an easier route if using food approved or GRAS ingredients. Druchta and Johnston (1997) published a scientific status summary on edible films that includes their safety and health issues. They reported that the process of determining acceptability of materials for edible polymer films follows procedures identical to those used in determining appropriateness of such materials for food formulation. The process is outlined as follows:

1. An edible polymer will be generally recognized as safe (GRAS) for use in edible films if the material has previously been determined as GRAS, and its use in an edible film is in accordance with current good manufacturing practices (food grade, prepared and handled as a food ingredient, and used in amounts no greater than necessary to perform its function), and within any limitations specified by the Food and Drug Administration (FDA).
2. If the edible polymer film uses a protein – such as wheat, soy, corn or any ingredient that could potentially contain an allergen – each such ingredient must be declared appropriately to the consumer, no matter how small the amount used, with the appropriate warning about the allergenic ingredient.
3. If the edible polymer film material used is not currently GRAS, but the manufacturer can demonstrate its safety, the manufacturer may either file a GRAS Affirmation Petition to the FDA or proceed to market the material without FDA concurrence (self-determination).
4. The manufacturer may not need to establish that use of an edible polymer in edible films is GRAS if the material had received pre-1958 FDA clearance and thus has "prior sanction."
5. Finally, if the material cannot be demonstrated to be GRAS or does not have "prior sanction," the manufacturer must submit a food additive petition to the FDA prior to its use in films and coatings.

If the intent is to make biodegradable polymer packaging film, the FDA regulates all materials proposed for food packaging to ensure they are safe for food contact under conditions of the intended use. As summarized by Druchta and Johnston (1997), there are several categories of biodegradable polymers acceptable for use as food packaging:

1. If a biodegradable polymer developed for use in food packaging is found to be GRAS by the developer (self-determination) or affirmed as GRAS by the FDA,

the polymer may be used in food packaging; if a GRAS polymer is combined with a food-grade synthetic polymer to enhance biodegradable character, the developer should supply the FDA with information that addresses whether the proposed use of the product can be considered a "good manufacturing practice." FDA requires information on the migration profile and identity of migrants from biodegradable packaging to food during typical storage conditions and times; the agency also requires information on environmental aspects of the biodegradable package use.

2. If a biodegradable polymer developed for use in food packaging is not GRAS, it may be used only after a food additive petition and environmental assessment are approved.

Most of polysaccharides are natural polymers; these offer the greatest opportunities as materials for biodegradable films and packages, since their safety, biodegradability and environmental compatibility are already assured. Even the chemically derived gum polymers – such as CMC, HPMC and PGA – that are food approved, are potential polymeric materials for use in biodegradable packaging. However, if these gum polymers require complexing or cross-linking with other chemicals to achieve rigidity, elasticity or water impermeability of their plastic packages, then their regulation changes and could follow the fate of non-edible, polymeric materials. Examples of these materials include cellophane, microbial polyesters (e.g., polyhydroxybutyrate/valerate co-polymers), polylactic acid, and combinations of starch with polyvinyl alcohol, where migration of toxic components to food could be a possibility. Hence, in this case, safety of these packaging films needs to be established first with the FDA. Any questionable substance that can reasonably be expected to migrate from the package into foods is classified as an indirect food additive, subject to FDA regulations per 21 CFR 170.39.

3.6 Future Trends

3.6.1 Novelty Films

With the versatility and established safety of most gum polymers, novel applications for gum films seem almost limitless. The concept of drug, vitamin and mineral delivery films is likely to expand and, while there are more and more novelty film products appearing in stores, other creative concepts are likely to evolve, including cheese, ketchup, or sour cream films, or various spice films that food scientists, product formulators or chefs could create. Convenience and microbial stability of dried forms of food films is the biggest selling point for developing these products. The market for edible films has already experienced noteworthy growth, from ~$1 million in 1999 to more than $100 million per year today. According to Tara McHugh of the U.S. Department of Agriculture, as quoted by

Klapthor,[2] breath mint and cold medication markets have been responsible for much of that growth. James Rossman, former maker of edible films and a current consultant, projects that retail sales of edible film, about $100 million a year today, are expected to hit at least $350 million by 2008 and $2 billion by 2012 (http://www.foodnavigator.com/Financial-Industry/Tate-Lyle-invests-in-edible-films-market).[3]

3.6.2 Biodegradable Films and Packages

Developing cost-effective biodegradable films and rigid/semi-rigid packages that match properties of synthetic plastics is still a task to be accomplished in the future. In 2005, a total of 28.9 tons of synthetic plastic materials were generated and 16.3 tons discarded as municipal solid waste (EPA 2006). Likewise, there have always been health concerns regarding residual monomers and components in plastics – including stabilizers, plasticizers and condensation components, such as bisphenol – that can migrate into food (Marsh and Bugusu 2007). However, bioplastics are currently much more expensive than most petroleum-based polymers, so substitution is likely to result in increased packaging costs. Cellophane, a biodegradable, cellulose-derived film, was invented a long time ago (in 1908), but has not really competed with plastics. Natureworks, LLC, owned by Cargill Inc., manufactures polylactide from corn sugar; and now Wal-Mart uses biopolymers by employing polylactide to package fresh-cut produce (Bastioli 2005). Research should focus on improving heat sealability and the printability of biodegradable films, in addition to making them cost-competitive with plastics.

3.6.3 Edible Films and Coatings

According to Lin and Zhao (2007), despite significant benefits from using edible coatings for extending product shelf life and enhancing the quality and microbial safety of fresh and minimally-processed fruits and vegetables, commercial applications on a broad range of edible coating of fruits and vegetables are still very limited. Of many improvements that needed for polysaccharides to be used more extensively as edible film and coating materials, most important is improving the water-resistance of gum films. Creation of nanocomposites, incorporation of lipid to the gum polymer matrix, or formation of nanolaminates can provide solutions to this challenge. Weiss et al. (2006), in their scientific status summary on materials in food nanotechnology, describe how a polysaccharide-clay nanocomposite can be formed. A carbohydrate is pumped together with clay layers, through a high shear cell, to produce a film that contains exfoliated clay layers.[4] The nanocomposite carbohydrate film has substantially

---

[2] http://www.am-fe.ift.org/cms/?pid=1000355]
[3] http://www.businessweek.com/magazine/content/06_31/ b3995059.htm
[4] http:www.pre.wur.nl/ UK/Research/Food+Structuring/ Microstructured/

reduced water vapor permeability. Similarly, chitosan containing exfoliated hydroxyapatite layers maintains functionality in humid environments, with good mechanical properties.

Incorporation of hydrophobic ingredients – such as lipids for improving the moisture barrier, while maintaining the desirable resistance to vapor, gas, or solute is another research area that can be pursued. Decher and Schlenoff (2003) created a nanolaminate based on deposition technique in which charged surfaces are coated with interfacial films that consist of multiple nanolayers of different materials. Nanolaminates can give food scientists advantages for the preparation of edible coatings and films over conventional technologies, because they can be more efficient moisture, lipid and gas barriers.

3.6.4 Gum Suppliers

Aqualon
   Products: Cellulose Derivatives, Guar
   http://aqualon.com
Colloides Naturels International
   Products: Gum arabic
   http://www.cniworld.com
CP Kelco
   Products: Carrageenan, cellulosics, gellan, pectin, xanthan
   http://www.cpkelco.com
Danisco
   Products: Pectin, LBG, guar gum, alginate, carrageenan, xanthan gum, cellulose gum
   http://danisco.com
Degussa
   Products: General Hydrocolloids and blends (alginates, agar, pectin, xanthan, carrageenans, galactomannan)
   http://www.texturantsystems.com
FMC biopolymer
   Products: Carrageenan, alginate, microcrystalline cellulose and konjac
   http://www.fmcbiopolymer.com/
Gum technology
   Products: General hydrocolloids and blends (Acacia, Agar, Alginates, Carrageenans, Cellulose Gum, Fenugreek, Gelatin, Guar Gum, Gum Blends, Inulin, Karaya, Konjac, Locust Bean Gum, Oat Fiber, Pectin, PGA, Pullulan, Tara Gum, Tragacanth, Xanthan Gum)
   http://www.gumtech.com/products/index.php
ISC [Importers Service Corporation]
   Products: Gum Arabic, Karaya, Ghatti, Tragacanth
   http://www.iscgums.com/product.htm

PL Thomas
    Products: General hydrocolloids and blends
    http://www.plthomas.com/gums_overview.htm
Roko
    Products: agar
    http://www.rokoagar.com
Tate and Lyle
    Products: Xanthan
    http://www.tateandlyle.co.uk
TIC Gums, Inc.
    Products: General Hydrocolloids and specialty blends (Gum Acacia, Agar,
        Alginates, Carrageenans, Cellulose Gum, Guar Gum, HPMC, Tara gum,
        Gum Ghatti, Inulin, Karaya, Konjac, Locust Bean Gum, Methylcellulose,
        Modified Gum Acacia, Pectin, PGA, Tara Gum, Tragacanth, Xanthan
        Gum, and specialty blends of gums, starches and emulsifiers)
    http://www.ticgums.com

Acknowledgements I would like to recognize the assistance of the following TIC Gums employees in preparing and testing the gum films and in editing the photos and figures shown in this chapter for print quality: Nick Pippen, Dan Grazaitis, Renrick Atkins, Frances Bowman and Maureen Akins.

References

Amanatidou A, Slump RA, Gorris LGM, Smid EJ (2000) High oxygen and high carbon dioxide modified atmospheres for shelf life extension of minimally processed carrots. J Food Sci 65:61–6

Arnold FW (1968) Infrared roasting of coated nutmeats. U.S. patent 3, 383, 220

Aspinall GO, Christensen JB (1965) Gum ghatti (Indian gum) Part IV. Acidic oligosaccharides from the gum. J Chem Soc 2673–2676 DOI: 10.1039/JR9650002673

Aspinall GO, Hirst EL, Wickstrom A (1955) Gum ghatti (Indian gum) Part I. The composition of the gum and the structure of two aldobiuronic acids derived from it. J Chem Soc 1160–1165 DOI: 10.1039/JR9550001160

Aspinall GO, Auret BJ, Hirst EL (1958a) Gum ghatti (Indian gum) Part II. The hydrolysis products obtained from the methylated degraded gum and the methylated gum. J Chem Soc 221–230 DOI: 10.1039/JR9580000221. 47

Aspinall GO, Auret BJ, Hirst EL (1958b) Gum ghatti (Indian gum) Part III. Neutral oligosaccharides formed on partial acid hydrolysis of the gum. J Chem Soc 4408–4414 DOI: 10.1039/JR9580004408

Aspinall GO, Bhavanadan VP, Christensen JB (1965) Gum ghatti (Indian gum) Part V. Degradation of the periodate-oxidized gum. J Chem Soc 2677–2684 DOI: 10.1039/JR9650002677

Banks NH (1985) Internal atmosphere modification in pro-long coated apples. Acta Hort 157:105

Bastioli C (2005) Handbook of biodegradable polymers. Toronto-Scarborough, Ontario, Canada, Chem Tec Publishing, 553 p

Bender H, Wallenfels K (1961) Investigations on pullulan II. Specific degradation by means of a bacterial enzyme. Biochem Z 334:79–95

Bernier B (1958) The production of polysaccharides by fungi active in the decomposition of wood and forest litter. Can J Microbiol 4:195–204

Bouveng HO, Kiessling H, Lindberg B, McKay J (1962) Polysaccharides elaborated by Pullularia pullulans. I. The neutral glucan synthesized from sucrose solutions. Acta Chem Scand 16:615–622

Bouveng HO, Kiessling H, Lindberg B, McKay J (1963) Polysaccharides elaborated by Pullularia pullulans. II. The partial acid hydrolysis of the neutral glucan synthesised from sucrose solutions. Acta Chem Scand 17:797–800

Bryan DS (1972) Prepared citrus fruit halves and method of making the same. U.S. patent 3, 707, 383

Carolan G, Catley BJ, McDougal FJ (1983) The location of tetrasaccharide units in pullulan. Carbohydr Res 114:237–243

Catley BJ (1970) Pullulan, a relationship between molecular weight and fine structure. FEBS Lett 10:190–193

Catley BJ, Whelan WJ (1971) Observations on the structure of pullulan. Arch Biochem Biophys 143:138–142

Catley BJ, Robyt JF, Whelan WJ (1966) A minor structural feature of pullulan. Biochem J 100:5P–8P

Chandramouli V, Kailasapathy K, Peiris P, Jones M (2004) An improved method of microencapsulation and its evaluation to protect *Lactobacillus* spp. in simulated gastric conditions. J Microbiol Methods 57:27–35

Chandrasekaran R, Radha A (1995) Molecular architectures and functional properties of gellan gum and related polysaccharides. Trends Food Sci 6:143–148

Chaplin M (2008) Water Structure and Science – Gum Arabic http://www.lsbu.ac.uk/water/hyarabic.html

Chen S, Nussinovitch A (2000) The role of xanthan gum in traditional coatings of easy peelers. Food Hydrocolloids 14(4):319–326(8)

Chien PJ, Sheu F, Yang FH (2007) Effects of edible chitosan coating on quality and shelf life of sliced mango fruit. J Food Eng 78:225–9

Choi JH, Cha DS, Park HJ (2001) The antimicrobial films based on Na-alginate and κ-carrageenan. In: IFT Annual Meeting, Food Packaging Division (74D). New Orleans, LA

Conca KR, Yang TCS (1993) Edible food barrier coatings. In: Ching C, Kaplan D, Thomas D (eds) Biodegradable Polymers and Packaging. Technomic Publishing Co. Inc, Lancaster, PA, pp 357–69

Cottrell IW, Kovacks P (1980) Alginates. In: Davidson RL (ed) Handbook of Water-Soluble Gums and Resins. McGraw-Hill, New York, NY, p 143

Daas PJH, Schols HA, de Jongh HHJ (2000) On the galactosyl distribution of commercial galactomannans. Carbohydr Res 329:609–619

Decher G, Schlenoff JB (2003) Multilayer thin films: sequential assembly of nanocomposite materials. Wiley VCH, Weinheim, Germany, p 543

DeLeenheer L, Hoebregs H (1994). Progress in the elucidation of the composition of chicory inulin. Starch 46:193

Donati I, Holtan S, Mørch YA, Borgogna M, Dentini M, Skjåk-Bræk (2005) New hypothesis on the role of alternating sequences in calcium-alginate gels. Biomacromolecules 6:1031–1040

Dow Chemical (2002) METHOCEL Cellulose Ethers Technical Handbook. http://www.dowchemical.com/methocel/ pharm/ resource/lit_gnl.htm

Draget KI (2000) Alginates. In: Phillips GO, Williams PA (eds) Handbook of Hydrocolloids. CRC , Boca Raton, FL pp 379–395

Druchta JM, Jonhston CD (1997) An Update on Edible Films-Scientific Status Summary. Food Technol 51(2):60, 62–63

Durango AM, Soares NF, Andrade NJ (2006) Microbiological evaluation of an edible antimicrobial coating on minimally processed carrots. Food Control 17:336–41

Ebringerova A, Hromadkova Z, Heinze T (2005) Hemicellulose. In: Heinze T (ed) Advances in Polymer Science [186] Polysaccharides-Structure Characterization and Use. Springer, Berlin, pp 1–68

Environmental Protection Agency, US (2006) Municipal solid waste in the United States: 2005 facts and figures. EPA530-R-06-011, Washington D.C., p 153

Evans AJ, Hood RL, Oakenfull DG, Sidhu GS (1992) Relationship between structure and function of dietary fibre: A comparative study of the effects of three galactomannans on cholesterol metabolism in the rat. Br J Nutr 68(1):217–229

Falshaw R, Bixler HJ, Johndro K (2001) Structure and performance of commercial kappa-2 carrageenan extracts I. Structure analysis. Food Hydrocolloids 15:441–452

Fisher LG, Wong P (1972) Method of forming an adherent coating on foods. U.S. patent 3, 676, 158

Giannouli P, Morris ER (2000) Cryogelation of xanthan. Food Hydrocolloids 17:495–501

Goodrum LJ, Patel A, Leykam JF, Kieliszewski MJ (2000) Gum arabic glycoprotein contains glycomodules of both extensin and arabinogalactan-glycoproteins. Phytochemistry 54:99–106

Goycoolea FM, Milas M, Rinaudo M (2001) Associative phenomena in galactomannan-deacetylated xanthan systems. Int J Biol Macromol 29:181–192

Han C, Lederer C, McDaniel M, Zhao Y (2004a) Sensory evaluation of fresh strawberries (*Fragaria ananassa*) coated with chitosan-based edible coatings. J Food Sci 70:S172–8

Han C, Zhao Y, Leonard SW, Traber MG (2004b) Edible coatings to improve storability and enhance nutritional value of fresh and frozen strawberries (*Fragaria ananassa*) and raspberries (*Rubus ideaus*). Postharvest Biol Technol 33:67–78

Hanna D, Shaw EHJ (1941) Chemical constitution of gum ghatti. Proc S Dakota Acad Sci 21:78

Hershko V, Nussinovitch A (1998) Relationships between hydrocolloid coating and mushroom structure. J Agric Food Chem 46:2988–97

Iijima H, Takeo K (2000) Microcrystalline cellulose: an overview. In: Handbook of Hydrocolloids, Phillips GO, Williams PA (eds) CRC/Woodhead Publishing, Boca Raton, p 664

Iverson CE, Ager SP (inventors) CH.sub.2O Incorporated (assignee)(2003) Method of coating food products and a coating composition. U.S. patent 6, 586, 029

Jansson PE, Keene L, Lindberg B (1975) Structure of the exocellular polysaccharide from *Xanthomonas campestris*, Carbohydr Res 45:275–82

Juang RS, Shao HJ (2002) A simplified equilibrium model for sorption of heavy metal ions from aqueous solutions on chitosan. Water Res 36(12):2999–3008

Kato K (1973) Isolation of oligosaccharides corresponding to the branching-point of Konjac Mannan. Agr Biol Chem 37(9):2045–2051

Kato K, Matsuda K (1969) Studies on the chemical stucture of konjac mannan. Part I. Isolation and characterization of oligosaccharides from the partial acid hydrolyzate of mannan. Agric Biol Chem 33, 1446–1453.

Kester JJ, Fennema OR (1986) Edible films and coatings: a review. Food Technol 40(12):47–59

King AH (1983) Brown seaweed extracts (alginates). In: Food Hydrocolloids. vol II Glicksman M (ed) CRC, Boca Raton FL, pp 115–88

Kirk-Othmer (1993) Cellulose ethers. In: Encyclopedia of Chemical Technology, 4th edn. vol 5, Kroschwitz J (ed). Wiley, New York, NY, pp 541–561

Krochta JM (1997) Edible protein films and coatings. In: Food Proteins and Their Applications. Damodaran S, Paraf A (eds) Marcel Dekker, New York, NY

Krumel KL, Lindsay TA (1976) Nonionic cellulose ethers. Food Technol 30(4):36–8, 40, 43

Labropoulos KC, Niesz DE, Danforth SC, Kevrekidis PG (2002) Dynamic rheology of agar gels: theory and experiment. Part I. Development of a rheological model. Carbohydr Polym 50:393–406

Le Cerf D, Irinei F, Muller G (1990) Solution properties of gum exudates from Steculia urens [karaya gum]. Carbohydr Polym 13(4):375–386

Leathers TD (2003) Biotechnological production and applications of pullulan. Appl Microbiol Biotechnol. 62:468–473

Lee JY, Park HJ, Lee CY, Choi WY (2003) Extending shelf-life of minimally processed apples with edible coatings and antibrowning agents. Lebens Wissen Technol 36:323–9

Li B, Xie B, Kennedy JF (2006) Studies on the molecular chain morphology of konjac glucomannan. Carbohydr Polym 64(4):510–515

Lin D, Zhao Y (2007) Innovations in the Development and Application of Edible Coatings for Fresh and Minimally Processed Fruits and Vegetables. Compr Rev Food Sci Food Safety 6(3):60–75

Lootens D, Capel F, Durand D, Nicolai T, Boulenguer P, Langendorff V (2003) Influence of pH, Ca concentration, temperature and amidation on the gelation of low methoxyl pectin. Food Hydrocolloids 17:237–244

Lowings PH, Cutts DF (1982) The preservation of fresh fruits and vegetables. In Proceedings of the Institute of Food Science and Technology Annual Symposium, Nottingham, UK, p 52

Maeda M, Shimahara H, Sugiyama N (1980) Detailed examination of the branched structure of konjac glucomannan. Agric Biol Chem 44:245–252

Maekaji K (1974) The mechanism of gelation of konjacmannan. Agric. Biol. Chem., 38, 315–321

Maekaji K (1978) Nippon Nogeikagakukaishi. 52:251, 485, 513

Maftoonazad N, Ramaswamy HS (2005) Postharvest shelf-life extension of avocados using methyl cellulose-based coating. Lebens Wissen Technol 38:617–24

Mason DF (1969) Oct 14. Fruit preservation process. U.S. patent 3, 472, 662

Mazza G, Qi H (1991) Control of after cooking darkening in potatoes with edible film-forming products and calcium chloride. J Agric Food Chem 39:2163–6

Marsh K, Bugusu B (2007). Food packaging-roles, materials and environmental issues. J Food Sci 72(3):39–55

May CD (2000) Pectins. In: Handbook of Hydrocolloids. Phillips GO and Williams, PA (eds) Woodhead Publishing Ltd, Cambridge, UK, pp. 219–229

Meheriuk M, Lau OL (1988) Effect of two polymeric coatings on fresh quality of 'Bartlett' and 'd'Anjou' pears. J Am Soc Hort Sci 113:222–6

Mei Y, Zhao Y, Yang J, Furr HC (2002) Using edible coating to enhance nutritional and sensory qualities of baby carrots. J Food Sci 67(5):1964–1968

Melton LD, Mindt L, Rees DA, Sanderson GR (1976) Covalent structure of the polysaccharide from Xanthomonas campestris: Evidence from partial hydrolysis studies. Carbohydr Res 46:245–57

Morgan BH (1971) Edible packaging update. Food Prod Devel 5:75–7, 108

Murray JCF (2000) Cellulosics. In: Handbook of Hydrocolloids. Phillips GO, Williams PA (eds) Woodhead Publishing Ltd, Cambridge, UK, pp 219–229

Naouli N (2006) Molecular weight analysis of linear and branched gums using triple detection technique. Application Report Number 0076-06 by Polymer Laboratories. (Unpublished)

Nieto M, Grazaitis D (2006) Edible Films with Unique Properties Using Gums. (Unpublished) Presented at IFT in Orlando, FL

Nisperos-Carriedo MO, Baldwin EA (1990) Edible coatings for fresh fruits and vegetables. In: Subtropical Technology Conference Proceedings. Lake Alfred, FL

Park SI, Zhao Y (2004) Incorporation of a high concentration of mineral or vitamin into-chitosan-based films. J Agric Food Chem 52:1933–9

Park S, Stan SD, Daeschel MA, Zhao Y (2005) Antifungal coatings on fresh strawberries (Fragaria ananassa) to control mold growth during cold storage. J Food Sci 70: M202–7

Pedroni VI, Schulz PC, Gschaider ME, Andreucetti N (2003) Chitosan structure in aqueous solution. Colloid Polym Sci 282(1):100–102

Petkowicz CLO, Reicher F, Mazeau K (1998) Conformational analysis of galactomannans: from oligomeric segments to polymeric chains. Carbohydr Polym 37:25–39

Pérez S, Mazeau K, Hervé du Penhoat C (2000) The three-dimensional structures of the pectic polysaccharides. Plant Physiol Biochem 38:37–55

Pérez S, Rodríguez-Carvajal MA, Doco T (2003) A complex plant cell wall polysaccharide: rhamnogalacturonan II. A structure in quest of a function, Biochimie 85:109–121

Ralet MC, Dronnet V, Buchholt HC, Thibault JF (2001) Enzymatically and chemically de-esterified lime pectins: characterisation, polyelectrolyte behaviour and calcium binding properties. Carbohydr Res 336:117–125

Ribeiro C, Vicente AA, Teixeira JA Miranda C (2007) Optimization of edible coating composition to retard strawberry fruit senescence. Postharvest Bio Technol 44:63–70

Risbud M, Hardikar A, Bhonde R (2000) Growth modulation of fibroblast by chitosan-polyvinyl pyrrolidone hydrogel: implications for wound management. J Biosci 25(1):147–159

Sacharow S (1972) Edible films. Packaging 43:6, 9

Sanderson GR (1981) Polysaccharides in foods. Food Technol 35:50–7, 83

Santerre CR, Leach TF, Cash JN (1989) The influence of the sucrose polyester, Sempfresh, on the storage of Michigan grown 'McIntosh' and 'Golden Delicious' apples. J Food Proc Preserv 13:293–305

Shahidi F, Arachchi JKV, Jeon Y (1999) Food applications of chitin and chitosans. Trends Food Sci Technol 10:37–51

Shimahara H, Suzuki H, Sugiyama N, Nishizawa K (1975a) Isolation and Characterization of Oligosaccharides from an Enzymic Hydrolysate of Konjac Glucomannan. Agri Biol Chem 39(2):293–299

Shimahara H, Suzuki H, Sugiyama N, Nishizawa K (1975b) Partial Purification of ß-Mannanases from the Konjac tubers and Their Substrate Specificity in Relation to the Structure of Konjac Glucomannan. Agri Biol Chem 39(2):301–312

Sowa W, Blackwood AC, Adams GA (1963) Neutral extracellular glucan of Pullularia pullulans (de Bary) Berkhout. Can J Chem 41:2314–2319

Srinivasa P, Baskaran R, Ramesh M, Harish Prashanth K, Tharanathan R (2002) Storage studies of mango packed using biodegradable chitosan film. Eur Food Res Technol 215(6):504–508

Stephen AM, Churms SC (1995) Gums and mucilages. In: Food Polysaccharides and Their Applications. Stephen AM (ed) Marcel Dekker, New York, NY, pp 377–425

Sugiyama N, Shimahara H, Andoh T, Takemoto M, Kamata T (1972) Molecular weights of Konjac Mannans of various sources. Agri Biol Chem 36(8):1381–1387

Sworn G (2000) Gellan gums. In: Handbook of Hydrocolloids. Phillips GO, Williams PA (eds) CRC/Woodhead Publishing, Boca Raton, pp 117–35

Takahashi R, Kusukabe I, Kusano S, Sakurai Y, Murakami K, Maekawa A, Suzuki T (1984) Structures of Glucomanno-oligosaccharides from the Hydrolytic Products of Konjac Glucomannan Produced by a β-Mannanase from Streptomyces sp. Agric Biol Chem 48, 2943–2950

Takigami S (2000) Konjac Mannan. In: Handbook of Hydrocolloids. Phillips GO, Williams PA (eds) CRC/Woodhead Publishing, Boca Raton pp 413–424

Tischer CA, Iacomini M, Wagner R, Gorin PAJ (2002) New structural features of the polysaccharide from gum ghatti (Anogeissus latifolia). Carbohydr Res 37:2205–2210

Tsujisaka Y, Mitsuhashi M (1993) Pullulan. In: Whistler RL, BeMiller JN (eds) Industrial Gums. Polysaccharides and Their Derivatives, 3rd edn. Academic, San Diego, CA, pp 447–460

Ueda S, Fujita K, Komatsu K, Nakashima Z (1963) Polysaccharide produced by the genus Pullularia. I. Production of polysaccharide by growing cells. Appl Microbiol 11:211–215

Vargas M, Albors A, Chiralt A, Gonzalez-Martinez C (2006) Quality of cold-stored strawberries as affected by chitosan-oleic acid edible coatings. Postharvest Biol Technol 41:164–71

Verbeken D, Dierckx S, Dewettinck K (2003) Exudate gums: occurence, production, and applications, Appl Microbiol Biotechnol 63:10–21

Wallenfels K, Bender H, Keilich G, Bechtler G (1961) On pullulan, the glucan of the slime coat of Pullularia pullulans. Angew Chem 73:245–246

Wallenfels K, Keilich G, Bechtler G, Freudenberger D (1965) Investigations on pullulan. IV. Resolution of structural problems using physical, chemical and enzymatic methods. Biochem Z 341:433–450

Weiss J, Takhistov P, McClements J (2006) Functional materials in food nanotechnology. J Food Sci 71(9):107–116

Wielinga WC, (2000) Galactomannans. In: Handbook of Hydrocolloids, Phillips GO, Williams PA (eds) CRC/Woodhead Publishing, Boca Raton, pp 413–423

Williams PA, Phillips GO (2000) Gum Arabic. In: Handbook of Hydrocolloids. Phillips GO, Williams PA (eds) CRC/Woodhead Publishing, Boca Raton, pp 155–168

Yuen S (1974) Pullulan and its applications. Process Biochem 9:7–9

Zhang D, Quantick PC (1998) Antifungal effects of chitosan coating on fresh strawberries and raspberries during storage. J Hort Sci Biotechnol 73:763–7

Chapter 4
Structure and Function of Starch-Based Edible Films and Coatings

Michael E. Kramer

4.1 Introduction

Edible films and coatings satisfy a variety of needs and meet specific product challenges for a large number of food applications. There is a general lack of agreement as to what constitutes a coating. A layer of seasoning on a snack or an oil spray applied to a cracker or a baked product, are examples of edible coatings. Further examples include soft, hard and chocolate panning in confectionery; application of carnauba wax to a gummy candy to preserve individual piece identity; application of icings or glazes to baked goods, use of caramel coatings for popcorn, and enrobing or dipping items in chocolate. Layers of barbeque sauce or fruit glaze on meats are coatings. Seasonings added to a chip, extruded corn collet or a rub applied to a chicken wing are often referred to as coatings. Tempura, battered and breaded fried appetizers are dependent upon coatings for their crunchy texture and eating quality. Egg wash layer added to yeast-leavened baked items for gloss is a protein-based aqueous edible coating. Liquid between air cells in foam could be described as a solute-stabilized edible film. The early Apollo astronauts ate foods coated with starch-based films to prevent the crumbs from becoming airborne and floating around the weightless environment of the cabin.

There are also examples of less traditional coatings and films, which are often freestanding or self-supporting. Edible packaging has been a topic of interest for many years, though few if any commercial examples exist. An invisible, edible coating would likely be more acceptable to consumers than petroleum-based plastic packaging. Edible coatings can make excessive packaging unnecessary, which is also perceived as a positive consumer benefit.

Freestanding films, such as breath strips and film drug delivery systems, are relatively new phenomena. Though these freestanding films are often formulated with the same kinds of ingredients as traditional edible coatings, there are additional

M.E. Kramer
Technical Services, Grain Processing Corporation, 1600 Oregon Street,
Muscatine, IA 52761, USA
e-mail: mike_kramer@grainprocessing.com

M.E. Embuscado and K.C. Huber (eds.), *Edible Films and Coatings for Food Applications*, 113
DOI 10.1007/978-0-387-92824-1_4, © Springer Science+Business Media, LLC 2009

demands on their performance. They have to possess strength and integrity to be removed from backing material and then undergo whatever processing is necessary to deliver the film to its intended use. One of the biggest challenges to freestanding films is to have individual layers remain separate and not adhere to each other or to curl once product is cut into individual pieces or sheets. Usually, attention must be paid to the packaging material to ensure that the water-soluble polymers used in these products do not absorb moisture, which would result in the individual pieces sticking together.

Use of starch and starch derivatives for producing edible coatings has a long history (Kester and Fennema 1986). Generally, such coatings have the advantage of lower cost compared to other alternative high tensile strength materials. Use of high amylose cornstarch improves the properties of the resulting films. Chemical substitution and acid hydrolysis of amylose-containing starches improve clarity and flexibility of coatings made from them. Glycerin and polyhydric alcohols are effective plasticizing agents, which improve film properties but do so at some expense of barrier characteristics. Applications and methods used for preparation of edible starch films and coatings are described in this chapter.

4.2 Starch Structure and Properties

Starch is the main polysaccharide energy storage material in the plant kingdom. It is a mixture of the predominantly linear α–(1→4) glucan or amylose (Fig. 4.1), and the highly branched, high molecular weight glucan or amylopectin (Fig. 4.2). Amylopectin is a glucan with α–(1→4) glycosidic linkages containing α–(1→6) branch points.

Of the two starch polymers, amylose is more closely associated with the ability to form films and coatings due to its predominant linear nature. Starch is both abundant in nature and readily available as an inexpensive commercial product. A film or coating can be made from any type of starch that contains amylose. Amylose derived from fractionated starch has also been used to make all-amylose films,

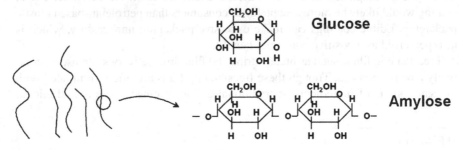

Fig. 4.1 Structure of amylose

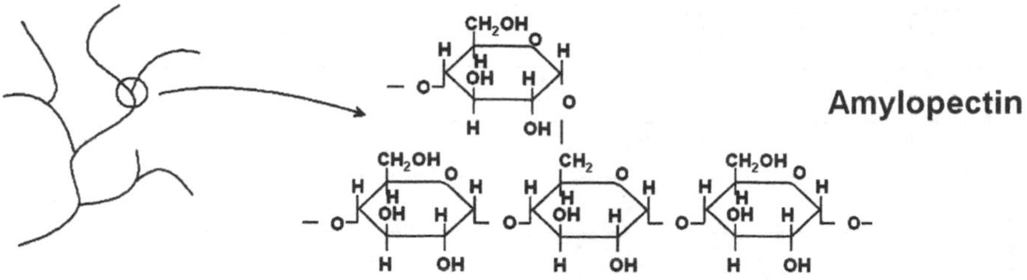

Fig. 4.2 Structure of amylopectin

resulting in several predictable properties that can be attributed to the amylose fraction of starch (Galliard 1987).

All starches with some amylose will form films when cast under the right conditions. Young (1984) cites examples from the 1960s using amylose-containing starch and fractionated amylose and amylopectin to form films. The presence of water as a plasticizer is identified here. Water generally decreases the tensile strength of the film by weakening intermolecular forces.

Amylose and amylopectin molecules are packaged into semi-crystalline aggregates and these aggregates are systematically packaged into starch granules. In these tightly packed starch granules, the starch molecules have little affinity for water and are not functional (Blanshard 1987). Native starch granules observed microscopically under plane-polarized light exhibit the characteristic polarization or "Maltese" cross, which is due to birefringence properties of the crystalline portion of the granule.

If starch granules remained in their inert state as they are found in nature, there would be little interest in starch for most applications. When heat is applied to native starch in the presence of sufficient water, granules begin to open up, swell and hydrate, which initiates the process of "gelatinization" or loss of granular and molecular order. The physical size of the granule swells to several times its original size, and native crystallinity is lost. Microscopic evaluation will demonstrate loss of the polarization cross.

Gelatinization occurs over a range of temperatures, and is dependent upon the type of starch and its modification. The most apparent change upon starch gelatinization is a dramatic increase in viscosity. Monitoring the viscosity through heating and gelatinization steps is used to characterize different starches. Instruments used to generate pasting profiles include the Rapid Visco Analyser (RVA), manufactured by Newport (Fig. 4.3) and Brabender Viscoamylograph. Gelatinized starch has swollen starch granules with some of the amylose leached into the surrounding inter-granular fluid. If the granule is heated further, the granules grow larger and larger until eventually they lose their integrity and identity, with amylose and amylopectin released to the surrounding liquid.

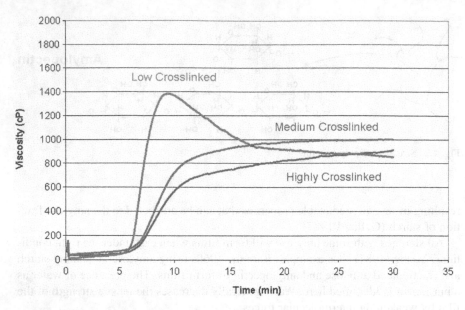

Fig. 4.3 RVA graphs of selected modified starches

4.3 Chemical Modification

In commercial food processing, there are demands placed upon starch, which cannot be met with native starch forms. To compensate for the various ways that native starches fail to deliver optimal performance, starches are chemically modified to enhance their properties or to create starch products with unique properties for various applications. The Food and Drug Administration limits the types of reactions and reagent levels that may be employed for the manufacture of food-grade modified starches. The Code of Federal Regulations defines "modified food starch, or food starch-modified" (21CFR172.892) specifically, and methods that deviate from those outlined may not be called food starch-modified (Anonymous 2007a). The starch manufacturer and food processor using these ingredients need not identify specific reactions or treatments used to produce modified food starches, provided the detailed definition is met.

4.3.1 Cross-Linked Starches

For the majority of food processing applications, fully swollen, intact granules are desired for maximum viscosity and optimum water stabilization. Fruits, vegetables and other high water content foods have a tendency to leach water into the food matrix continuously in the first 3–4 days after processing and fully intact granules

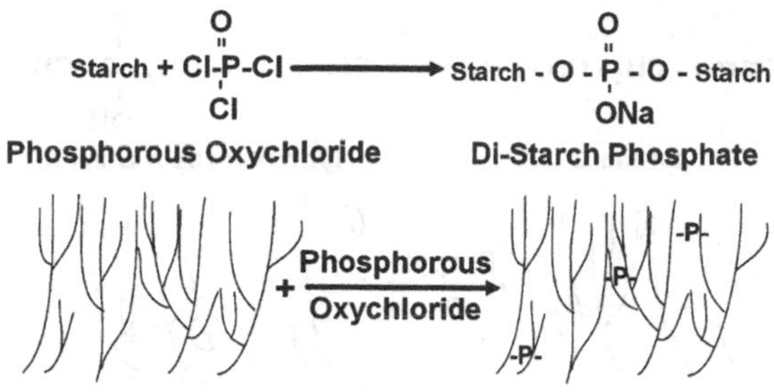

Fig. 4.4 Cross-linking using phosphorous oxychloride

will continue to absorb this freed water to maintain constant viscosity and texture. Many types of modified food starch products are chemically cross-linked using phosphorous oxychloride, sodium trimetaphosphate or adipic anhydride to prevent fragmentation of starch granules (Fig. 4.4). These chemical bonds reinforce the native granule structure, which prevents early swelling and cookout, as well as unwelcome granule disruption.

Specially modified starches designed for formation of films and coatings are usually not designed to maintain granule integrity. Full functionality of a coating starch is achieved only when the majority of starch granules are fully disrupted and amylose and amylopectin polymers are fully dispersed in the aqueous medium. For this reason, starches specially modified to possess useful coating attributes are not typically cross-linked.

4.3.2 Substituted Starches

In addition to cross linking, starches are often substituted with various reagents to modify then properties and functionality. Substitution reagents include succinic anhydride, acetic anhydride, and propylene oxide (Fig. 4.5). These chemical entities covalently attach to carbohydrate polymers inside of the starch granule.

There are several positive attributes contributed by substitution modifications. The first is that introduced functional groups help to open up the granule, increasing its affinity for water. This results in lower starch gelatinization temperatures and better hydration. Another advantage is that the resulting starch gel is much less firm and exhibits better clarity. In native starches, the amylose portion of starch can re-associate with time in a process referred to as retrogradation. As retrogradation proceeds, the gel gradually becomes more opaque as water is expelled out. Steric hindrance caused by substituted groups on the polymer chains prevents re-associations, so that retrogradation and weeping do not readily occur. This advantage also makes the gel more resistant to breakdown during successive freeze-thaw cycles, which generally cause unmodified native starch gels to lose water and thus, their functionality.

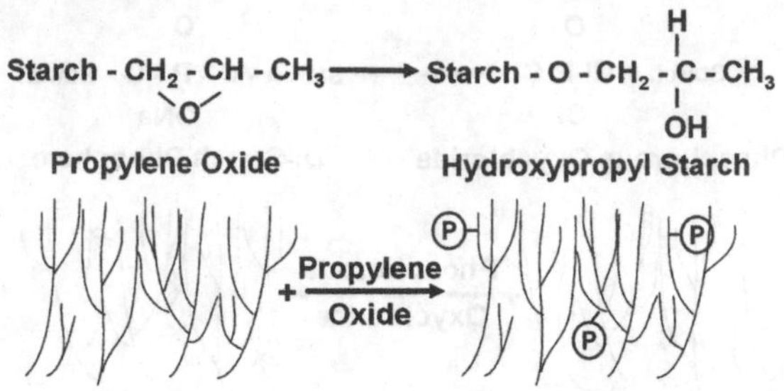

Starch - CH$_2$ - CH - CH$_3$ ⟶ Starch - O - CH$_2$ - $\overset{\overset{\text{H}}{|}}{\underset{\underset{\text{OH}}{|}}{\text{C}}}$ - CH$_3$

Propylene Oxide   **Hydroxypropyl Starch**

Fig. 4.5 Substitution reaction using propylene oxide

It has been known since the 1960s that partial etherification with propylene oxide will result in better starch-based films. Hydroxypropylated starches generally produce films, which are clearer and more flexible compared to native starch films. High amylose starches derivatized with propylene oxide were studied in the mid-1960s (Jokay et al. 1967).

4.3.3 Oxidized Starch

Food starches are oxidized to improve their adhesion properties in coating applications such as batters. Sodium chlorite, calcium hypochlorite and sodium hypochlorite are all approved starch oxidizing reagents for use in batter applications. Deep fried food coatings such as tempura coatings, beer battered coatings and other coating types for meats, fish and vegetables often contain oxidized starches. French fries are frequently coated with a batter to improve their eating quality by retaining crispness after frying or as a way of delivering seasoning to the surface. An ideal batter coating firmly adheres to the surface of the food to be coated throughout the freezing and frying processes. When used, breadcrumbs need to be retained on the food surface also. Moisture permeability can be a desirable property for coatings, because water vapor needs to be allowed to escape from food during cooking without blowing off pieces of the coating into the fryer.

4.3.4 Acid-Hydrolyzed Starches

Another type of chemical modification, which has found use in designing starch products specifically for films and coatings, is acid modification. Acid-treated or "thin boiling" starch is produced by treating granular starch with a mineral acid to partially depolymerize starch molecules to a desired end point (Fig. 4.6). Acid-modified

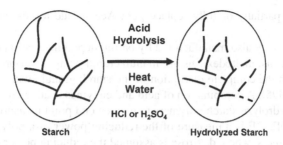

Fig. 4.6 Chemical modification of starch – acid hydrolysis

starch is usually used in applications where there is a high level of solids present in addition to maintaining a relatively low hot paste viscosity during processing, but where a firm or rigid setback is desired after processing and cool down. Examples include imitation cheese, jelly candies, processed meats, and extruded snacks and cereals. Acid modification or thinning serve the purpose of partially debranching the amylopectin portion of starch located within granule amorphous regions, so that behavior of the treated starch becomes more amylose-like (due to an increase in the linear component). Experience with dextrins and maltodextrins indicates that shorter polymer length results in better film formation when compared to native forms of starch.

4.3.5   Dextrins and Maltodextrins

Dextrins are often used to make various types of coatings on everything from pharmaceutical products, to confections, to edible glues and sealants. The traditional method of manufacture of dextrins is in a dry, heated environment where glucose polymers partially depolymerize and then repolymerize in linkages different from the α–$(1\rightarrow4)$ and α–$(1\rightarrow6)$ glycosidic bonds found exclusively in native starches (Kennedy and Fischer 1984). Acid may also be employed and variables include starch type, amount/type of acid, moisture content during heating, and time/temperature of heating (Nisperos-Carriedo 1994). These dextrin products can be soluble in cold water, and have fair film formation properties and good adhesive properties. Dextrins have found use in confectionery products as a coating and as an alternative to gum acacia. In panned candies, a layer of gum acacia solution is often utilized as a barrier between a high fat candy component, such as chocolate or nut butter, and a sugar shell. An aqueous coating of dextrin is also applied to the outside of chocolate prior to coating with a solvent-based coating such as edible shellacs, often referred to as confectioners' glaze. The undercoating here prevents solvent flavors from migrating from the coating into the chocolate thus spoiling the flavor. Tapioca-based and corn-based dextrins have been used successfully in these kinds of candy applications to partially or fully replace gum Acacia due to cost and availability considerations.

Maltodextrins are also manufactured by partial depolymerization of starches. In contrast to dextrins, maltodextrins are manufactured using a wet process that usually begins with the complete disruption of the starch granule with the use of heat (Anonymous 1995). A combination of acid and enzymes is added to a native starch dispersion to hydrolyze starch polymers to a given end point measured by dextrose equivalent or DE. DE is a measure of the reducing power of a polymeric mixture of glucose polymers, where dextrose is assigned the reducing power of 100. Starch polymers have a single reducing end, and depolymerization brings about a drop in molecular weight, which results in more reducing ends and an accompanying rise in DE. Under the definition provided in 21CFR184.1444, maltodextrins are polymers of D-glucose linked primarily by α–$(1\rightarrow4)$ glycosidic bonds, possessing a DE of less than 20 (Anonymous 2007b).

Maltodextrins are water-soluble at concentrations up to 70% (w/v), and have good film forming properties. Being of intermediate molecular weight, they form weak coatings on their own with good oxygen barrier properties. Murray and Luft (1973) studied the use of maltodextrin as a functional coating on fruit slices and nutmeats to extend shelf life.

4.4   High-Amylose Cornstarch

High-amylose starches are produced from genetic mutants of corn (maize) that have higher than typical amylose to amylopectin ratios. Starches with amylose contents of 50, 70 or even 90% are sometimes used to produce films. High-amylose starches generally have the disadvantage of being highly crystalline, requiring high temperatures to achieve full gelatinization. To cook out high- amylose starch granules to full dispersion requires cooking temperatures well above 100°C, which can be only achieved at above atmospheric pressures or with direct injection of steam.

High-amylose starches have found widespread use in applications where firm gels and some loss of water are desirable, such as jellied candies. These high-amylose starches are also used to make films and coatings, and were used in coatings prepared for Apollo space flights.

Films made from purified amylose isolated from conventional corn starch were also evaluated (Young 1984). Lourdin et al. (1995) studied the effect of amylose to amylopectin ratio using a combination of native starches and starch fractions. They found that tensile strength and elongation increased with increasing percentages of amylose. There are more recent examples in the patent literature where the use of high amylose starches have been used to extend the shelf life of deep fried foods (textural quality) after they have left the fryer (US patent 5976607). Battered vegetables and French fries are often coated with a high-amylose starch batter so that they will remain crisp for additional minutes under a heat lamp. High-amylose and hydroxypropylated modified food starches have been found to be effective for these types of foods. Pagella et al. (2002) also found that high-amylose starches produced the most commercially acceptable freestanding films by criteria set forth in their study. Films made from high-amylose rice and pea starches were found to have excellent oxygen barrier properties, better than those of protein-based alternatives (Mehyar and Han 2004).

4.5 Additives

4.5.1 Polyols

A number of additives that can be included in the formulation of a coating solution which will alter the properties of the resulting coating layer or freestanding film to improve performance. Generally, plasticizers are added to these polymers to improve the overall flexibility and extensibility of the resulting film or coating. In the realm of edible films, polyhydric alcohols are the most common types of plasticizers for edible films, including propylene glycol, glycerin, sorbitol and other polyols. There are many publications where various types of polyols were studied for their effects on the resulting films (Arvanitoyannis et al. 1996, 1997; Psomiadou et al. 1996; Arvanitoyannis and Biliaderis 1998; Parra et al. 2004; Fama et al. 2005; Rodriguez et al. 2006). In most studies, starch-based films are plasticized most effectively with the use of glycerin, which has specific advantages of being relatively non-sweet, inexpensive, and having been granted GRAS (generally recognized as safe) status by the FDA.

Strength, pliability and flexibility of a film or coating can be measured using tensile testing techniques. A strip of pre-cast film of specific known dimensions, in either rectangular or bar bell shape, is pulled apart under a known force using Texture Analyzer or Instron-type testing instruments (ASTM D2370) Anonymous Method #D2370. A strip of pre-cast film of specific known dimension is pulled apart at a carefully defined speed, while resistance is recorded over time. Peak force needed to break the film, as shown by a sudden drop in force, is recorded as a measure of the overall film strength. The extent or distance that the film is stretched prior to the failure point is the extensibility, which is often expressed as a percentage of the length of the film piece tested. This percent extensibility is used as a measure of flexibility, pliability and resiliency of the film or coating. In general, starch-based films are very strong when compared to those of other edible polymers, but usually score less favorably on the measurement of extensibility (Donhow and Fennema 1994; Chang et al. 2000). The strength of the film has also been demonstrated to be dependent upon moisture content of the film, which is in turn influenced by the relative humidity of the environment in which it is stored (Chang et al. 2000; Bertruzzi et al. 2003).

Plasticizing polymers improves pliability and flexibility of the film or coating. Addition of glycerin or other plasticizing agents makes the resulting film more pliable. Further addition of plasticizing agents will result in rubbery, sticky films. Thus, it is necessary to optimize the amount of plasticizing agent used in a formulation to attain desired film properties. However, there are some applications, such as vegetarian soft capsules, where a rubbery, extensible sheet is considered desirable.

4.5.2 Sugars

Sugars and other low molecular weight materials, when added to starch-based films, should also act as effective plasticizers (Zhang and Han 2006a). Generally this effect is true up to a point, but sugars generally make resulting films more glass-like and brittle. Shamekh et al. (2002) obtained better results with films made from starch hydrolysates when low molecular weight sugars were removed. In some coating applications (e.g. desserts), cracking as a result of brittleness does not constitute a complete failure, and sugars can be incorporated up to the level that is necessary to achieve desired sweetness. In more demanding applications where an oxygen barrier is desired, cracks are unacceptable, such as in tablet coatings for pharmaceutical products. Very little sugar can be tolerated in these starch-based edible films.

4.5.3 Maltodextrins

Maltodextrins have a number of advantages as additives to edible films. They can slightly reduce the brittleness of starch films through a plasticization and humectancy effect due to their relatively lower molecular weight and slight hygroscopic properties. The range of available DE for maltodextrins allows the coating developer to dial in the desired property by choosing the appropriate DE to add to the coating formulation. Obviously, in a tack coating to adhere spherical particulates to a food surface stickiness will be much more welcome than in an invisible coating intended to extend shelf life of a high fat snack.

The most important function that maltodextrin can serve is to increase the solids of the coating solution to achieve faster drying. It is typical that 10–20% (w/v) solids of maltodextrin can be added to the solution before excessive viscosity, sweetness or undesirable flavors are observed. Although it is possible to raise solids levels to values much higher than 20% with addition of higher DE maltodextrins or corn syrups, this strategy is often found to be counter productive. Excessively high solids content for coating solutions is often found to prolong drying compared to those in the 20–40% range. This observation is probably due to the increase in hygroscopicity of smaller molecular weight components and water binding capability of larger, starch-like fractions (Zhang and Han 2006b)

4.5.4 Detackifiers

A number of different products have been proposed as detackifiers for addition to film forming starch-based polymer solutions to decrease overall stickiness of the resulting surface. Polysorbates, lecithin and other surfactants have been used to decrease overall tackiness of starch-based films and coatings.

One reason to reduce tackiness of films is to prevent multiple sheets of freestanding films or flat surfaces of coated materials from adhering to each other. In addition to surfactant additives, small amounts of fine insoluble particulates have been successfully used to decrease tackiness of a film or coated surface.

4.5.5 Humectants

Glycerin is used as a plasticizer or humectant to maintain an adequate moisture level for a continuous film casting. Keeping the film hydrated assures adequate flexibility and resiliency. When starch-based edible films are subjected to relative humidity environments below 20–25%, they can experience cracking. Use of glycerin and polyols can lower tolerances to 10–15% relative humidity. When a film is made from highly soluble polysaccharides such as gum arabic, high relative humidity conditions will bring about dissolution of the film.

One of the easiest families of functional ingredients to incorporate into freestanding films and coatings is color. FD&C colorants are commonly used at such low levels that they have a minimal effect on the integrity of the coating. Exceptions to this rule are coatings with tight tolerances for performance, such as pharmaceutical tablets, and those where an opacifier, like titanium dioxide, would need to be added as an undercoat.

4.5.6 Lipids

Addition of lipids to a coating solution can improve barrier properties of the resulting coating. This phenomenon has been demonstrated through the improvement of water vapor transmission and oxygen barrier properties of films and coatings (Garcia et al. 2000). Addition of high melting point waxes has been shown to be particularly effective in improving moisture barrier properties of films (Han et al. 2006). Addition of approximately 20% (based upon polymer weight) edible oil to a starch-based coating solution is generally found to extend shelf life of the coating layer in high-oil foods such as nutmeats.

4.5.7 Emulsifiers and Wetting Agents

It is sometimes beneficial to add a wetting agent or surfactant to the coating solution to improve coating efficiency. When applying an aqueous solution to a fatty surface such as nuts, chocolate or compound coatings, it is usually advantageous to add a little lecithin or other emulsifiers to the coating solution to lower surface tension and allow the surface to be coated to wet. If the surface to which the coating will be applied is not sufficiently wetted by coating solution, the resulting coating layer will not tightly adhere to the food surface. Subsurface bubbles can ruin the appearance of the product. These bubbles will make a transparent coating opaque or alter the hue of a colored coating.

4.6 Methods of Application and Manufacture

Coatings can be applied to food surfaces by just about any method. Starch solutions have been applied using dipping, spraying, and ladling directly onto a tumbling mass of product. For meat products, a deluge machine has been used. This machine functions as a waterfall apparatus, and coats the product surface much the way that a chocolate enrober does. In each example, the liquid dispersion of polymer is applied to the food surface to be coated, and water is allowed to evaporate off until the coating is deposited. The coating will be in equilibrium with the environment of the package and also with the food itself. Coatings on low moisture foods will be hard, non tacky and transparent. Coatings on higher moisture materials will be more pliable and slightly tacky. High moisture foods such as meats will have a semi-solid paste like consistency.

4.6.1 Free Films

One family of coatings that has received a lot of interest in the last 10 years is that of freestanding films, which are intended to be consumed without any backing material or center. Challenges of producing an acceptable breath-freshening film, for example, includes making the film with enough integrity that it can be cut, packaged and remain dispensable throughout product shelf life. These criteria must be accomplished without unacceptably extending the desired time of film dissolution and flavor delivery on the tongue or roof of the mouth. Dissolution in the mouth is largely a function of the nature of polymers used and thickness of the piece. Thicker pieces will have better strength, but might be unacceptable in the mouth or have an undesirably long dissolution time.

Freestanding films can be made in the laboratory on a variety of surfaces. The method most often used in our lab is to affix sheets of polyethylene food wrap onto the bench top surface with adhesive tape. A thin film of starch dispersion is then applied using a Gardner draw down bar (Fig. 4.7) or bird applicator. After water is allowed to evaporate, the film can then be peeled from the surface where it is cast and cut to whatever use is required. Surfaces for casting include plastic, glass, Plexiglas and various types of semi-flexible plastic sheets used for cutting boards. A textured surface or an especially flat, shiny surface can be imparted to the film depending upon the surface used. Backing material should be selected based on the ease with which the film can be removed. Some materials, such as glass, can have a tendency to hold onto the film without releasing it upon drying. Other materials, such as siliconized paper, might release too readily, which could result in ripped sheets and small pieces as the film separates from the backing too quickly and unevenly. Flexible materials have obvious advantages for removal of edible film.

Commercially, films are typically deposited, dried and removed on a continuous basis, usually on a rotating belt. The length of the belt, speed, temperature and air velocity can all be altered to ensure good drying at the appropriate point in the process. Belts made of silicone and stainless steel are often used for the drying step.

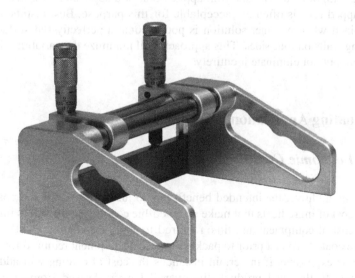

Fig. 4.7 Paul Gardner universal applicator (Paul N. Gardner Company, Inc., Pompano Beach, FL)

4.7   Methods for Measuring Effectiveness

One of the challenges of measuring any desirable property of a coating is that most testing protocols are designed for freestanding films. Much of the instrumentation and tests are based upon non-edible plastic packaging materials. A freestanding film is, by definition, not equivalent to a coating applied to a surface. Some researchers have tried to control the difference between coatings and freestanding films by applying the test polymers to a neutral backing material, such as paper and measuring oxygen and water vapor transmission rates on the coated paper in comparison to uncoated paper (Hurtado et al. 2001).

The differences between a true coating and a freestanding film are particularly large when measuring fat barrier properties. Fat migration tests in which a given quantity of oil is dropped upon a piece of pre-cast film to measure migration rates to some collection surface, like a filter paper, are somewhat artificial. This is a very different environment than the same peanut oil, for example, embedded in nut mass with an edible coating of gum acacia, dextrin or modified food starch used to prevent its migration over a period of months into a chocolate coating surrounding it.

There is also the challenge of casting a film with defined dimensions. This is true whether transmission rate or tensile strength is to be measured. Instruments and test procedures often require a particularly large piece of film. For inflexible materials, it can be difficult to produce an intact freestanding film of sufficient size. Instrumentation is usually designed with non-edible packaging in mind. Larger sheets of film can often be cut to the width and length desired for testing, but thickness can be a problem. The common method of casting the film by pouring or spreading with draw down bars, film applicators, bird applicators, formed rods, or wire wrapped rods is often not acceptable for this purpose. Best results are usually obtained when polymer solution is poured onto a perfectly flat surface with restraining walls on four sides. This approach will minimize the variability in the Z dimension, but not eliminate it entirely.

4.8   Coating Applications

4.8.1   Economic Considerations

As with any additive, the intended benefit has to justify the additional cost of its use. The cost of ingredients that make up an edible coating are often less important than the cost of equipment and time required to apply the coating and the cost of drying the coated product prior to packaging. Most equipment required for coating processes is expensive. If in certain instances the cost of coating will add a substantial cost to the final product, the overall benefit derived from coating the product must outweigh this additional cost. This way, consumer value of the coated product is maintained.

4.8.2 Coatings to Enhance Appearance

An example of a coating used primarily for visual appeal, such as a shiny surface, is the application of shellac-based coatings to surfaces of chocolate-panned items. Chocolate-covered peanuts, malted milk balls and jellybeans commonly have a shine coat added to their surfaces (Fig. 4.8). In the case of chocolate items, shellac provides the added benefit of preventing cocoa butter from staining packaging. A thin shellac layer also protects individual pieces from rubbing against each other, giving undesirable scuff marks to the chocolate.

There are a number of challenges to replacing a solvent-based ingredient, such as shellac, with an aqueous-based coating. It might be necessary to add a surfactant to the coating solution to allow the food surface to be coated while the coating is being deposited by evaporation. For coating sugar confectionery, it is a real challenge to add an aqueous coating solution, containing perhaps 80% water, to a substrate that is water-soluble. This obstacle can be overcome by either putting down a pre-coat of a water-insoluble material, such as wax, prior to application of the aqueous coating or by the use of aqueous coating techniques employed by the pharmaceutical industry such as dipping or spraying.

Aqueous-dispersed, starch-based films are usually applied as a solution. The water of the coating solution is then evaporated off to deposit the coating. Another disadvantage that aqueous systems have over some solvent-based systems is that the coating itself remains as water-soluble as the original starch used in the coating solution.

Fig. 4.8 Malted milk balls (this figure is not referred to from in the text)

Starch-based coatings do not undergo a cross-linking or curing step to make them more water-insoluble. This can limit moisture barrier properties of the film, and restrict use of the film in higher humidity environments. This limitation can also be considered a positive attribute in various applications where the coating is to be washed off once it has performed its protective function, perhaps at some intermediate step in a process.

Perhaps the biggest challenge comes from the fact that water is less volatile than ethanol, which is used to disperse and deposit shellac and many protein-based coatings. The presence of water typically leads to longer drying times, which can contribute to greater expense.

4.8.3 Flavored Coatings

Flavored coatings can be divided up into basic categories. A coating might be of a similar flavor as the item to be coated, thus, reinforcing the overall flavor impact. Coatings might also be flavored with a contrasting flavor for a novelty effect. If a coating is to be applied to extend shelf life, it might be necessary to flavor the coating to diminish delay in the initial release of flavor when first placed in the mouth.

In pharmaceutical tablets and nutritional supplements, it is often desirable to apply a coating to mask or disguise an unpleasant flavor. This goal is usually accomplished by a combination of the masking properties of the polymer, a pleasant flavor to distract or with the addition of flavor masking agents.

4.8.4 Oxygen Barriers

Starch-based coatings usually compare very favorably to other coating alternatives in regard to oxygen barrier properties. By creating a barrier between atmospheric oxygen and the food, an extension of shelf life can be demonstrated (Phan et al. 2005). One obvious type of food that might gain advantages from such a coating layer is nutmeats. Nutmeats have many characteristics that make them ideal candidates for coating. Because of their high oil content, oxidative rancidity is the limiting factor in their shelf life. They also have a sufficiently high monetary value to justify whatever costs might be associated with coating. Lastly, their need for individual piece identity is enhanced with coating. There are some challenges in coating nuts. It is important that the coating step does not allow nuts to become bruised or crushed, such that they begin to express oils to their surfaces. Although nuts are usually coated with an aqueous solution that includes emulsifiers, once oil flows to the nut surface, coating pretty much becomes impossible. Walnuts and macadamia nuts have been found to be particularly problematic. Walnuts are very fragile and a small amount of tumbling can easily bruise the nuts. Macadamias are very hard, but also very high in fat.

It is possible to coat either of these two nut varieties, provided care is taken to conduct the coating as gently and as quickly as possible. Nuts can also be a challenge for aqueous coatings, as it is very important that the water content of nuts remain very low for both desired textural properties and overall shelf life.

One area in which the use of starch-based edible films and coatings has enjoyed extensive practice is the process whereby volatile, oil soluble flavors and fragrances are encapsulated into a glassy carbohydrate-based matrix. Fragrance and flavor oils are often highly susceptible to oxidation. Use of various gum, starch and malto-dextrin products to construct a carbohydrate-based capsule provides an effective barrier for greatly extended shelf life.

4.8.5   Moisture Barriers

Water-soluble polymers might not at first be thought of as a choice for impeding moisture migration from a high moisture component of a food to that of a low moisture content. Starch-based coatings have been demonstrated to at least slow mois-ture migration (Hurtado et al. 2001; Chung and Lai 2005). Whether or not the reduction in moisture transfer is commercially significant is dependent upon the shelf life of the food and the gradient between the two components. Arvanitoyannis and Biliaderis (1998) found that addition of sodium caseinate to a starch-based film lowered the water vapor transmission rate relative to that of a pure starch film, but did so at the expense of the physical attributes of the film.

In many chocolate confectionery applications, it is necessary to form an oil barrier to prevent non-cocoa butter oils from entering the chocolate, causing the chocolate to lose its temper and to bloom. Oils with low melting points can migrate to the surfaces of chocolates and re-crystallize, forming a greasy grayish white layer known as fat bloom (Minifie 1989). Coconut and nut oils can be kept isolated from chocolate in such products with the use of edible coatings. Another advantage of coating high-fat confectionery products with a sealant coat on the outer surface is to avoid a scuffed appearance and to inhibit migration of mobile fat components to the product surface, leading to staining of the package material. Figure 4.8 shows coated malted milk balls to protect the product.

Coatings can be formulated with antioxidant ingredients to improve the over-all protection provided by the film. Murray and Luft (1973) used SO_2 in their maltodextrin coatings to further enhance color preservation of dried apple and apricot slices.

4.8.6   Tack and Adhesive Coatings

One relatively straightforward application for edible films is as a tack or adhesive coating in which the wet solution is used as an edible glue to adhere one food component to another. A very simple example might be an oil spray applied to a snack food, followed by coating with a seasoning blend. In this example, there is no drying step, and seasonings are weakly adhered to the snack food surface. Where it is necessary to improve adhesion or fix particulates, it is necessary to use a polymer-based adhesive coating material. In such instances, maltodextrins are added to improve adhesion. Inclusion of a high DE maltodextrin within the formulation will increase adhesion and tackiness of the coating. This tackiness is accompanied by an increase in pliability and hygroscopicity of the coating.

An example of a tack coating is adherence of a powdered seasoning blend to a product surface. In this scenario, seasoning should be fixed fast to the surface of the snack food to prevent rub off during distribution and/or to minimize the tendency of seasoning to adhere to the consumer's skin or to stain the fingers. For some snack and bakery applications, larger pieces, such as poppy, caraway or sesame seeds or grains like wheat berries or oatmeal, might be adhered to product surfaces. These pieces can be held in place with an edible tack coating. There are other examples where components need to be fixed to each other. Examples include granola clusters as well as granola bars. Ready-to-eat cereal flakes might be coated with pieces of whole grain, with or without additional flavorings applied to product surfaces.

For applications in which large particulates need to be held fast onto a larger product piece (as in Fig. 4.9), best results are usually obtained with a three-stage application method. In a rotating drum or pan or on a belt, the larger piece is first wetted with the film forming solution. Next, the smaller particulate is added, and product is mixed thoroughly. Lastly, an additional layer of tack coating is applied on top as a seal coating. In these tack-type coating applications, it is usually only necessary to apply 0.2–0.5% of coating on a final dry weight gain basis to achieve desired results. This range is large, because both the density of the item to be coated and the size of the piece (surface area) influence the amount of tack material needed.

Low moisture snack foods are not ideal substrates for use of a coating solution containing up to 80% water. Extruded corn collets and wheat-based baked snacks such as crackers are very prone to staling and deformation at higher moisture levels. It is possible to apply starch-based coatings to these snack foods for shine, protection or to attach particulates. Good results can be achieved through application of a minimal amount of material to achieve the desired effect and quick dehydration of the coating before deleterious effects can begin to occur. Use of a high molecular weight coating material to assure adequate moisture control is also important. For particularly sensitive materials, a side vented, perforated walled or aqueous film coating device might be necessary.

4.8.7   Anti-Microbial Coatings

Coatings intended to delay or prevent growth of microorganisms have been proposed for many years (Fama et al. 2006). For many foods, deleterious, microbial growth is largely a surface phenomenon. Thus, limiting bacterial growth on the

Fig. 4.9 Spices and particulates on bread

surface of food is anticipated to improve overall shelf life. Even tiny breaches in the coating could be a large concern for these kinds of products. There is plenty of interest in these types of coatings in the industry, and there is evidence that they can be very effective (Jagannath et al. 2006).

4.8.8 Coatings to Alter or Enhance Texture

In sugar-shelled or hard-panned candies, a molded piece of chocolate, peanut butter or other confection is placed in a rotating candy pan and coated with multiple applications of highly-concentrated sugar solution. The panning action ensures that solution is applied evenly and that a hard sugar shell results upon crystallization. Dextrins and starch-based coatings are sometimes included in final layers of coatings to produce speckled and mottled surface effects. In some cases, a starch-based polymer is added to more conventional sugar shelling to alter the bite or strength of the coating. Glassy polymers, such as high molecular weight maltodextrins and starch hydrolysis products, have a more resilient and flexible texture, which makes them less brittle in the mouth. Rough uneven coatings or less crystalline and brittle coatings are typically considered to be faults, if they were not intentional. However, unusual flavors, colors, and textures are what drive novelty in confectionery products.

In some applications, it is advantageous to coat pieces to improve product flow, maintain superior piece identity or to prevent clumping and agglomeration of different components during distribution and shelf life. Near transparent coatings that prevent clumping in granola and trail mix coatings are one example of this type of application.

4.8.9 Flavor Delivery

A successful flavor delivery film has to possess the strength and stability of any freestanding film while delivering its load of flavor components. Flavor compounds are often not water-miscible, which can present various challenges to water dispersible polymers such as starch. Oil-based flavors, such as citrus and other volatile oils, can be immobilized in the viscous film-forming liquid in droplets small enough to prevent coalescence. Surfactants and emulsifying agents are only occasionally necessary for these applications. Emulsion breakdown and phase separation are prevented through homogenization or high shear mixing to assure suspension of small droplets and an adequately high solution viscosity to slow oil droplet mobility. Through these meane, coalescence and creaming are minimized. The temporary emulsion need only be stable for the time it takes for water to evaporate off, leaving behind the film with the oil droplets encased within the resulting glassy matrix.

Many flavor ingredients such as cocoa are water insoluble, and have to be evenly dispersed throughout the polymer matrix, without being disruptive to the integrity of the film. As with delivery of any "active" component, the minimum level of flavor needed to deliver the desired impact must be balanced against whatever negative contribution the insoluble component will have on strength and resiliency of the film.

4.8.10 Drug Delivery

Drug delivery through edible films has advantages over more conventional drug delivery devices such as liquid, capsules, or tablets. Even without water, this type of medication can be easily ingested. Edible film delivering medication can provide an obvious advantage for people who have trouble swallowing, suffer from dysphagia or for patients who reject other easy to swallow medications. Small children and household pets also fit into this category in which swallowing can be difficult. Topical application of films to a well-targeted area of the oral cavity is also an improvement over other delivery techniques. An example here would be a topical anesthetic for pain relief.

4.9   Conclusion

The concept of starch-based edible films is a technology that has been recognized to have significant potential for providing a variety of solutions to the food industry. The real advantages of starch are its relatively low cost because it is abundant in nature and it can be subjected to a number of modifications (chemical and genetic) to alter and improve its performance. In many successful applications of starch-based coatings, other polymers can be added to augment and improve performance characteristics. These could include gums to prevent loss of moisture, or maltodextrins to change gloss and tack characteristics or cellulosic or protein-based ingredients to improve their performance and functionality.

References

Anonymous (1995) MALTRIN® "maltodextrin and corn syrup solids for food formulations" Grain Processing Corporation
Anonymous (2007a) 21Code of Federal Regulations 172.892
Anonymous (2007b) 21Code of Federal Regulations 184.1444
Anonymous Method #D2370 American Society of Testing Materials
Arvanitoyannis I, Billaderis CG (1998) Physical properties of polyol-plasticized edible films made from sodium caseinate and soluble starch blends. Food Chemistry 62(3): 333–342
Arvanitoyannis I, Psomiadou E, Nakayama A (1996) Edible films made from sodium caseinate, starches, sugars or glycerol, Part 1. Carbohydrate Polymers 31: 179–192
Arvanitoyannis I, Psomiadou E, Nakayama A, Aiba S, Yamamoto N (1997) Edible films made from gelatin, soluble starch and polyols, Part 3. Food Chemistry 60(4): 593–604
Bertruzzi MA, Armada M, Gottifredi JC (2003) Thermodynamic analysis of water vapour sorption of edible starch-based films. Food Science and Technology International 9(2): 115–121
Blanshard JMV (1987) Starch granule structure and function: a physiochemical approach. In Galliard T (ed) Starch: Properties and Potential, 2nd Edn. Wiley, New York, NY, 16–54
Chang YP, Cheah PB, Seow CC (2000) Plasticizing-antiplasticizing effects of water on physical properties of tapioca starch films in the glassy state. Journal of Food Science 65(3): 445–451
Donhow IG, Fennema O, (1994) "Edible films and coatings: characteristic, formation, definitions, and testing methods. In Krochta J, Baldwin E, Nisperos-Carriedo M (eds) Edible Coatings and Films to Improve Food Quality, Technomic, Lancaster, 1–24
Fama L, Rojas A, Goyanes S, Gerschenson L (2005) "Mechanical properties of tapioca-starch edible films containing sorbates." Swiss Society of Food Science and Technology. Elsevier LWT 38: 631–639
Fama L, Florese SK, Gerschenson L, Goyanes S (2006) Physical characterization of cassava starch biofilms with special reference to dynamic mechanical properties at low temperatures. Carbohydrate Polymers 66: 8–15

Galliard T (1987) Starch availability and utilization. In Galliard T (ed) Starch: Properties and Potential, 2nd Edn. Wiley, New York, NY, 1–15

Garcia MA, Martino MN, Zaritzky NE (2000) Lipid addition to improve barrier properties of edible starch-based films and coatings. Journal of Food Science 65(6): 941–947.

Han JH, Seo GH, Park IM, Kim GN, Lee DS (2006) Physical and mechanical properties of pea starch edible films containing beeswax emulsions. Journal of Food Science 71(6): E290–E296

Higgins C, Qian J and Williams K (1999) Water dispersible coating composition for fat-fried foods US Patent 5976607

Hurtado ML, Estevez AM, Barbosa-Canovas G (2001) Physical characterization of a potato starch edible coating used in walnut storage. Proceeding of the 4th International Conference on Postharvest, Ben-Arie R, Philosoph-Hadas S, (eds) Acta Hort. 553, ISHS, 627–628

Jagannath JH, Radhika M, Nanjappa C, Murati HS, Bawa AS (2006) Antimicrobial, mechanical, barrier and thermal properties of starch-casein based, neem (Melia azardirachta) extract containing film. Journal of Applied Polymer Science 101: 3948–3954

Jokay L, Nelson GE, Powell EL (1967) Development of edible amylaceous coatings for foods. Food Technology 21: 1064–1066

Kennedy HM, Fischer AC (1984) Starch and dextrins in prepared adhesives. In Whistler R, BeMiller JN, Paschall EF, (eds) Starch Chemistry and Technology, 2nd Edn. E. Academic, New York, NY, 593–610

Kester JJ, Fennema OR (1986)Edible films and coatings a review. Food Technology 1986: 47–59

Lourdin D, Dell Valle G, Colonna P (1995) Influence of amylose content on starch films and foams. Carbohydrate Polymers 27: 261–270

Mehyar GF, Han JH (2004) Physical and mechanical properties of high amylose rice and pea starch films as affected by relative humidity and plasticizer. Journal of Food Science 69(9): E449–454

Minifie B (1989) Chocolate, Cocoa and Confectionery, 3rd Edn. Van Nostrand Reinhold, New York, NY, p 109

Murray DG, Luft LR (1973) Low-DE corn starch hydrolysates. Food Technology 27: 32–40

Nisperos-Carriedo M. (1994) Edible coatings and films based on polysaccharides. In Krochta J, Baldwin E, Nisperos-Carriedo M, (eds) Edible Coatings and Films to Improve Food Quality, Technomic, Landcaster, 305–335

Pagella C, Spigno G, DeFaveri DM (2002) Characterization of starch-based edible coatings. Trans IChemE 80(Part C): 193–198

Parra DF, Tadini CC, Ponce PAB, Lugao AB (2004) Mechanical properties and water vapor transmission in some blends of cassava starch edible films. Carbohydrate Polymers 58: 475–481

Phan TD, Debeaufort F, Luu D, Voilley A (2005) Functional properties of edible agar-based and starch-based films for food preservation. Journal of Agriculture Food Chemistry 53: 973–981

Psomiadou E, Arvanitoyannis I, Yamamoto N (1996) Edible films made from natural resouces; microcrystalline cellulose (MCC), Methylcellulose (MC) and corn starch and polyols – Part 2. Carbohydrate Polymers 31: 193–204

Rodriguez M, Oses J, Ziani K, Mate JI (2006) Combined effect of plasticizers and surfactants on physical properties of starch-based edible films. Food Research International 39: 840–846

Shamekh S, Myllarinen P, Poutanen K, Forseell P (2002) Film formation properties of potato starch hydrolysates. Starch/Starke 54: 20–24

Young A (1984) Fractionation of starch. In Whistler R, BeMiller J, Paschall E, (eds) Starch Chemistry and Technology, 2nd Edn. Academic, New York, NY, 249–284

Zhang Y, Han JH (2006a) Mechanical and thermal characteristics of pea starch films plasticized with monosaccharides and polyols. Journal of Food Science 71(2): E109–E118

Zhang Y, Han JH (2006b) Plasticization of pea starch films with monosaccharides and polyols. Journal of Food Science 71(6): E253–E261

Chapter 5
Lipid-Based Edible Films and Coatings

Frédéric Debeaufort and Andrée Voilley

5.1 Introduction

The quality of food products depends on their organoleptic, nutritional, and micro-biological properties, all of which are subject to dynamic changes during storage and distribution. Such changes are mainly due to interactions between foods and their surrounding environment or to migration between different components within a composite food.

In the last 20 years, there have been over 45 edible packaging patents. During 2006 alone, there were 174 scientific papers focused on edible packaging. Most work covered in these papers deals with water vapour transfer. However, there are other potential applications. For example, edible packaging can be used to encapsulate flavour and aroma compounds, antioxidants, antimicrobial agents, pigments, ions that prevent browning reactions, or nutritional substances such as vitamins.

Typically, bio-based polymers, or biopolymers, are obtained from renewable resources. Figure 5.1 shows the different categories of bio-based materials. Polysaccharides such as starch, cellulose and chitosan; proteins such as casein and gluten; and lipids such as waxes are used in manufacture of these types of packaging materials. While polymers extracted from biomass are biodegradable, polymers synthesized from bio-derived monomers have to be tested for biodegradability. In addition, to facilitate use of such packaging materials in the industry, biopolymers have to be capable of being processed under the same conditions as petroleum-based plastics.

When a packaging material such as a film, a thin layer, or a coating is an integral part of a food and is eaten with the food, it qualifies as "edible" packaging (Guilbert

F. Debeaufort (✉)
Universitary Institute of Technology, Department of Bioengineering,
Université de Bourgogne, 7 bd Dr Petitjean, BP 17867, F-21078, Dijon, France
e-mail: frederic.debeaufort@u-bourgogne.fr

A. Voilley
EMMA Department, Université de Bourgogne – ENSBANA,1 esplanade Erasme,
F-21000, Dijon, France
e-mail: voilley@u-bourgogne.fr

M.E. Embuscado and K.C. Huber (eds.), *Edible Films and Coatings for Food Applications*,   135
DOI 10.1007/978-0-387-92824-1_5, © Springer Science+Business Media, LLC 2009

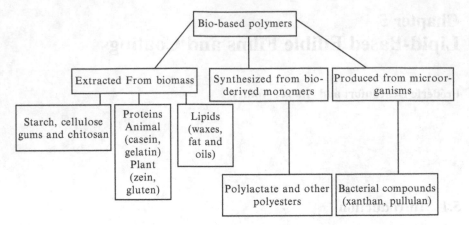

Fig. 5.1 Different types of biomaterials

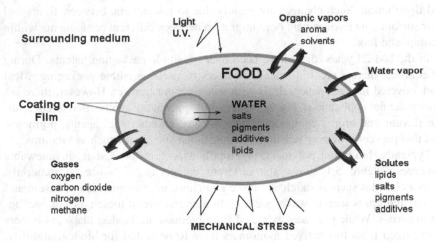

Fig. 5.2 Transfers that can be potentially controlled by edible barriers

and Cuq 1998). Coatings are either applied to or formed directly on foods, while films, on the other hand, are self-supporting structures that can be used to wrap food products. They are located either on the food surface or as thin layers between different components of a food product. An example of the latter would be a film placed between the fruit and the dough in a pie to provide protection against transfer and mechanical stress (Fig. 5.2).

Because they are both a packaging and a food component, edible films and coatings have to fulfill the following specific requirements:

– Good sensory attributes (neutral or favourable to the product to be coated)
– High barrier and mechanical efficiencies

- Biochemical, physicochemical, and microbial stability
- Non-toxicity and safety
- Simple technology
- Non-polluting nature
- Low raw material and processing cost

5.2 Formulation of Edible Films and Coatings

Because there are many different types of edible packaging, composition of edible barriers is very diverse. There are no specific regulations that deal with formulation and application of edible films and coatings. Since they are eaten with the food to which they are applied, all food ingredients and additives allowed by Codex Alimentarius, US Food and Drug Administration, or European Union regulations can be potentially used in their formulation. These regulations allow an almost infinite choice of components for formulation of films and coatings. Choice of which component to use, however, largely depends on the targeted product objectives and on technological and sensory constraints. The scientific and industrial patent literature over the last 20 years reveals a broad range of substances cited as film and coating constituents. More recently, new film/coating formulations have been based on vegetable or fruit purees, including those of bananas, apples, and zucchini. Krochta and Demulder-Johnston (1997) have provided a fairly comprehensive list of potential applications of film-forming substances.

The most commonly used hydrophobic film-forming barrier materials include, in decreasing order of barrier efficiency:

- Waxes
- Lacs/lacquers
- Fatty acids and alcohols
- Acetylated glycerides
- Cocoa-based compounds and their derivatives

Decreasing efficiency can be explained by the chemical composition of the molecules, such as presence of polar components, hydrocarbon chain length, and number of unsaturation or acetylation. For components having a common chemical nature, an increasing chain length modifies barrier properties because molecular polarity decreases, and this does not favour water solubility of the film (McHugh and Krochta 1994b, c).

5.2.1 Lipid-Based Edible Films

Edible barriers based on hydrophobic substances such as lipids were developed specifically for limiting moisture migration within foods. These hydrophobic substances are effective barriers against moisture migration because of their apolar nature (Morillon et al. 2002). A list of hydrophobic substances used as components of edible films and coatings, based on research and patents published in the last 15 years, is provided in Table 5.1.

A very wide range of compounds could be used in formulation of edible packaging. The decision on what to use will depend mostly on the target application. Many studies have investigated use of coatings on fresh fruits and vegetables to control desiccation. Substances most often used for this application are waxes. In enology, wine bottle corks may be coated with beeswax and paraffin wax to prevent soaking of the cork and release of flavours from the cork to the wine. This coating method offers an alternative to use of plastic corks in wine bottles. Cocoa butter and cocoa-based films are widely used in confectionery and biscuit industries. Other films are used in pizza and dough products to prevent moisture migration and in crispy and dry cereal-based products to prevent loss of flavour and water from fat or aqueous stuffing and filling. Most food-grade lacs and varnishes are applied in the pharmaceutical industry, as a protective layer for active substances. In the food industry, these are used to improve surface appearance (color, sheen, and gloss) and reduce surface stickiness. Few studies have investigated effects of applying lacs and varnishes of natural origin to foods. Nevertheless, many food additive suppliers have proposed using them as coating agents. Emulsifiers and surface active agents are sometimes used as barriers to gases and/or moisture on fish or meat products. They may also be added to a coating formulation to improve surface adherence between the coating and food article to be coated.

Table 5.1 Hydrophobic substances used as film former or barrier compounds

Oils, fats, shortening, margarine	Animal and vegetable native oils and fats (peanut, coconut, palm, palm kernel oils, cocoa, milk butters, lard, tallow, etc.)
	Fractionated, concentrated and/or reconstituted oils and fats (fatty acids, mono-, di-, and triglycerides, cocoa butter substitutes, etc.)
	Hydrogenated and/or transesterified oils (margarine, shortenings, etc.)
Waxes	Natural vegetable and animal waxes: candelilla, carnauba, jojoba bees, whales
	Non-natural waxes: paraffins, mineral, microcrystalline, oxidized or non-oxidized polyethylene
Natural resins	Asafoetida, Benjoin, Chicle, Guarana, Myrrhe, Olibanum (incense), Opoponax, Sandaraque, Styrax (Turkey) resins
Essential oils and liquorices	Camphor, mint, and citrus fruit essential oils
	Liquorices and pure glycyrrhizin
Emulsifiers and surface active agents	Lecithins, mono- and diglycerides, mono- and diglyceride esters
	Fatty sucrose esters, fatty alcohols, fatty acids

5.2.2 Composite Edible Films

Pure lipids can be combined with hydrocolloids, such as proteins, starches or celluloses, and their derivatives, either by incorporating lipids into the hydrocolloid film-forming solution (emulsion technique) or by depositing lipid layers onto the surface of the pre-formed hydrocolloid film to obtain a bilayer (Kamper and Fennema 1985; Krochta and De Mulder Johnston 1997). Multi-component films have been extensively reviewed by Wu et al. (2002) and Debeaufort et al. (2002). Addition of non-lipid compounds or particles such as sugar crystals, fibres, and proteins as dispersed components in fat materials permits formation of fat dispersions, such as chocolates (Fig. 5.3). These composite films take advantage of the functional properties of each component of the film to provide both barrier and mechanical properties. The resulting water barrier efficiency of bilayered films is often of the same order of magnitude as that of pure lipid (Debeaufort et al. 2002), and is much higher than that of emulsion-based films (Debeaufort et al. 1993; Martin-Polo et al. 1992a, b; McHugh and Krochta 1994a, b). However, the hydrocolloid layer is hydrophilic, and tends to absorb water when the film is in direct contact with a moist phase. Furthermore, additional processing steps for film making (casting and drying) make bilayer films difficult to use in high-speed commercial production.

The most cited substances used to form composite films are cellulose ethers, but pectinate, chitosan, starch, alginates, and carrageenans have been used in other composite films (Karbowiak et al. 2006, 2007). These substances are often combined with stearic or palmitic acids, beeswax, acetylated monoglycerides, and lecithins (Wu et al. 2002).

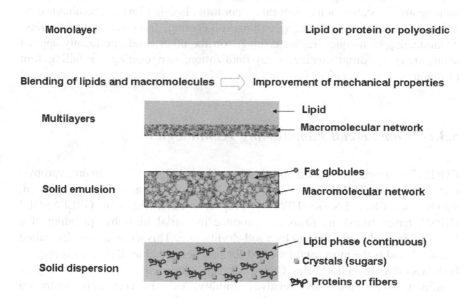

Fig. 5.3 Possible structures of lipid-based composite films

5.2.3 Additives to Improve Functionality

Other substances have often been added to film and coating formulations for improving their functional characteristics, such as enhancing film-forming ability of solutions, suspensions, or emulsions; promoting adherence of coating to the support; or controlling flow and spread properties of coating solutions, suspensions, and emulsions. Emulsifiers and texturizing agents have been used to stabilize dispersed systems; acid and alkaline substances have been used to improve protein solubilization; and tanning agents such as sulfites, aldehydes, and ascorbic acid have been used to increase reticulation or grafting of biopolymers, such as proteins. Moreover, some antioxidants (e.g., ascorbic acid), antimicrobial agents, or enzymes (e.g., transglutaminase) can promote intra- or intermolecular bonds. Enzymes are also used to increase solubilization of biopolymers, whereas some salts are used to induce polymer gelation (e.g., alginates or pectinates) (Guilbert and Cuq 1998).

5.3 Film and Coating Technologies

There are many technologies that can be used to make edible films and coatings. The choice of a specific process depends on the nature of film or coating constituents and on the intended shape of the barrier layer. Polysaccharides and proteins have to be modified using processes such as polymerization, gelation, coagulation, and co-acervation. Lipids-based coatings are more often obtained from either melting and crystallization or solvent evaporation. Edible films can be obtained by extrusion, co-extrusion, spreading, casting, roll coating, drum coating, pan coating, or laminating techniques. Edible coatings, on the other hand, are mainly applied using spraying, drum coating, spray-fluidization, pan coating, or falling film techniques.

5.3.1 Commercial Film-Making Technologies

COGIN® (carrageenans-based films), Chris-Kraft Polymer Inc. (hydroxypropyl-methylcellulose (HPMC)-based films), BioEnvelop® (cellulose derivative-based, starch-based, films), ENAK® (HPMC-, carrageenan-based films), and GREENSOL® (HPMC-based films) are firms that produce industrial films by spreading of a film-forming solution followed by a roll-drying step. This technique is also called "casting" and is the most useful method for producing edible films and coatings at both laboratory and pilot scales. On the lab scale, spreading thicknesses and drying conditions (temperature and relative humidity) may be accurately controlled (Debeaufort et al. 2002).

Extrusion technologies are often used for industrial production of films, tubes, poaches, and casings. Such is the case with those produced by companies like SOUSSANA® (gelatin-based tubes and collagen-based films); ENAK (alginate- and carrageenan-based films); and GREENSOL-MONOSOL® (hydroxypropyl methylcellulose- and methylcellulose-based films).

5.3.2 Coating Technologies

Coating application consists of applying a liquid or a powder ingredient onto a base product. Table 5.2 gives some examples of technology, products and principles.

Surface properties play a key role in the success of coating application. Application of coatings generally requires a four-step process:

1. Deposition of coating material (solution, suspension, emulsion or powder) on the surface of the product to be coated through spraying, brushing, spreading, or casting.
2. Adhesion of coating material (solution, suspension, emulsion or powder) to the food surface.
3. Coalescence (film-forming step) of the coating on the food surface.
4. Stabilization of the continuous coating layer on its support or food product through co-acervation by drying, cooling, heating, or coagulation.

Table 5.2 Coating processes and product applications

Term	Industry	Process	Principle
Enrobing	Chocolate	Enrobe	Cover a confectionery centre with tempered chocolate
Encapsulation	Food ingredient	Various systems	Formulate and create by physical or chemical reactions a protective matrix or layer containing an active ingredient in view or preservation or controlled release.
Panning	Confectionery	Pan	Build successive layers (a few microns at a time) of sugar syrup or chocolate with alternative drying/cooling periods
Seed coating	Seed	Drum, fluidized bed	Form a protective layer of a polymer film containing the active pesticides the seed requires in the storage and early stage of the germination
Seasoning	Snack	Drum	Disperse a coarse or finely ground herbs and spices mix onto snacks, croutons, chips, etc.
Breading	Prepared meals	Duster, coater	Meat chunks, prepared cheese, vegetable pieces can be pre-dusted, coated with a batter, and covered with breading, a fine or coarse ingredient obtained from bread or a processed dough

To achieve success in coating operations, specific parameters dealing with the base
product to be coated, processing equipment, and the coating formulation have to
be taken into account, as illustrated in Fig. 5.4.

Choice of the coating system or apparatus depends mainly on size and
shape of the product to be coated and on the desired thickness of the coating
(Fig. 5.5).

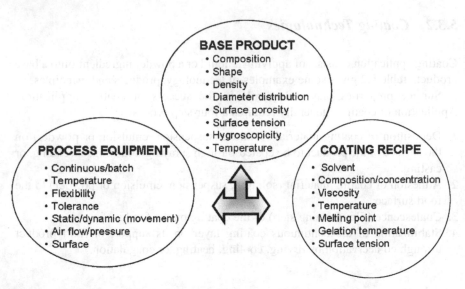

Fig. 5.4 Parameters to be controlled in a coating operation

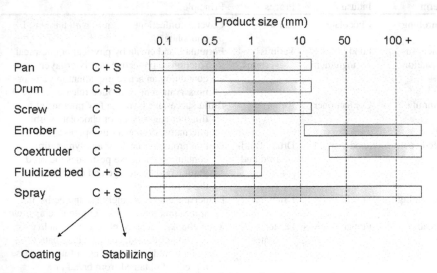

Fig. 5.5 Coating techniques based on the size of the product to be coated

5.3.2.1 Enrobing

Enrobing involves application of a thick coating layer by dipping the product to be coated in solution batter or in molten lipid (e.g., a chocolate-based coating). Coating of fresh or frozen products with a batter and/or breading can enhance palatability, add flavor to an otherwise bland product and reduce moisture loss and oil absorption during frying.

5.3.2.2 Pan Coating

Pan coating is used to apply either thin or thick layers onto hard, almost spherical particles in a batch process. For example, coating confectionery centres (peanuts and almonds) with gum Arabic provides a uniform base layer for further coating and a hydrophilic/lipophilic surface, prevents moisture and fat migration and allows incorporation of additional flavour.

5.3.2.3 Drum Coating

Drum coating is often the best technique for applying either a thin or a thick layer onto hard or solid foods in a continuous process (e.g., nuts). Oiling and salting of nuts enhances palatability, adds flavor, and delays lipid oxidation in peanuts. Adding a chocolate coating to cornflake cereals enhances palatability, adds flavor and delays moisture absorption (Fig. 5.6).

5.3.2.4 Screw Coating

A screw coater allows application of a thin layer of coating onto a solid and firm food material in a continuous process. For example, it is often used to deposit thin coatings on sticky particles, such as cheese shreds and pieces, to improve anti-caking

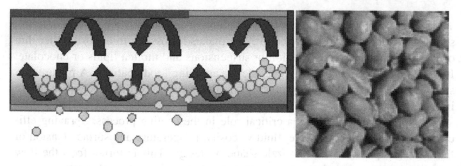

Fig. 5.6 Drum coating of peanuts

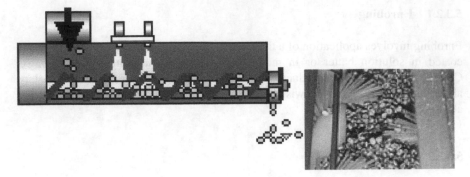

Fig. 5.7 A screw coater used in coating particles of cheese

properties and prevent agglomeration of particles when product is stored in flexible packaging (Fig. 5.7).

5.3.2.5　Fluidized-Bed Coating

Fluidized-bed coating is a technique used to apply a very thin layer onto dry particles of very low density and/or small size. Agglomeration of the powder promoted by the fluidized-bed coating technique enhances dispersion and solubility of the coating material. Powder is fluidized with hot air and sprayed simultaneously with a binder liquid. This process causes particle adhesion, agglomeration and drying of agglomerates. It is applicable to both batch and continuous operations. The process of coating seeds with a pesticide slurry is often done by this technique. Seeds are treated by coating them with a polymer containing the desired pesticide/fungicide combination needed for a successful germination period. The process increases resistance of seeds to both mechanical breakage and biological attack.

5.3.2.6　Spray Coating

The spray-coating technique can be used alone or in combination with pan, drum, screw and fluidized-bed coaters. Spraying makes it possible to deposit either thin or thick layers of aqueous solution or suspensions and molten lipids or chocolate. It is the most commonly used technique for applying food coatings. A picture of a spray-coating apparatus and examples of products coated by this process are shown in Fig. 5.8.

The spraying nozzle plays a critical role in the coating process. Spraying efficiency depends on the pressure, fluid viscosity, temperature and surface tension of the coating liquid, as well as nozzle shape or design. This in turn affects the flow rate, the size of the droplets, spraying distance and angle, and overlap rate.

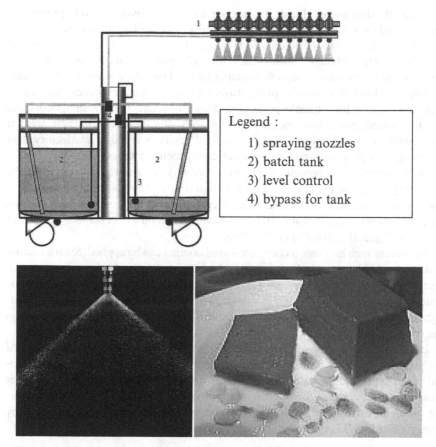

Fig. 5.8 A spray coater and a coated chocolate cake

5.4   Barrier Properties of Edible Packaging

5.4.1   Control of Mass Transfers Within Foods

Food preservation helps food maintain its original properties/qualities up to the point of consumption. It is accomplished by protecting the food article against conditions of its surrounding environment and controlling migration and mass transfer, including water loss. Indeed, product quality depends on the organoleptic, microbiological, nutritional and functional properties. These properties are subject to change during storage and distribution, mainly due to moisture migration either from the food to the environment or between different parts within a composite heterogeneous product.

Many physical and/or chemical processes have been developed to improve food stability and to prolong shelf-life. However, only food packaging makes it possible to control mass transfer between a food material and its surrounding environment (Kadoya 1990). Although plastics can effectively protect a food product from its surrounding environment, they often cannot be used to reduce mass transfer within a composite food. For instance, pizza crusts rapidly lose their crispiness because of moisture transfer from tomato paste. The dough of a tart, filled with apricots and laid on custard cream, loses its crispy texture 8 h after cooking (Fig. 5.9). Even if nutritional and microbiological quality of food products remains satisfactory, loss of key sensory properties represents significant product deterioration.

In heterogeneous composite foods, moisture migration always occurs from the wet compartment to the dry one, either as liquid or as vapour, even when the wet compartment contains solid water, as is the case with frozen products. For example, in an ice-cream cone, moisture transfer occurs as liquid (1) or vapour (2,3,4), as shown in Fig. 5.10 (Morillon et al. 2002).

To prevent such transfers, a chocolate-based coating can be applied. Such a coating makes it possible to control migration of liquid water from the cryo-concentrated sucrose solution of the ice cream (1), as well as water vapour transfer within the cone (2 and 3) and from the external environment (4), because it guarantees an air-tight contact with the paper/aluminium material typically used as external packaging for such products.

In all cases, no matter what the state of water, product temperature, or nature of the product, different regions within a food tend to balance their chemical potential often expressed as water activity. Thus, the driving force that governs water transfer is not the water content differential but the water activity or water chemical potential differential. This phenomenon is well described and predicted by Henry's and Fick's laws.

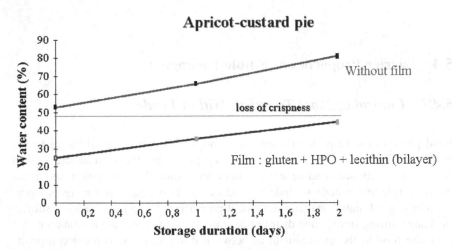

Fig. 5.9 Changes in water content in the dough of a pie filled with apricot and custard

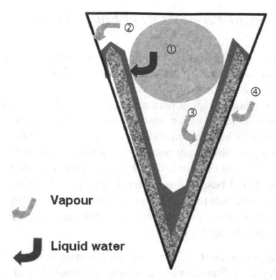

Fig. 5.10 Schematic representation of moisture transfers occurring in an ice-cream cone

Moisture transfer within and throughout a food product depends on an extrinsic parameter, water activity differential; and on two intrinsic parameters, solubility coefficient (S), which represents product affinity for water (thermodynamics), and diffusivity coefficient (D), which deals with water mobility (kinetics) in the system. Permeability is usually expressed as $P = DS$. On the other hand, moisture transfer between two compartments with different hydration levels depends on both thermodynamic and kinetic parameters. The primary means for reducing mass flow rests in controlling these two parameters.

First, the product formulation should be changed in order to balance, as much as possible, water activity of different compartments within a composite food. It is possible to either increase or decrease water activity by decreasing the content of solutes and dry matter, or by adding water activity depressors (e.g., polyols), respectively. However, this remedy induces changes in the final product composition, which can often be unfavourable to sensory quality. In general, it has been found that such transfers can be prevented by this means when maximum water activity differential is less than 0.1 and water activity of the food ranges between 0.3 and 0.75.

Second, kinetics of migration can be modified by applying a barrier layer at the interface between two compartments with differential moisture levels. These barriers are applied as thin layers, films, and coatings, which qualify as packaging material when they provide selectivity against transfer (Debeaufort et al. 1998). Theoretically, these barriers can be applied to food independent of temperature or moisture differentials between the different regions within a composite food. In fact, their efficiency is due to numerous factors (extrinsic or intrinsic), all of which vary according to functional properties desired, the coating or film formulation, application process, and nature of the targeted food.

5.4.2   Water Vapour Permeability of Edible Films

To protect products against moisture transfer, most efficient barriers are composed of hydrophobic substances. They are usually classified according to their water vapour permeability, as described in Fig. 5.11.

Table 5.3 provides water vapour permeability values of some lipid-based edible films as a function of composition and structure. Some lipids and food-grade hydrophobic substances have permeability values close to those of plastic films, such as low-density polyethylene or polyvinyl chloride. Permeability of lipid materials having a high solid fat content is usually much lower than that of liquid lipids. However, most solid lipid-based films are brittle when used alone; so they often are combined with hydrocolloids to form bilayer or emulsion films. Their structure and barrier properties depend on the preparation technique, with the most common preparation techniques being the following:

- Dipping the support in molten lipid to obtain a homogenous fat layer.
- Casting and drying a film-forming emulsion in which fats are dispersed in a hydrocolloid aqueous solution to obtain an emulsified edible film.
- Depositing a lipid layer previously molten or solubilized in an adequate solvent onto a hydrocolloid-based film, which is used as a mechanical support to obtain bilayer or multilayer films.

Barrier efficiency of bilayer films is of the same order of magnitude as that of the pure lipid or plastic films, and is much lower than that of emulsion-based films. Permeability of the latter is often very close to that of protein- or polysaccharide-based edible films, as shown in Table 5.4. Water vapour permeability values can vary over a wide range, according to many parameters related to external conditions (such as temperature, relative humidity, and pressure) and composition and structure of the film.

Composition and structure of edible films affect the water transfer mechanism, and therefore their barrier performance, just as much as physicochemical properties of the substance going through the barrier (e.g., concentration or moisture affinity). Both edible films and coated or wrapped food products are exposed to various factors such as temperatures during processing, transport, and storage. So, knowledge of the behaviour of the barrier layer, as a function of physical and

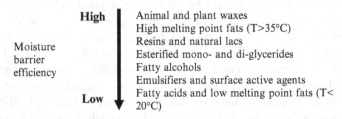

Fig. 5.11 Classification of hydrophobic substances according water vapour permeability

Table 5.3 Water vapour permeability (WVP in 10^{-11} g m^{-1} s^{-1} Pa^{-1}) of hydrophobic substances used in edible films

Composition	Temperature (°C)	RH gradient (%)	WVP
Paraffin wax	25	0–100	0.02
Candelilla wax	25	0–100	0.02
Carnauba wax + shellac	30	0–92	0.18
Carnauba wax + glycerol monostearate	25	22–100	35
Carnauba wax + glycerol monostearate	25	22–65	1.36
Microcrystalline wax	25	0–100	0.03
Beeswax	25	0–100	0.06
Capric acid	23	12–56	0.38
Myristic acid	23	12–56	3.47
Palmitic acid	23	12–56	0.65
Stearic acid	23	12–56	0.22
Acetyl acyl glycerol	25	0–100	22 to 148
Shellac	30	0–84	0.36 to 0.77
Shellac	30	0–100	0.42 to 1.03
Gumlac dewaxed	25	22–100	2.37
Gumlac dewaxed	25	22–75	1.51
Triolein	25	22–84	12.1
Hydrogenated cotton oil	26.7	0–100	0.13
Hydrogenated palm oil	25	0–85	227
Hydrogenated peanut oil	25	0–100	390
Native peanut oil	25	22–44	13.8
Cocoa butter substitute	25	22–84	0.6
Non-tempered cocoa butter	26.7	0–75	23.5
Tempered cocoa butter	26.7	0–75	4.9
Tempered cocoa butter	26.7	0–100	26.8
Milk chocolate	26.7	22–75	1.12
Milk chocolate	26.7	0–100	89.2
Dark chocolate	20	0–81	0.24

chemical parameters, is needed for improving the formulation of edible packaging. Many factors affect functional properties of edible films. These factors can be categorized into two groups (1) those concerning composition, and (2) those dealing with structure.

5.4.3 Effect of Film Composition

5.4.3.1 Chemical Nature and Concentration of Ingredients

Lipid-based edible films have a low affinity for water, which explains why they have low moisture permeability. However, each hydrophobic substance has its own physicochemical properties. Thus, edible films that are based on lipids exhibit variable behaviours against moisture transfer. Polarity of lipid constituents has to be considered, such as the distribution of electrostatic potential on molecules,

Table 5.4 Water vapour permeabilities (WVP in 10^{-11} g m^{-1} s^{-1} Pa^{-1}) of composite edible films

Composition	Temperature (°C)	RH gradient (%)	WVP
Bilayer films			
Methylcellulose/paraffin wax	25	22–84	0.2–0.4
Methylcellulose/paraffin oil	25	22–84	2.4
Methylcellulose/beeswax	25	0–100	0.058
Methylcellulose/carnauba wax	25	0–100	0.033
Methylcellulose/candelilla wax	25	0–100	0.018
Methylcellulose/triolein	25	22–84	7.6
Methylcellulose/hydrogenated palm oil	25	22–84	4.9
Hydroxypropylmethyl cellulose/stearic acid	27	0–97	0.12
Carnauba wax + glycerol monostearate de glycérol/dewaxed gum lac	25	22–75	4.29
Carnauba wax + glycerol monostearate/dewaxed gum lac	25	22–65	0.69
Emulsion-based films			
Methylcellulose + PEG400 + behenic acid	23	12–56	7.7
Methylcellulose + triolein	25	22–84	14.4
Methylcellulose + hydrogenated palm oil	25	22–84	13.2
Methylcellulose + PEG400 + myristic acid	23	12–56	3.5
Carboxymethylcellulose + butter oil + pectine	25	22–84	15.1
Hydroxypropyl cellulose + PEG400 + acetylated monoglycerides	21	0–85	8.2
Gluten + acegtylated monoglycerides	23	0–11	5.6–6.6
Gluten + oleic acid	30	0–100	7.9
Gluten + soy lecithin	30	0–100	10.5
Gluten + paraffin wax	25	22–84	1.7
Gluten + paraffin oil	25	22–84	5.1
Gluten + trioleine	25	22–84	9.7
Gluten + hydrogenated palm oil	25	22–84	7.4
Sodium caseinate + acetylated monoglycerides	25	0–100	18.3–42.5
Sodium caseinate + lauric acid	25	0–92	11
Sodium Caseinate + beeswax	25	0–100	11.1–42.5
Whey protein + palmitic acid	25	0–90	22.2
Whey protein + stearilic alcohol	25	0–86	53.6
Whey protein + beeswax	25	0–90	23.9–47.8

which depends on the chemical group, aliphatic chain length, and presence of unsaturation (Callegarin et al. 1997).

Waxes are the most efficient substances for reducing moisture permeability, owing to their high hydrophobicity caused by their high content of long-chain fatty alcohols and alkanes. The most efficient substance is beeswax, followed by stearyl alcohol, acetyl acyl glycerols, alkanes (hexatriacontane), triglycerides (e.g., tristearin), and fatty acids (e.g., stearic acid). This classification can be explained by molecular hydrophobicity, which defines the degree of interaction with water. Stearyl alcohol is a good barrier compared to triglycerides or fatty acids because its hydroxyl group has less affinity for water than carbonyl and carboxyl groups. Considering only polarity, alkanes would be predicted to be the best moisture barriers, though this was not observed to be the case for example hexatriacontane because of another contributing factor, structure.

For components having the same chemical nature, chain length influences moisture barrier properties. McHugh and Krochta (1994b, c) found that moisture barrier efficiency of fatty alcohols and fatty acids increases with the carbon atom number (from 14 to 18), because the relative proportion of the apolar part of the molecule increases, minimizing water solubility and moisture transfer. Many authors also have shown that among carboxylic acids, stearic and palmitic acids present the lowest water vapour permeabilities. Nevertheless, when the carbon chain contains more than 18 atoms, films containing arachidonic or behenic acids have higher moisture permeabilities. This observation is explained by the fact that very long chains induce a heterogeneous structure of the polymer network, decreasing moisture barrier properties (Koelsch and Labuza 1992). These results are shown in Fig. 5.12.

Unsaturated fatty acids are less efficient in controlling moisture migration, because they are more polar and exhibit different crystallization tendencies than saturated lipids. Indeed, films containing stearic or palmitic acids have been shown to better reduce the desiccation of oranges than coatings composed of oleic acid (Hagenmaier and Baker 1997). Conformation of unsaturated fatty acids is important as well, because their chemical structures (e.g., molecular conformation) are different. Besides physical state, crystallization changes with the degree of unsaturation, and consequently modifies film density and permeability.

Concentration of hydrophobic substances is also a key factor in barrier efficiency. Water vapour permeability of films composed of stearic acid and HPMC depends on fatty acid concentration, and strongly decreases at concentration values higher than 14–30% (Fig. 5.13). Moreover, the physical structure of all ingredients is also responsible for barrier properties of composite, lipid-based edible films.

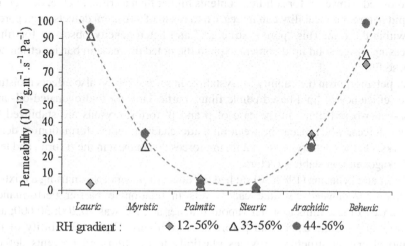

Fig. 5.12 Influence of the chain length of fatty acids on water vapour permeability of emulsion-based edible films composed of methylcellulose and lipids

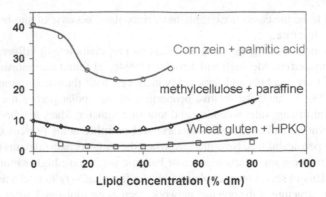

Fig. 5.13 Influence of lipid content on water vapour transfer in emulsion-based films with different film compositions (*HPKO* hydrogenated palm kernel oil)

5.4.3.2 Effect of Physical State and Structure of Ingredients

The solid or liquid state of lipid compounds strongly influences the moisture barrier efficiency of a film. Permeability of a hydrogenated cottonseed-oil-based film increases 300 times when the liquid fraction of lipid varies from 0 to 40%. Martin Polo et al. (1992a) showed that permeability of paraffin wax increases by 100 times when the paraffin oil content increases from 75 to 100%. Several authors have demonstrated that an increase in solid fat content, especially between 0 and 30%, improves barrier efficiency, as shown in Fig. 5.14. The solid structure of fats is denser and limits diffusion of water. Moreover, solubility of water in solid lipids is also reduced. However, for solid fat contents higher than a critical value, depending on lipid nature, permeability can increase as a result of structural defects (e.g., porosity) within the film. This "porous structure" and heterogeneity observed for films possessing a high solid lipid content explain the noted decrease in barrier efficiency of these films.

Fat polymorphism (i.e., ability to crystallize in several forms) also affects moisture barrier efficiency of lipid-based edible films, particularly for hydrogenated-oil- and cocoa-butter-based films. In the case of β and β' forms, crystals are stabilized by London forces, which occur between aliphatic chains to allow formation of dense networks. On the contrary, these stabilizing forces do not exist in the α form, and lead to a hexagonal, non-stable structure.

Kester and Fennema (1989a, c) studied the effect of polymorphism using a mixture of fully hydrogenated soy and canola oils in relation to barrier performance. Resistance of these films to water vapour transfer at 25°C was 40,200; 30,600; and 38,700 s m^{-1} for the α, β', and β forms, respectively. Higher flexibility of the hexagonal crystal structure provides plasticity to the film and prevents defects from forming during crystallization. This in turn improves moisture permeability.

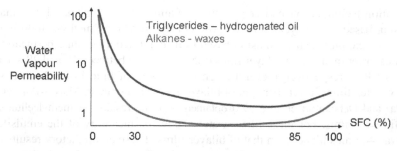

Fig. 5.14 Influence of solid fat content (SFC) of lipid-based barrier layer (bilayers) on the water vapour transfer

Crystallization in the β form, when compared to crystallization in the β' form, increases mass transfer resistance of films. This effect is mainly explained by the greater density of β crystals.

Waxes serve as good barriers against moisture transfer, but their efficiencies cannot be related to polymorphism or crystal size. Indeed, of the waxes, carnauba wax provides the most heterogeneous, porous surface, whereas microcrystalline wax exhibits the smoothest surface. However, films composed of carnauba or microcrystalline waxes have neither the best nor worst moisture barrier performance. Their permeabilities are probably more affected by their affinity for water.

Kester and Fennema (1989b) studied the effect of ageing on moisture permeability of stearyl alcohol-based films in which size and shape of crystals were modified without changing polymorphism. Stearyl alcohol crystallizes in a orthorhombic form (β'), and moisture transfer resistance of coatings stored at 48°C for 0, 14, and 35 days increased from 11 to 50% during this time. The mechanism responsible for improvement in barrier properties with time is not fully described and explained, but it seems that two parameters are important (1) the decrease in structural defects when recrystallization occurs at a low rate, and (2) the increase of solid fat content as liquid lipid crystallizes.

The chemical nature of film ingredients contributes to moisture barrier efficiency, because it partially affects affinity of the film for water. While physical structure of the lipids determines film structure, and thus moisture permeability, film structure is another parameter that needs to be considered.

5.4.4 Influence of Film Structure on Barrier Properties

5.4.4.1 Role of Preparation Technique

Edible films containing lipids may be obtained using different techniques which determine barrier structure. Several authors have studied the influence of film preparation technique on barrier properties. Kamper and Fennema (1984) made emulsion-based and bilayer films composed of hydroxypropylmethylcellulose and stearic and palmitic acids. Emulsion-based films exhibited a moisture permeability 40 times lower than that of bilayer films (0.5×10^{-12} and between 19 and 28×10^{-12} g m^{-1} s^{-1} Pa^{-1}, respectively), even though the amount of lipid in emulsion-based films was ten times lower than that of bilayer films. In contrast, Martin-Polo et al. (1992a) and Debeaufort et al. (1993) obtained opposite results for methylcellulose–paraffin wax films. For these films, water vapour permeability of the emulsified film was 40 times higher than that of bilayer films. These contradictory results are likely explained by considering film preparation conditions. Though the initial emulsion prepared by Kamper and Fennema (1984) was homogeneous, a phase separation was observed during drying, which led to an apparent bilayer structure for the emulsified film. Debeaufort et al. (1993), using scanning electron microscopy, revealed the irregular, rough, and heterogeneous surface of emulsified films and, contrastingly, the smooth and homogeneous surface of bilayer films. A more homogeneous distribution of the lipid layer at the film surface was found to be more efficient for controlling water transfer.

The most efficient moisture barriers are generally obtained when the film structure contains a continuous lipid layer, obtained either from a bilayer technique or after the destabilization of the emulsion during drying for an emulsion-based film. In chocolate, the continuous phase is composed of a lipid, as a result of conching and tempering. Dispersed hydrophilic particles (sucrose crystals, cocoa powder) do not significantly affect the water vapour permeability, except where there is high relative humidity, for which film structure is modified.

Because of their poor mechanical properties, lipids are often used in conjunction with a support, which can sometimes be considered as a model food to be coated. However, this support can strongly influence film properties by modifying the distribution of hydrophobic substances. If the support is porous, the barrier is less efficient because the surface to be protected is much greater and a rough surface makes application of a continuous and regular coating layer more difficult.

Adhesion of a film to a food product is a physical phenomenon that can be controlled by the overlapping of rough surfaces; by wettability forces due to interfacial forces between materials, where at least one is liquid; by electrostatic forces between charged polymers; or by chemical bonds when a reaction occurs. Barron (1977) showed that chocolate coating, whose adhesion to a biscuit is very good, tends to crack because the latter swells when moisture is absorbed. In this case, too strong of an adhesion between the coating and support is not always an advantage. Interaction occurring between the support and coating has to be taken into account because it can modify coating barrier properties, either due to a partial absorption of the barrier layer by the support or due to a heterogeneous lipid distribution. Indeed, a barrier coating can be efficient only if it forms a homogeneous and continuous layer on the entire surface of the food product to be protected. When cracks appear, all barrier properties are lost. Continuity and homogeneity of the barrier also strongly depend on film thickness.

5.4.4.2   Influence of the Film Thickness

Film thickness can modify film properties, such as barrier performance. According to the following equation, WVTR (water vapour transfer rate) corresponds to the amount of water vapour (Δm) transferred through a film area (A), during a defined time (Δt), where permeability (*WVTR*) is normalized by thickness (x) and partial vapour pressure gradient (Δp):

$$\text{WVTR} = \frac{\Delta m}{A \Delta t} (\text{g m}^{-2}\,\text{s}^{-1}) \text{ and } P = \text{WVTR} \frac{x}{\Delta p} (\text{g m}^{-1}\,\text{s}^{-1}\,\text{Pa}^{-1}).$$

An increase in film's permeability with thickness indicates water affinity, which could be due to the presence of hydrophilic compounds in the formulation. It seems, then, that the influence of film thickness varies with the nature of both the lipid and other film components.

Mass transfer resistance of lipid compounds against gas migration, such as oxygen, is mainly due to structure; the more dense (e.g., compact) the crystals and the more homogeneously they are distributed and oriented, the more difficult the gas diffusion and the lower the permeability. In the case of water migration, barrier properties are more complex. Affinity for water is often as important as the film structure itself. Thus, transfer of water molecules within films and coatings depends both on diffusivity (kinetic factor) and sorption (thermodynamic factor) phenomena, even in the case of hydrophobic substances.

5.4.5   Effect of Properties of Penetrant Molecules

In the case of dense materials as opposed to porous ones, properties of the penetrant, such as its concentration, polarity, or physical state, affect not only its interactions with film components but also kinetic phenomenon of transfer. Size and shape of the diffusing molecules influence the diffusivity, whereas polarity and its ability to condense influence the sorption. Water is a polar compound having a small size that tends to favour both sorption and diffusivity in dense materials. This ability makes these materials more permeable to moisture than to non-condensable gases. In this chapter, only water transfer will be considered.

5.4.5.1   Effect of Water Concentration Gradient

The driving force for moisture transfer is the vapour pressure or the water activity difference between the two sides of the film. The higher the gradient, the greater the transfer rate, if there is no interaction between water and the film. However, permeability remains constant, as observed with paraffin wax films, hydrogenated cottonseed oil, or dark chocolate, as long as humidity on the high side of the gradient is lower than 80%. Other behaviours were observed: particularly an increase in permeability of paraffin oil coatings, cocoa liquor and butter, milk chocolate, and acetylstearyl glycerols. Water vapour permeability of milk chocolate increases from 10.7×10^{-12} to 892×10^{-12} g m^{-1} s^{-1} Pa^{-1} when the humidity gradient varies from 22–75% to 0–100%. This phenomenon is attributed to moisture sorption by hydrophilic particles (milk powder, sucrose crystals, and cocoa powder) within the coating, which leads to a loss of film integrity.

It is well known that permeability of hydrophilic films depends both on the relative humidity (RH) difference and on the absolute humidity values. For instance, for the same RH gradient, permeability increases with an increasing vapour pressure value. Tiemsra and Tiemsra (1974) established a direct relationship between water vapour transfer rate and humidity for films composed of peanut oil, peanut butter, or cocoa liquor.

$$\mathrm{WVTR} = \frac{k a_{w1} \Delta a_w}{x}.$$

In this equation, WVTR is the water vapour transfer rate (g m^{-2} s^{-1}), a_{w1} is the higher water activity value of the gradient Δa_w, x is the film thickness (m), and k is an adjusted mass transfer coefficient (g m^{-1} s^{-1}). This model would be more interesting if the k parameter could be correlated to a film's physicochemical and/or structural characteristics. In the case of hydrophilic films, the permeability increase at high relative humidity levels is often due to swelling and plasticization of the polymer network by moisture sorption. This creates a less dense structure in which chain extremities are more mobile. In this scenario, water diffusivity and water permeation become easier.

Even for hydrophobic substances, which theoretically have a low affinity for moisture, water affects the barrier efficiency of edible films and coatings when they contain hydrophilic particles or polar groups, as in chocolate or cocoa-based coatings. Most published reports deal with transfer of water vapour. However, films and coatings are often used for highly hydrated products in which water is in a liquid state, such as gels. Water in contact with the barrier layer can also be in a solid state, as in the case of frozen foods. Therefore, it is interesting to consider the effect of the physical state of water on its transfer.

5.4.5.2 Effect of Physical State of Water on Barrier Performances

For edible films and coatings, a few authors consider the direct contact between the barrier layer and the moist compartment. Morillon et al. (1998) studied the effect of the physical state of water, liquid or vapour, on its migration through lipid-based edible films and through hydrophilic and hydrophobic synthetic films. They reported that water permeability of lipid–sugar-based edible films is the same for both water physical states as long as water activity is lower than 0.85 (Fig. 5.15).

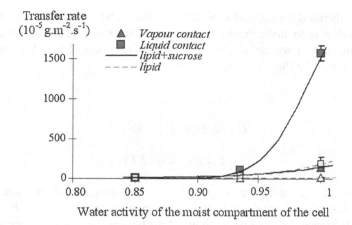

Fig. 5.15 Effect of the state of water (liquid or vapour) on water transfer rate through cocoa-based edible barriers

At higher water activities, transfer of liquid water through lipid-based edible films and hydrophilic plastic films is much greater than transfer of vapour water. Two phenomena may occur (1) a partial solubilization of the sucrose of the film, and/or (2) a swelling of the matrix in hydrophilic films.

When the penetrant is partly in a solid state, as water is in frozen foods, its transfer through edible or plastic films has been scarcely studied. The discontinuity of solubility, diffusivity, or permeability of some molecules when the state of the diffusing molecule changes is commonly called the "Schroeder Paradox" (Vallieres et al. 2006), and is often observed for hydrophilic polymers over their melting temperature ranges. This phenomenon seems to be due to a modification of energy required for deformation of molecules when they diffuse, and becomes very weak above their melting points. Moreover, all molecules that diffuse in and through dense polymers are in a liquid state. So when there is no state change, such as melting or condensation, less energy is required for transfer. This phenomenon could explain why a liquid penetrant in contact with the membrane often favours moisture transfer.

The influence of a penetrant's physical state, and especially water transfer through dense films and coatings, is a field that has been poorly considered to date, but is important to better understand the mechanism of moisture transfer in composite foods. Nevertheless, the physical state and structure of edible barriers also strongly depends on temperature.

5.4.5.3 Influence of Temperature

Food products can be exposed to a wide range of temperature variations during processing, and are often stored at sub-ambient temperatures. Temperature tremendously affects all thermodynamic and kinetic phenomena and, thus, moisture transfer. As long as there is no modification of the film or coating structure due to temperature, permeability, diffusivity, and solubility coefficients vary according to the Arrhenius law, as described by the following equation.

$$P = P_0 \exp(-E_{a,P} / RT),$$

$$D = D_0 \exp(-E_{a,D} / RT),$$

$$S = S_0 \exp(-\Delta H_S / RT).$$

In this equation, $E_{a,P}$ and $E_{a,D}$ are the apparent activation energies of the permeation and diffusion phenomena (kJ mol^{-1}); ΔH_S is the enthalpy of sorption (kJ mol^{-1}); P_0, D_0, and S_0 are the reference values of permeability, diffusivity and solubility, respectively; R is the ideal gas constant (8.31 kJ mol^{-1} K^{-1}); and T is the temperature (K).

From the definition of permeability, which is the product of solubility and diffusivity coefficients, the relationship between energies is described by the following equation.

$$E_{a,P} = E_{a,D} + \Delta H_S.$$

In this equation, ΔH_S is always a negative value for water vapour, which means that solubility decreases as temperature increases, whereas $E_{a,D}$ is positive and diffusivity rises with temperature. According to the transfer, which is mainly driven either by kinetic or thermodynamic phenomena, permeability increases or decreases as temperature increases. For instance, water vapour permeability of polyethylene waxes increases as temperature varies from 25 to 40°C; activation energy is between 5 and 15 kJ mol^{-1}. This observation indicates that moisture transfer seems to mainly depend on diffusion. Kester and Fennema (1989b) reported an increase in moisture transfer as temperature decreased, which can be explained by a greater moisture sorption by the filter paper used as a mechanical support. Other authors have not shown any effect of temperature on moisture transfer, because transfer rate variations are directly related to the vapour pressure of water, as observed for hydrogenated cottonseed oil and cocoa butter over a temperature range of 3–26.7°C, or between 21 and 27°C for purified acetylstearyl glycerol.

Over a wider range of temperatures, some authors have observed (1) a decrease in moisture permeability with temperature, and (2) an increase in permeability for low temperatures. Indeed, this phenomenon was also observed for plastic packaging, edible films composed of fatty acid or fatty acid–beeswax mixtures and chocolate- or cocoa-based coatings (Fig. 5.16).

This increase in moisture permeability observed by several authors can be due to a constant or an increasing water vapour transfer rate, in spite of the temperature decrease. For lipid-based films, such as chocolate or cocoa butter, temperature modifies the solid fat content (SFC), and thus affects both structure and barrier performance. Indeed, water vapour transfer rate of chocolate decreases between 26 and 20°C, but remains constant from 20 to 10°C, which is explained by the fact

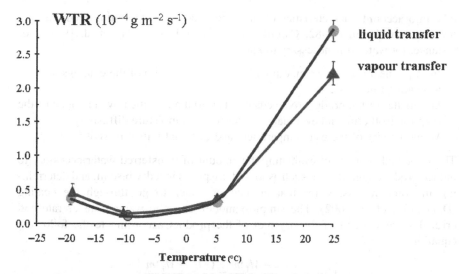

Fig. 5.16 Effect of temperature on liquid and vapour water transfer rate (WTR) through cocoa-based films

that SFC increases from 80% at 26°C to 90% at 20°C. For both plastic and edible packaging materials, authors explain that this behaviour is due to phase transitions; the appearance of defects, such as holes or cracks caused by temperature dilatation or contraction; or by a greater moisture sorption at low temperatures. This particular behaviour of permeability as temperature decreases was also reported by Morillon et al. (1999) for hydrophilic and hydrophobic films. However, these authors conclude that the driving force for moisture transfer is not vapour pressure difference, but rather chemical potential. They further calculated a "corrected permeability" that always decreases as temperature declines.

Temperatures below 0°C allow better preservation of foods, particularly by inhibition of mold growth. But chemical and physicochemical reactions still occur, though they are significantly slowed down by sub-freezing temperatures. The effect of a temperature below 0°C on moisture transfer has been hardly studied in the field of edible barriers. Indeed, the few works published on moisture transfer at such temperatures only consider the amount of water amount gained or lost by the food. This approach does not take into account real barrier performance or permeation phenomena occurring within edible films.

5.4.6 Prediction of the Shelf-Life of Coated Products

Initially developed for prediction of shelf-life of products packaged in plastic flexible or semi-rigid packaging, several models have been adapted to coated foods by considering edible films and coatings as conventional packaging. These models take into account characteristics of the food product, surrounding medium, and barrier layer (Labuza 1982; Cardoso and Labuza 1983; Hong et al. 1991). For instance, for water it is necessary to know

- Film permeability, or water vapour transfer rate (WTR) of the coatings, as well as coating thickness.
- Dry matter of the product to be coated, its initial and critical water contents, the sorption isotherm, and eventually its density and moisture diffusivity.
- Water activity of the wet compartment and eventually its diffusivity.

These methods consist of evaluating the amount of transferred water necessary to induce product degradation, such as loss of crispness in a dry biscuit, and determining the time necessary for that amount of water to go through the coating (Debeaufort et al. 2002). The simplest model uses the water transfer rate and critical water content of the dry part of the product according to the following equation.

$$\text{Time} = \frac{(M_C - M_i)m}{\text{WTR } A} = \frac{(M_C - M_i)m}{PA\,\Delta p}x.$$

In this equation, WTR is the water transfer rate (g m^{-2} s^{-1}); A is the surface exposed to the transfer; x is the coating thickness (m); m is the dry matter of the product (g); M_i and M_c are the initial and critical water contents, respectively; and Δp is the water vapour partial pressure differential across the barrier.

However, this model assumes that the water activity differential between the two faces of the coating remains constant over storage, which never occurs. The progressive change in water activity in the dry compartment (e.g., a cereal-based biscuit) has to be taken into account from the sorption isotherm modelled as linear between the initial and critical water contents for predictions to be in accordance with reality (Labuza and Contreras-Medellin 1981). Parameters used in this model are given in Fig. 5.17, where M_i, M_c, and M_e are the initial, critical, and equilibrium water contents, respectively; aw_i, aw_c, aw_e are the water activities; and b_s is the slope of the linear-considered sorption isotherm between the initial and critical moisture content points.

In an attempt to consider water activity variation in the dry compartment, Labuza and Contreras-Medellin (1981) and Labuza (1982) suggested the following equation.

$$\ln \frac{M_i - M_e}{M_c - M_e} = \frac{PAp^0}{xmb}\,\text{time}.$$

In this equation, P is the coating permeability (g m^{-1} s^{-1} Pa^{-1}), A is the exposed surface (m^2), p^0 is the saturated vapour pressure at the experiment temperature, x is the coating thickness (m), m is the dry matter of the product (g), and b is the slope of the sorption isotherm of the dry compartment ($g_{water}/g_{product}$). This model was used by Biquet and Labuza (1988) for chocolate coatings applied between agar gel and microcrystalline cellulose. Morillon et al. (1998) also compared the estimated shelf-life calculated from this latter model and experimental values determined from

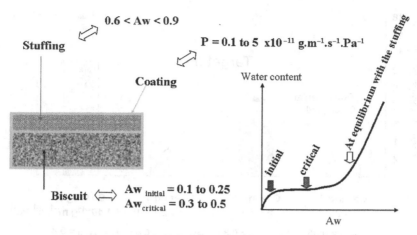

Fig. 5.17 Schematic representation of shelf-life prediction for a model composite food (stuffed biscuit)

sensorial experiments for a wafer coated with a cocoa-based coating. Figure 5.18 provides the estimated (bars) and measured (triangles) shelf-life values for two different coatings whose permeability is either 1 or 2.5×10^{-11} g m^{-1} s^{-1} Pa^{-1}, and two wafers whose critical water activity varies from 0.35 to 0.45 before losing crispiness.

Shelf-life varied from 14 days to more than 18 months on the basis of coating efficiency and wafer sensitivity to water. Very few results on real products have been published, and one of the main difficulties is to know the coating permeability in real conditions of the application. More recently, Roca et al. (2006) and Bourlieu et al. (2007) have developed newer models that take into account moisture diffusivity in each compartment (dry and wet and in the edible barriers), based on sorption kinetics, isotherm data, and water vapour permeability of the edible film.

5.4.7 Oxygen and Carbon Dioxide Permeabilities of Edible Films and Coatings

Barrier properties of edible films and coatings against oxygen and carbon dioxide are given in Table 5.5. Gas barrier properties of hydrocolloid-based edible films are very important for fresh fruit and vegetable applications. In dry conditions, permeabilities of these edible films and coatings are very low, down to 3,000 times lower than those of plastic polymers. However, though oxygen permeability of plastic films (except EVOH or Nylon) remains constant irrespective of relative humidity, that of edible films and coatings usually increases significantly in the presence of moisture. This behaviour is most often caused by plasticization of the biopolymer network by

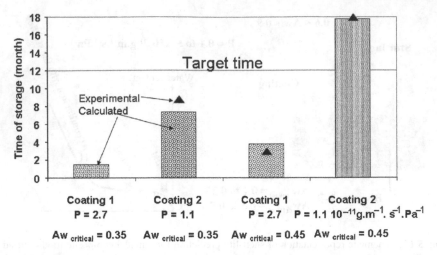

Fig. 5.18 Shelf-life of a wafer in ice cream, estimated from the Hong et al. (1991) model and sensory analyses (crispness)

Table 5.5 Oxygen (P_{O_2}) and carbon dioxide (P_{CO_2}) permeabilities of edible films (10^{-15} g m^{-1} s^{-1} Pa^{-1})

Film	P_{O_2}	P_{CO_2}	T (°C)	a_w
Methylcellulose	8.352	1315.60	30	0
Beeswax	7.680	–	25	0
Carnauba wax	1.296	–	25	0
Corn zein	0.560	9.50	38	0
Wheat gluten	0.016	0.31	25	0
Chitosan	0.009	–	25	0
Pectin	21.440	937.20	25	0.96
Wheat gluten	20.640	1,614.80	25	0.95
Starch	17.360	–	25	1
Chitosan	7.552	352.44	25	0.93

water, which induces greater solubility and diffusivity of gases in the biopolymer matrix. Lipids are not as well suited as proteins or polysaccharides because of their strong affinity for oxygen. Moreover, some lipid compounds (mainly unsaturated fatty acids) are oxygen sensitive, and can undergo rancidity, causing off-flavor development.

5.5 Mechanical Properties

As mentioned earlier, materials used to make edible or biopolymer film packaging include polysaccharides, proteins, and lipids, or combinations of these compounds for composite films. The goal is to combine advantages associated with different components to obtain optimal properties. Mainly, polysaccharides and proteins provide structural support, while lipids are used to create an effective water barrier. Moreover, several studies have shown the influence of the physical state and polymorphism of lipids on functional properties of composite films. The way in which lipid is added (dispersed in the hydrocolloid aqueous solution to obtain an emulsion in a single packaging layer or cast on an existing packaging to form a bilayer film) influences film mechanical properties. The food industry has focused on emulsified films, which require only one step in manufacture, as opposed to the two or more steps required for bilayer and multilayer films. The nature of the emulsifier used for lipid dispersion controls film properties. Drying temperature and time have a large impact on film physical properties. Each of these experimental conditions associated with film formulation and formation needs to be taken into account and studied to evaluate impact of each parameter on film properties.

Bravin et al. (2006) investigated the effect of different types of surfactants (glycerol monostearate (GMS), Tween 60 and 80) on tensile strength (TS) and percentage of elongation at break (E) of an emulsified edible film composed of cornstarch and methylcellulose (MC), containing cocoa butter or soybean oil as lipid compounds. Emulsified films were prepared by MC or starch solubilization in water/alcohol or water, respectively, at temperatures higher than 75°C before being mixed together. Lipid was then added before the mixture was spread onto glass plates using a thin layer chromatography spreader. Films were dried at 25°C, at a relative humidity of 40% for 15 h. All films were maintained for least 10 days at 53% RH before mechanical properties were measured. Table 5.6 shows TS and E values of cocoa butter films in relation to lipid and emulsifier contents. Values showed substantial variation, depending on the film formulation used.

Table 5.6 Tensile strength (TS) and elongation (E) of edible film as affected by cocoa butter/lipid content (L), emulsifier (EM) and emulsifier type (GMS, Tween 60 and 80)

	EM (%)	TS (MPa)	
L (%)		10	20
GMS	10	11.9	9.1
	30	10.9	8.5
Tween 60	10	9.9	8.6
	30	8.3	6.7
Tween 80	10	20.6	8.6
	30	18,2	10.7
		E (%)	
L (%)		10	20
GMS	10	28.6	13.6
	30	17.6	11.6
Tween 60	10	28.1	19.0
	30	20.1	19.7
Tween 80	10	4.3	6.0
	30	4.6	3.7

The TS data are in the same range as those observed for LDPE films. Increasing the total lipid level reduced TS, which correlated with development of a global heterogeneous film structure after lipid was added (Fig. 5.19). Images of film sections containing 20% butter revealed formation of a less oriented, less compact structure, with evident internal cavities and discontinuities. Use of oil (liquid state) at a 10% addition level limited the adverse impact of lipid addition on film mechanical properties (Table 5.7).

Phan et al. (2002a, b) investigated the effect of drying temperature on emulsified films made from arabinoxylans, hydrogenated palm kernel oil (HPKO), and two sucrose esters, SP 10 and SP 70 (emulsifiers that are soluble in oil and water, respectively). Film was prepared by solubilizing the polysaccharides in water at 75°C before addition of glycerol at 15%. Lipid addition at a rate of 30% on a dry basis was carried out. The emulsifier mixture was dispersed prior to addition to the arabinoxylan film-forming solution. The solution was spread onto glass plates for film drying. Temperatures of 30, 40, and 80°C were employed for drying. The values for the mechanical properties of the films as a function of the drying temperature are presented in Figs. 5.20 and 5.21. Tensile strength (S) and percentage of elongation at break (E) for films containing lipid ranged from 0.8 to 3 times lower than those without lipid addition. Despite the liquid fat choice, no pseudo-plasticizer effect or lubricant was observed, and both S and E decreased.

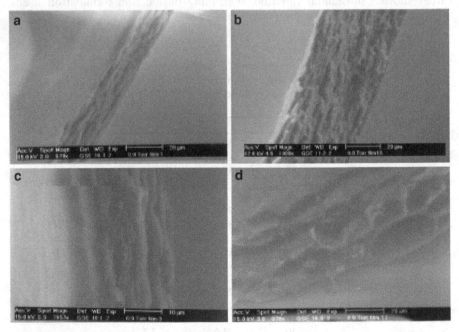

Fig. 5.19 Scanning electron microscopy cross-section photomicrographs of 10% soybean oil films (**a, b**) and 20% cocoa butter films (**c, d**)

Table 5.7 Tensile strength (TS) and elongation (E) of edible film as affected by soybean oil/lipid content (L), emulsifier (EM) and emulsifier type (GMS, Tween 60 and 80) (Bravin et al. 2004)

	EM (%)	TS (MPa)	
L (%)		10	20
GMS	10	11.3	10.4
	30	11.1	9.2
Tween 60	10	26.2	19.9
	30	14.8	12.9
Tween 80	10	22.8	15.7
	30	24.4	19.6
		E (%)	
L (%)		10	20
GMS	10	15.0	6.0
	30	11.6	5.1
Tween 60	10	6.2	7.6
	30	4.4	4.1
Tween 80	10	5.5	4.6
	30	5.0	4.6

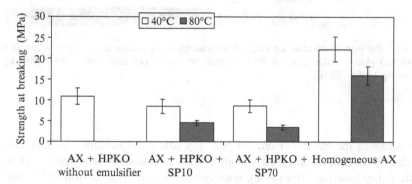

Fig. 5.20 Strength (S) at breaking point for arabinoxylan-based edible films as a function of drying temperature (Phan et al. 2002b) (*AX* arabinoxylans, *HPKO* hydrogenated palm oil kernel, *SP10* sucroester 10, *SP70* sucroester 70)

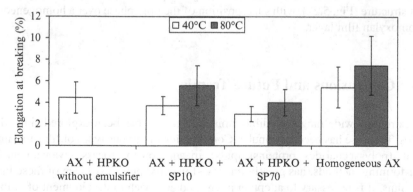

Fig. 5.21 Elongation (E) at breaking point for arabinoxylan-based edible films as a function of drying temperature (Phan et al. 2002b) (*AX* arabinoxylans, *HPKO* hydrogenated palm oil kernel, *SP10* sucroester 10, *SP70* sucroester 70)

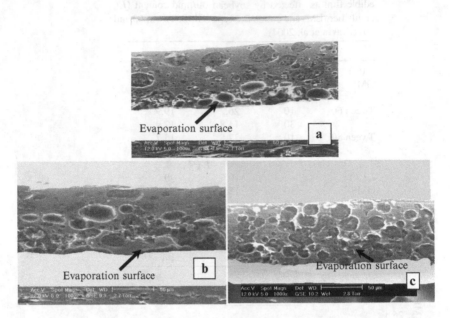

Fig. 5.22 Environmental scanning electron microscopy cross-section micrographs of arabinoxy-lan–HPKO–glycerol films dried at 80°C without emulsifier (**a**), containing 2.5% SP10 (**b**) or containing 2.5% SP70 (**c**)

The nature of the emulsifier did not play any role in improvement of mechanical resistance, but temperature greatly modified film physical performance. S decreased while E increased as film drying temperature was increased from 40 to 80°C. At higher temperatures, the solvent evaporated more rapidly, and the films dried more quickly. In addition, number of interactions between arabinoxylan chains was reduced; the network was less organized and more stretchable, which tended to increase E. Moreover, the increase in drying temperature promoted a bilayer-like film structure (Fig. 5.22), with superposition of the lipid phase over a homogeneous arabinoxylan film layer.

5.6   Conclusions and Future Trends

An extremely wide range of edible moisture barriers has been explored since the 1950s. Their use has become popular. Combination of various types of film-forming agents (polysaccharides, proteins, lipids), in addition to process advancement in film-forming methods, has greatly improved water vapour resistance of these bar-rier films. It is necessary to adopt an integrated approach in development of edible barriers to combine regulatory, nutritional, organoleptic and technical requirements

(Bourlieu et al. 2007). Indeed, most published patents not only disclose barrier composition but also the food product onto which it has to be applied and the appropriate technique for applying the barrier. This integrated approach should also be adopted in research to study determination of barrier efficiency for specific barrier applications in a given food product.

The field of ready-to-eat food products is still developing and provides a significant challenge in terms of mass transfer control. In particular, edible moisture barriers may provide an interesting answer to consumer demand for composite products with good nutritional value and stable organoleptic properties.

References

Barron LF (1977) The expansion of wafer and its relation to the cracking of chocolate and "bakers chocolate" coatings. *J. Food Technol*, 12:73–84

Biquet B, Labuza TP (1988) Evaluation of the moisture permeability characteristics of chocolate films as an edible moisture barrier. *J. Food Sci*, 53:989–998

Bourlieu C, Guillard V, Powell H, Valles-Pamies B, Guilbert S, Gontard N (2007) Modelling and control of moisture transfers in high, intermediate and low composite food. *J. Food Chem*, 106(4):1350–1358

Bravin B, Peressini D, Sensidoni A (2006) Development and application of polysaccharide–lipid edible coating to extend shelf-life of dry bakery products. *J. Food Eng*, 76(3):280–290

Callegarin F, Quezada Gallo JA, Debeaufort F, Voilley A (1997) Lipids and biopackaging. *J. Am. Oil Chem. Soc*, 74:1183–1192

Cardoso G, Labuza TP (1983) Prediction of moisture gain and loss for packaged pasta subjected to a sine wave temperature/humidity environment. *J. Food Technol*, 18:587–606

Debeaufort F, Martin-Polo M, Voilley A (1993) Polarity, homogeneity and structure affect water vapour permeability of model edible films. *J. Food Sci*, 58(2):426–434

Debeaufort F, Quezada-Gallo JA, Voilley A (1998) Edible films and coatings: tomorrow's packagings: a review. *Crit. Rev. Food Sci*, 38(4):299–313

Debeaufort F, Voilley A, Guilbert S (2002) Les procédés de stabilisation des produits par des films "barrière" In Lorient D, Simatos D, Le Meste M (Eds.), *"Propriétés de l'eau dans les aliments,"* Tec & Doc, Lavoisier, Paris, pp. 549–600

Guilbert S, Cuq B (1998) Les films et enrobages comestibles. In Multon JL et Bureau G (Eds.) *L'emballage des denrées alimentaires de grande consommation*, Tec & Doc, Lavoisier, Paris, pp. 472–530

Hagenmaier RD, Baker RA (1997) Edible coatings from morpholine-free wax microemulsions. *J. Agric. Food Chem*, 45:349–352

Hong YC, Koelsch CM, Labuza TP (1991) Using the L number to predict the efficacity of moisture barrier properties of edible coating materials. *J. Food Proc. Preserv*, 15:45–62

Kadoya T. (1990) Food Packaging. Academic, New York, NY

Kamper SL, Fennema O (1984) Water vapour permeability of an edible, fatty acid, bilayer Film. *J. Food Sci*, 49:1482–1485

Kamper SL, Fennema O (1985) Use of edible film to maintain water vapour gradients in food. *J. Food Sci*, 50:382–384

Karbowiak T, Hervet H, Leger L, Champion D, Debeaufort F., Voilley A (2006), Effect of plasticizers (water and glycerol) on the diffusion of a small molecule in iota carrageenan biopolymer films for edible coating application. *Biomacromolecules*, 7:2011–2019

Karbowiak T, Debeaufort F, Champion D., Voilley A (2007) Influence of thermal process on structure and functional properties of edible films. *Food Hydrocoll*, 21(5–6):879–888

Kester JJ, and Fennema O (1989a) The influence of polymorphic form on oxygen and water vapour transmission through lipid films. *JAOCS*, 66:1147–1153

Kester JJ, Fennema O (1989b) Tempering influence on oxygen and water vapour transmission through a stearyl alcohol film. *JAOCS*, 66:1154–1157

Kester JJ, Fennema O (1989c) Resistance of lipid films to water vapour transmission. *JAOCS*, 66:1139–1146

Koelsch CM, Labuza TP (1992) Functional, physical and morphological properties of methyl cellulose and fatty acid-based edible barriers. *Lebensm. Wiss. Technol*, 25:404–411

Krochta JM, DeMulder-Johnston C (1997) Edible and Biodegradable Polymer Films: Challenges and Opportunities. *Food Technol*, 51:61–74

Labuza TP (1982) Moisture gain and loss in packaged foods. *Food Technol*, 36:92–97

Labuza TP, Contreras-Medellin R (1981) Prediction of moisture protection requirements for foods. *Cereal Food World*, 26:335–343

Martin-Polo MO, Mauguin C, Voilley A (1992a) Hydrophobic Films and Their Efficiency against Moisture Transfer. I. Influence of the film preparation technique. *J. Agric. Food Chem*, 40:407–412

Martin-Polo M, Voilley A, Blond G, Colas B, Mesnier M, Floquet N (1992b) Hydrophobic films and their efficiency against moisture transfer. 2. Influence of the physical state. *J. Agric. Food Chem*, 40:413–418

McHugh TH, and Krochta JM (1994a) Milk-protein-based edible films and coatings. *Food Technol*, 48(1):97–103

McHugh TH, Krochta JM (1994b) Permeability properties of edible films. In: Krochta JM, Baldwin EA, Nisperos-Carriedo M. *Edible Films and Coatings to Improve Food Quality*. Technomic Publishing Company, Lancaster, USA

McHugh TH, Krochta JM (1994c) Water vapour permeability properties of edible whey protein–lipid emulsions films. *JAOCS*, 71:307–311

Morillon V; Debeaufort F, Capelle M, Blond G, and Voilley A (1998) Effect of the water properties on the moisture barrier efficiency of lipid-sugar based edible coatings. ISOPOW VII – *Water Management in the Design and Distribution Of Quality Foods*. May 30–June 4, Helsinki, Finland

Morillon V, Debeaufort F, Blond G.A., Voilley (1999) Temperature influence on the moisture transfer through synthetic films. *J. Membr. Sci*, 4377:1–9

Morillon V, Debeaufort F, Blond G, Capelle M, Voilley A (2002) Factors affecting the moisture permeability of lipid-based edible films: a review. *Crit. Rev. Food Sci. Nutr*, 42(1):67–89

Phan TD, Debeaufort F, Peroval C, Despre D, Courthaudon JL, Voilley A (2002a) Arabinoxylan-lipid-based edible films: 3 influence of drying temperature on emulsion stability and on film structure and functional properties. *J. Agric. Food Chem*, 50(8):2423–2428

Phan TD, Peroval C, Debeaufort F, Despre D, Courthaudon JL, Voilley A (2002b) Arabinoxylan-lipid-based edible films: 2 influence of sucroester nature on the emulsion structure and film properties. *J. Agric. Food Chem*, 50(2):266–272

Roca E, Guillard V, Guilbert S and Gontard N (2006) Moisture migration in a cereal composite food at high water activity: Effects of initial porosity and fat content. *J. Cereal Sci*, 43 (2):144–151

Tiemsra PJ, Tiemsra JP (1974) Moisture transmission through peanut oil films. *Peanut Sci*, 1:47–50

Vallieres C, Winkelmann D, Roizard D, Favre E, Scharfer P, Kind M (2006) On Schroeder's paradox. *J. Membr. Sci*, 278(1–2):357–364

Wu Y, Weller CL, Hamouz F, Cuppett SL and Schnepf M (2002) Development and application of multicomponent edible coatings and films: A review. *Adv. Food Nutr. Res*, 44:347–394

Chapter 6
Characterization of Starch and Composite Edible Films and Coatings

María A. García, Adriana Pinotti, Miriam N. Martino, and Noemí E. Zaritzky

6.1 Introduction

Starch-based and composite edible films and coatings can enhance food quality, safety and stability. They can control mass transfer between components within a product, as well as between product and environment. They can improve performance of the product through the addition of antioxidants, antimicrobial agents, and other food additives. Unique advantages of edible films and coatings can lead to the development of new products, such as individual packaging for particular foods, carriers for various food additives, and nutrient supplements. Film materials and their properties have been reviewed extensively in this book and previously (Guilbert 1986; Kester and Fennema 1986; Krochta and De Mulder-Johnson 1997).

Composite films can be formulated to combine the advantages of each component. Biopolymers, such as proteins and polysaccharides, provide the supporting matrix for most composite films, and generally offer good barrier properties to gases, with hydrocolloid components providing a selective barrier to oxygen and carbon dioxide (Guilbert 1986; Kester and Fennema 1986; Drake et al. 1987, 1991; Baldwin 1994; Wong et al. 1992; Baldwin et al. 1997). Lipids provide a good barrier to water vapour (Nisperos-Carriedo 1994; Baldwin et al. 1997), while plasticizers are necessary to enhance flexibility and improve film's mechanical properties.

Composition, microstructure and physical properties of biopolymeric films determine their possible applications. Controlling the film formulation allows tailoring of the mechanical and barrier properties of these materials, improving the efficiency of preservation for packaged foods. The study of film microstructure and interactions between film components provides insight into both fundamental aspects of material science and practical technologies for possible applications. Most methods used in characterization of films in the solid state are based on

M.A. García (✉), A. Pinotti, M.N. Martino, and N.E. Zaritzky
Centro de Investigación y Desarrollo en Criotecnología de Alimentos (CIDCA),
CONICET, Universidad Nacional de La Plata (UNLP), 47 y 116 La Plata (1900), Argentina
e-mail: magarcia@quimica.unlp.edu.ar

M.E. Embuscado and K.C. Huber (eds.), *Edible Films and Coatings for Food Applications*, 169
DOI 10.1007/978-0-387-92824-1_6, © Springer Science+Business Media, LLC 2009

detection of structural and thermodynamic properties involving the crystalline-amorphous structure. X-ray diffraction is probably the most important technique for observing structural properties of crystalline solid materials, polymers and undoubtedly food materials (Gontard et al. 1993; Krochta and De Mulder-Johnson 1997). Thermodynamic changes can be evaluated by several calorimetric techniques and by thermomechanical analysis (Cherian et al. 1995; Roos 1995; Galietta et al. 1998; García et al. 2000a; Sobral et al. 2001; Mali et al. 2002; Neto et al. 2005). Knowledge of thermodynamic state, molecular mobility and phase transitions are important, since they affect film barrier properties, which define performance of films under conditions of common use and abuse.

The objectives of this chapter are to describe: (a) properties of different starch and composite films and their relation to functionality, with special emphasis on film formulations based on starch, chitosan and methylcellulose; (b) common methods of characterizing composite films and coatings and (c) applications of these biomaterials.

6.2 Formulation of Films and Coatings

Hydrocolloids, proteins, and their derivatives represent the major components of film formulations, and form the primary supporting matrix of films. Functional, organoleptic, nutritional and mechanical properties of an edible film can be modified by addition of other ingredients in minor amounts. A main requirement for film formulation components is that they be "generally recognized as safe" (GRAS) substances, and be used in accordance with good manufacturing practices.

In general, protein films are poor moisture barriers due to the hydrophilic nature of most proteins, but are good barriers to gases such as oxygen and carbon dioxide (Cuq et al. 1998; McHugh and Krochta 1994a). Casein, collagen, zein, wheat gluten, whey proteins, gelatin and soy proteins have been extensively investigated due to their availability and mechanical resistance as components of edible films (McHugh and Krochta 1994a; Were et al. 1999; Oh et al. 2004; Sohail et al. 2006). Edible films and coatings discussed in this chapter are summarized in Table 6.1.

Table 6.1 Film formulations

Film matrix	Components
Starch	Gelatinized through thermal and alkaline treatment
	Starch source (% amylose content): corn starch
	native (25), amylomaize (65) and yam starch (30)
	Starch (2–4.5 g/100 g)
	Plasticizer: glycerol or sorbitol – 0–20 g/L)
Starch and lipid	Lipid: sunflower oil (0–8 g/L)
Methylcellulose and chitosan (MC–CH)	MC to CH ratio – 0:100, 25:75, 50:50, 75:25, 100:0
Corn starch (thermal gelatinized) and	CS (3.5–6 g/100 g)
chitosan (CS–CH)	Plasticizer glycerol, 0–50 g/100 g)
	CS to CH ratio – 0:100, 50:50 and 100:0

Most commonly used polysaccharides are cellulose derivatives, chitosan, starch, alginate, carrageenan and pectin due to their good film-forming properties. Starches are abundant in nature, commercially available, inexpensive and are totally biodegradable. Thus, they are attractive raw materials for the development of completely degradable products for specific market needs. Among the cellulose derivatives, methylcellulose (MC) has the best film-making properties, high solubility and efficient oxygen and lipid barrier properties (Park et al. 1993; Donhowe and Fennema 1993a, b; Nisperos-Carriedo 1994).

Chitin and its deacetylated product, chitosan, have received a lot of interest in agriculture, biomedicine, biotechnology and the food industry due to their unique properties, biodegradability and bioactivity (Muzzarelli et al. 1988; Kumar 2000; Tharanathan and Kittur 2003). Based on its antifungal, mechanical and oxygen barrier properties (Chen et al. 1996; Caner et al. 1998), chitosan film is a promising active packaging film material (Vermeiren et al. 1999). Production of chitosan from crustacean shells, a byproduct of the seafood industry, is economically feasible (Kumar 2000). However it lacks universal approval for food applications.

Composite biopolymer films can bring about improved mechanical and physical properties if components are structurally compatible. Thus, blending a starch or methylcellulose with chitosan can produce films with interesting properties and novel applications.

Plasticizers are the other components of edible films. These are low molecular weight compounds that are added to soften the rigid structure of films. Plasticizers improve mechanical, barrier and physical properties of biopolymer films (Banker 1966). They must be compatible with film-forming polymers, and reduce intermolecular forces and increase mobility of polymer chains (Donhowe and Fennema 1993c, 1994). Hydrophilic compounds such as polyols (glycerol, sorbitol) and polyethylene glycol are commonly used as plasticizers in hydrophilic film formulations (Gontard et al. 1993). Lipophilic compounds, such as vegetable oils, lecithin and, to a lesser extent, fatty acids, may also act as emulsifiers and plasticizers (Kester and Fennema 1986; Cuppet 1994; Donhowe and Fennema 1994; Hernandez 1994).

Food additives such as antioxidants, antimicrobial agents and nutrients, can be incorporated in film formulations to achieve specific functionalities. This concept of "active films" is a very promising application as it creates new avenues for designing packaging materials.

One of the most popular and oldest techniques to protect specific fruits and vegetables is application of natural wax and lipid coatings. Coatings are intended to protect product against dehydration, attack by fungi, and abrasion during processing and to improve product appearance (Hardenburg 1967; Paull and Chen 1989; Drake and Nelson 1990; Drake et al. 1991; Baldwin 1994; Hagenmaier and Baker 1994; Baldwin et al. 1997). In some products, the oily appearance limits their use (Baker et al. 1994; Baldwin et al. 1997).

Composite starch suspensions can be used as coatings or for preparation of edible films. The most common method of food coating involves spraying or immersion. Coating integrity is a critical factor that depends on, surface tension, adhesion to food substrate and flexibility of the coating. Unplasticized matrices are often brittle and rigid once dried, due to strong interactions between polymer chains that favor crystal development. This brings formation of cracks or and peeling off, especially for irregularly shaped fruits. Addition of a plasticizer can remedy this problem, since it can improve coating flexibility by reducing interactions between polymer chains. However the amount of plasticizer needs to be optimized since its amount can negatively alter the barrier and mechanical properties of a coating. García et al. (1998a, b, 1999, 2000b, 2001) described how they optimized plasticizer in composite coatings. For starch-based formulations, glycerol and sorbitol are common plasticizers used with typical concentrations in coating solutions ranging between 0 and 50 g/L. Maximum plasticizer concentration is limited by surface migration. For example, the maximum amount of glycerol suitable for these starch-based formulations was 20 g/L, while higher concentrations resulted in plasticizer migration to the surface, leading to a sticky coating (García et al. 1998a). For sorbitol, a concentration greater than 20 g/L can led to inconveniently long drying times.

Sunflower oil was included in coating formulations at concentrations between 0 and 8 g/L to enhance water barrier properties. Similar to plasticizers, lipid materials also may migrate to coating surface, depending on coating microstructure. Starch-based coatings containing less than 5 g/L sunflower oil did not exhibit oil migration (García et al. 2001). Wong et al. (1992) reported that similar lipid concentrations did not result in oil migration in chitosan-fatty acid films. However, a marked lipid migration was reported for methylcellulose films (Greener and Fennema 1989; Vojdani and Torres 1989a, b, 1990). Lipid concentration has been optimized for barrier properties by García et al. (2001).

Apart from sunflower oil, a composite coating may also include active ingredients such as antimicrobial agents and antioxidants. Active components remain at the surfaces of food products, where they are needed. The coating matrix limits diffusion of active components into the core of the food and thus, minimizes the amount of additive needed for the product (García et al. 2001).

Another important application of coatings is to reduce oil uptake of products during deep fat frying. Excessive fat in the diet has been linked to coronary heart disease. Thus, coatings applied to food before frying can help in reducing fat absorption. Hydrocolloids, especially cellulose derivatives such as methylcellulose, can be used to minimize oil content of fried foods.

The description of coating formulations used in published works (García et al. 1998a, b, 1999, 2000b, 2001, 2002, 2004a,b) is shown in Table 6.2.

Table 6.2 Coating formulations

Coating matrix	Components
Starch	Starch source (% amylose): corn starch native (25), amylomaize (65)
	Plasticizer – glycerol or sorbitol (0–20 g/L)
Starch and lipid	Lipid - sunflower oil (0–8 g/L)
Starch and antimicrobial agent	Active agents – potassium sorbate and citric acid
Methylcellulose (MC)	MC (0.5–1.5 g/100 g)
	Plasticizer – sorbitol (0–1 g/100 g)

6.3 Preparation and Characterization

Composite films are prepared by using 2 or more hydrocolloids. This requires dissolution of biopolymer molecules. Some polysaccharides, like cellulose derivatives, require solubilization at a higher temperature, while others require dissolution in a pH-regulated medium (e.g., chitosan).

Preparation of starch for the preparation of films requires gelatinization of native starch granules either by cold (e.g., alkaline treatment with NaOH) or thermal treatments. The most common technique to gelatinize starch is through thermal treatment (Mali et al. 2002). However, alkaline dispersion can also be used to alter the size distribution of starch polymers, that brings about a mild degree of starch hydrolysis or depolymerisation (García et al. 2000a; Romero-Bastida et al. 2005).

In general, preparation of protein-based films requires dissolution of protein in an appropriate solvent, e.g., zein, which is a prolamin, requires dissolution in ethanol (Ghanbarzadeh et al. 2007). Other examples of alcohol-soluble proteins are wheat gluten and fish myofibrillar proteins. Hydration of protein facilitates its solubilization (Sobral et al. 2001).

Cross-linking of proteins with lactic acid, tannic acid, ionized calcium or transglutaminase has been shown to increase resistance of films to water vapour and gas. In addition, preparation of films from aqueous emulsions of proteins and lipids or by lamination of a protein film with lipid can also decrease water vapour permeability (WVP). Lipids are frequently incorporated to reduce water vapour permeability, due to their hydrophobic character. Addition of lipids to a hydrocolloids matrix can be carried out by overcasting, leading to bilayer films, or by emulsion techniques (Baldwin et al. 1997; Shellhammer and Krochta 1997). In the latter case, the suspension can be used to obtain either coatings or films. With solid lipids, such as carnauba wax or saturated fats, the common procedure is to incorporate them via bilayer films, whereas the emulsion technique is preferred for liquid lipids, such as edible oils (Greener and Fennema 1989; Gontard et al. 1994; Hagenmaier and Baker 1994; Shellhammer and Krochta 1997). Emulsion preparation parameters are important factors for obtaining homogenous films and coatings, and will influence performance. For example, in the case of starch-based films, a high stirring velocity produced extensive foaming that could not be eliminated with a vacuum and thus produced low quality coating (García et al. 2001). Another important factor is the lipid to aqueous phase ratio. If the critical ratio is exceeded, migration of lipid may occur.

Plasticizers are often added once biopolymer solubilization has been achieved. Other functional additives (e.g., emulsifiers, antimicrobial or antioxidant agents) are incorporated into formulations after addition of the plasticizer. It must be stressed that each individual additive may require a particular condition for proper and effective incorporation.

As described previously, composite suspensions can be used to obtain either coatings or films. In the case of coatings, surface tension and rheological behavior of the suspension are important factors that will affect suspension spreadability and coating adhesion. Plasticizer and lipid addition, in the case of a starch coating suspension, decrease surface tension (Table 6.3), providing better adhesion to foodstuffs. The surface of vegetables already has low surface tension for natural protection. However, this natural advantage is a drawback for aqueous coating applications (Hershko and Nussinovitch 1998). Modifying the coating formulation to include plasticizers and/or lipids helps decrease surface tension to facilitate adhesion of coating to foodstuffs (Table 6.3). The similarity between liquid and solid film in values of dispersive and polar components influences spreadability of the liquid (Hershko and Nussinovitch 1998). Ghanbarzadeh et al. (2007) reported that sugar addition increases surface tension of zein films, which was attributed to an increase in polar forces and interactions within the film matrix.

Understanding the rheological behavior of film or coating suspensions is critical for process scale-up to ensure that processing requirements and machinability issues can be properly addressed. Rheological properties of composite starch suspensions exhibit pseudoplastic behavior ($n < 1$), while the Power law rheological model provided a satisfactory behavior fit enough to experimental data ($r^2 > 0.96$) (García et al. 2006). Plasticizer and lipid addition to corn and amylomaize starch suspensions decreased the flow behavior index and increased the consistency index. Apparent viscosity of corn starch suspensions decreased with both plasticizer and lipid addition (Table 6.3). Measurement of viscosity is important. Spraying requires a low viscosity, while immersion requires a higher viscosity coating solution.

In formulation of composite biopolymer films, it is important to characterize the miscibility of biopolymers and interactions that may occur between them, since these attributes ultimately influence film microstructure. For example, chitosan-starch suspensions show pseudoplastic flow behavior similar to those obtained with chitosan suspensions (Fig. 6.1). The presence of chitosan in the suspension decreased peak viscosity of the suspension. When chitosan was blended with methylcellulose, film solutions also showed pseudoplastic behavior ($n < 1$). Single component solutions of either MC or CH showed similar values for all rheological parameters; however, composite film solutions had higher apparent viscosities and

Table 6.3 Characterization of composite film suspensions

Formulation	Surface tension (dyne/cm)	Apparent viscosity η_{ap} at $\lambda = 692,48$ s^{-1} (mPa s)
Methylcellulose (MC)	87.00 ± 0.210	102.49 ± 0.83
Chitosan (CH)	72.17 ± 0.710	112.73 ± 0.89
CH:MC (75:25)	87.50 ± 0.690	140.68 ± 1.06
CH:MC (50:50)	81.07 ± 0.320	134.57 ± 1.19
CH:MC (25:75)	70.38 ± 0.470	127.11 ± 0.75
Starch-based (w/o additive)	60.68 ± 0.120	12.85 ± 0.063
Starch-based (w/ glycerol, G)	59.60 ± 0.764	11.95 ± 0.182
Starch-based (w/G + sunflower oil, SO)	59.08 ± 0.679	11.31 ± 0.092
Starch-based (w/sorbitol, S)	51.44 ± 0.082	12.59 ± 0.176
Starch-based (w/S + SO)	54.02 ± 0.266	9.64 ± 0.182

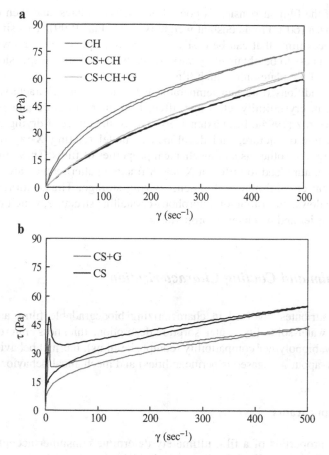

Fig. 6.1 Rheological behavior of film suspensions formulated with (**a**) chitosan (CH), corn starch–chitosan (CS + CH) and corn starch–chitosan–glycerol (CS + G + CH); (**b**) corn starch (CS) and corn starch with glycerol (CS + G) (cornstarch at 5%)

consistency indexes than those of single component solutions by themselves. These results point to a likely synergistic effect between constituents of the film solution, suggesting a high compatibility and high degree of interaction between the two polymer types. Besides, apparent viscosity of film solutions increased with CH concentration (Table 6.3).

6.4 Film Preparation

Films can be prepared by casting, extrusion or lamination, which are similar processes used in the synthetic polymer industry (Stepto 1987). Casting is a common, small-scale production method used to obtain biodegradable films. In this technique, a portion of the film suspension is poured onto acrylic plates, and then dried in a ventilated oven (60°C) to a constant weight (García et al. 1999). This simple technique produces films that can be easily removed from plates, and allows films of variable thickness to be obtained by varying the weight of film suspension applied and the area of the plate onto which films are cast.

Drying conditions (rate and temperature) determine film characteristics (e.g., water content, crystallinity, etc.) that affect film microstructure and properties. Bader and Göritz (1994a, b, c) extensively studied the effect of drying on amylomaize film microstructure, and developed a model relating X-ray diffraction patterns, sorption isotherms and mechanical properties of films. They stressed that drying temperature lead to different X-ray diffraction pattern which are associated with different crystalline developments during storage. Thus, tailoring drying conditions allows controlling of amorphous-crystalline structure, which is strongly related to barrier and mechanical properties.

6.4.1   Film and Coating Characterization

The main attributes involved in characterizing biodegradable films are: optical properties, water-solubility, water sorption/desorption, thickness, microstructure, crystallinity, biopolymer compatibility (composite films), thermal behavior, barrier properties (vapour and gaseous permeabilities) and mechanical behavior.

6.4.1.1   Appearance

The visual properties of a film ultimately determine consumer acceptability of the packaged products. Both subjective and objective techniques are used to characterize optical properties of films, with surface color commonly measured by a colorimeter and opacity determined by spectrophotometry. Polysaccharide films are typically colorless, though those of CH may exhibit a slightly yellow appearance. However, the intensity of the yellowness was negligible in comparison to that of whey protein-based films (Trezza and Krochta 2000a, b). Polysaccharide films are free of many of the problems associated with protein and lipid films, such as Maillard and oxidation reactions (Trezza and Krochta 2000a, b).

Film opacity is a critical property to consider if the film is to be used at a food surface. It can be measured using the method proposed by Gontard et al. (1992), in which the film spectrum is recorded over the visible range. Opacity is estimated as the area under the absorption curve (Au × nm). Transparent films are characterized by low values for area under the absorption curve measurements. Table 6.4 shows that CH films were the most transparent, and CS films the most opaque, increasing opacity with increasing starch concentration. Nevertheless, opacity values for starch-based films were still lower than those reported for wheat gluten films obtained under different solubilization conditions (Gontard et al. 1992). Addition

Table 6.4 Optical properties of corn starch, chitosan and composite films

Film composition	Film opacity (Au × nm)	Film color	
		Color differences (ΔE)	Chromaticity b*
Corn starch[a] (CS)	138.0 ± 0.8	0.83 ± 0.16	2.20 ± 0.25
CS with glycerol	109.6 ± 0.9	0.60 ± 0.10	2.23 ± 0.08
CS with chitosan	95.9 ± 1.05	1.74 ± 0.15	2.94 ± 0.15
CS with glycerol and chitosan	88.7 ± 0.7	2.16 ± 0.19	3.68 ± 0.21
Chitosan (CH)	18.9 ± 1.1	3.79 ± 0.50	5.04 ± 0.32

[a] *CS concentration*: 5 g/100 g film suspension

Table 6.5 Film solubility in water at 100°C of CH–MC films

Film composition	Film solubility in water (%)
MC	98.4 ± 2.2
CH:MC (25:75)	40.7 ± 6.3
CH:MC (50:50)	27.7 ± 1.6
CH:MC (75:25)	14.0 ± 2.1
CH	9.3 ± 0.9

of glycerol reduced starch film opacity, while blending corn starch with CH generated films with reduced opacity, but increased yellowness.

6.4.1.2 Water Solubility

Water solubility is an important factor in determining possible applications for composite biopolymer films. MC films are completely soluble in water, while CH films have lower water-solubility values. Composite MC–CH films had intermediate water-solubility, which decreased with increasing chitosan concentration (Table 6.5). Thus, by adjusting CH concentration in the film formulation, water-solubility of the composite film can be changed to meet requirements of a specific application. A similar trend in water solubility was observed for CH-starch composite films.

6.4.1.3 Water Sorption

Water sorption isotherms are useful for determination of film stability under various conditions, as many film constituents, especially hydrocolloids, are sensitive to relative humidity and temperature. Though sorption isotherms can be measured by several techniques, a simple procedure is described by Spiess and Wolf (1993).

Sorption isotherms for yam starch films (with and without glycerol) and native yam starch were compared at 25°C (Fig. 6.2). All isotherms showed a similar

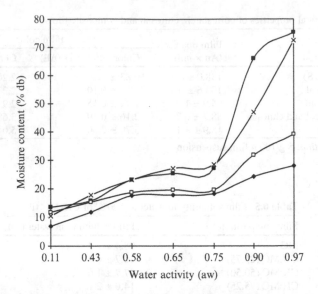

Fig. 6.2 Water sorption isotherms of yam starch films and native yam starch at 25°C. (*filled diamond*) yam starch (Film formulation: 4.0% yam starch (*open square*) without glycerol, (*times symbol*) with 1.3% glycerol, and (*filled square*) with 2.0% glycerol)

sigmoidal shape (type II isotherm). When samples were conditioned at water activity (a_w) values greater than 0.43, the plasticized starch film showed higher equilibrium moisture contents than both the native yam starch powder and the non-plasticized starch film. The hygroscopic characteristics of glycerol-plasticized starch films contributed to an increase in hydrophilicity of the film.

Controlled films showed higher equilibrium moisture content than native yam starch at $a_w > 0.75$. This result could be attributed to the fact that during film production, gelatinization led to a starch molecular reorganization that increased water absorption capacity of the unplasticized film compared to native starch. Unplasticized films showed similar behavior to those reported by other researchers working with high-amylose corn starch (Bader and Göritz 1994b) and tapioca starch films (Chang et al. 2000).

6.4.1.4 Thickness of Films

Film thickness can be measured with a digital film thickness gauge or by scanning electron microscopy (SEM), the first method being simpler and faster. Correlation between film thickness measured by SEM and a digital film thickness gauge was established for CH-starch films (García et al. 2006). These results are presented in Table 6.6 with a correlation coefficient of 0.9953.

Table 6.6 Film thickness of corn starch, chitosan and composite films

Film composition	Film thickness (μm)	
	Digital thickness gauge	SEM
Corn starch[a] (CS)	63.1 ± 1.7	69.2 ± 1.9
CS with glycerol (G)	44.4 ± 0.9	47.3 ± 1.9
CS with chitosan (CH)	51.5 ± 7.8	56.1 ± 11.3
CS with G and CH	52.2 ± 1.4	57.0 ± 2.1
CH	15.2 ± 1.8	13.9 ± 2.9

[a]*CS concentration*: 5 g/100 g film suspension

Barrier and mechanical properties depend on film thickness. A wide range of film thickness values has been reported for composite biopolymer films in the literature, indicating its dependence on both film composition and processing parameters.

6.4.1.5 Microstructure

Properties of composite films and coatings depend on several factors such as: the ratio of crystalline to amorphous zones, polymeric chain mobility, and specific interactions between functional groups of polymers and the permeant substance within amorphous zones. Common techniques used to elucidate film structure include SEM, X-ray diffraction, differential scanning calorimetry (DSC), thermo-mechanical analysis (TMA), dynamic mechanical analysis (DMA) and Fourier transform infrared spectroscopy (FTIR).

6.4.1.6 Scanning Electron Microscopy

SEM may be used to evaluate film homogeneity, layer structure, pores and cracks, surface smoothness and thickness. For example, cross sections of unplasticized corn starch film showed a multi-laminar structure, while those for films containing a plasticizer showed a more compact structure, regardless of the plasticizer used (Fig. 6.3, Table 6.6). SEM images of plasticized films containing lipid exhibited smooth surfaces and a compact structure, indicative of a homogeneous dispersion of lipids within the film matrix.

When corn starch was blended with chitosan, film surfaces appeared smooth (without pores or cracks) and homogeneous (no obvious phase separation), even for formulations that did not include a plasticizer (Fig. 6.4c, Table 6.6). The compact, homogeneous matrix of CS–CH films is an indicator of structural integrity and, consequently, good mechanical properties such as high resistance and elongation at break are expected. Similarly, in the case of MC–CH composite films, a compact structure with no pores or cracks was observed by SEM, even though film formulation did not include a plasticizer (Fig. 6.5). The composite

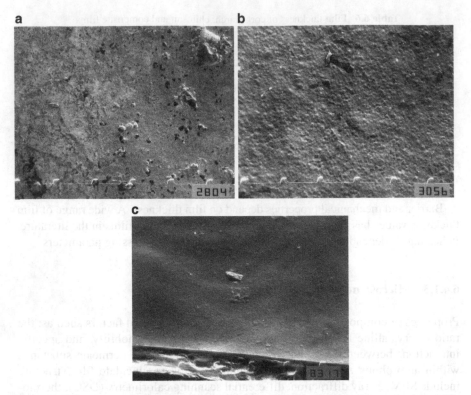

Fig. 6.3 SEM micrographs of the surface of corn starch-based films (**a**) without plasticizer, (**b**) with 20 g/L glycerol and (**c**) with 20 g/L sorbitol and 2 g/L sunflower oil (Magnification: 100 μm between marks)

film showed multilayered structure (Fig. 6.5c), though both CH and MC are compatible polymers that could limit chain mobility.

When dealing with coatings, characterization of microstructure provides important insight into performance of the different formulations. Light microscopy is used for evaluation of coating adhesion, thickness and uniformity, while estimation of coating integrity is conducted by SEM. Improved visualization of starch-based coatings using light microscropy may be achieved via staining with iodine solution. Coating thicknesses varied between 40 and 50 μm for all tested formulations (García et al. 1998a, b).

6.4.1.7 Infrared Spectroscopy

Fourier transform infrared spectroscopy (FTIR) is a useful technique to supplement microstructural characterization of composite films, since it may be used to evaluate interactions between film components. Figure 6.6a shows the IR spectra for films of MC, CH and a 50:50 CH–MC blend. The positions of peaks of the single component spectrum is similar to those described by different authors

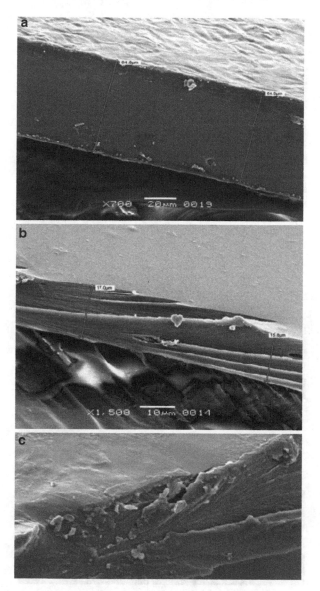

Fig. 6.4 SEM micrographs of the cross-sections of (**a**) corn starch-based films (**b**) chitosan films and (**c**) a film obtained by blending corn starch and chitosan

(Chen et al. 2003; Sionkowska et al. 2004; Wu et al. 2005; Zaccaron et al. 2005). Spectra of compo-site films show characteristic peaks of both polymers, and also small shifts in the positions of the $1{,}500{-}1{,}700$ cm^{-1} bands, which are related to amino and carbonyl groups. Figure 6.6b shows the effect of chitosan concentration in CH–MC blends with an increasing CH concentration, while spectra of

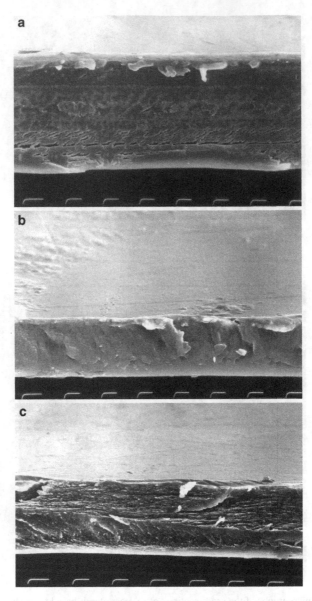

Fig. 6.5 SEM micrographs of the cross-section of (**a**) methylcellulose films, (**b**) chitosan films and (**c**) a film obtained by blending methylcellulose and chitosan (Magnification: 10 μm between marks)

composite films approach that of the CH film. Further, composite films showed a broad band between 2,985 and 3,600 cm^{-1} corresponding to the N–H and –OH superposition of chitosan (Sionkowska et al. 2004), while MC film showed only the characteristic –OH stretching at 3,457 cm^{-1} (Zaccaron et al. 2005).

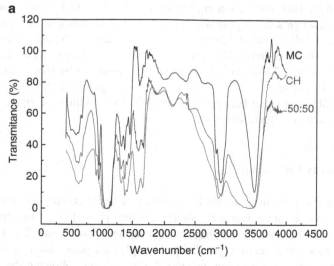

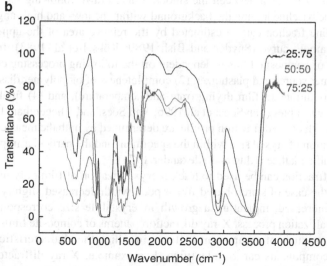

Fig. 6.6 FTIR characterization of: (**a**) MC, CH and 50:50 films and (**b**) composite films with different CH:MC ratios

According to Wanchoo and Sharma (2003), hydrogen bonding or other interactions between chemical groups of dissimilar polymers should theoretically cause a shift in the peak position of participating groups. In composite MC–CH films, this behavior is observed for –OH stretching, since this peak shifted from 3,463 cm⁻¹ in MC films toward 3,450 cm⁻¹ in CH-MC composite mixtures with increasing

CH proportion. This observation could indicate that the –OH group movement is compromised when another polymer is present. A shift was also observed in CH amino group ($-NH_2$) band at 1,560 cm^{-1} as the amount of MC decreased in composite films. However, other characteristic peaks, including those at 1,411 cm^{-1} (OH vibrations) and 2,904 cm^{-1} (C–H stretching) assigned by Pawlak and Mucha (2003) and Zaccaron et al. (2005) did not show a significant shift. These results suggest a mild interaction between CH and MC that might be attributed to their similar chemical and linear structures.

6.4.1.8 X-ray Diffraction

X-ray diffraction patterns of edible films, either of single component or composite materials, show an amorphous-crystalline structure characterized by sharp peaks associated with the crystalline diffraction and an amorphous zone. The higher the amorphous zone, the lower the crystallinity. Amorphous regions within the sample can be estimated by the area between the smooth curve drawn following the scattering hump and the baseline joining the background within the low- and high-angle points. The crystalline fraction can be estimated by the relative area of the upper regions above the smooth curve (Snyder and Bish 1989; Köksel et al. 1983). In general, crystallinity of composite films is dependent on the following processing conditions: (1) biopolymer source and plasticizer, (2) completeness of biopolymer dissolution in water, (3) conditions of film drying (rate and temperature), and (4) final moisture content of the samples (Van Soest et al. 1996; Van Soest and Vliegenthart 1997). The shape and width of the diffraction profile are determined by both the mean crystal size (and distribution of crystal size within the specimen) and the particular imperfections of the crystalline lattice (Klug and Alexander 1974).

X-ray diffraction can be used to track recrystallization of film polymers during storage. In the case of starch-based films, peak width decreased slightly and peak intensities increased, indicating a growth in crystallite size corresponding to a slow recrystallization process. X-ray diffraction patterns of composite films generally represent a mixture of component features in which the characteristic peaks of individual components can be identified. For example, X-ray diffractograms of composite MC–CH films are similar to those obtained for MC alone, although peak intensity at $2\theta = 8°$ (which is distinctive of MC films) decreased with an increasing CH concentration, suggesting that CH interferes with MC ordering (Fig. 6.7). Similar results were reported for CH-gelatin films by Chen et al. (2003), who attributed the decreased crystallinity of the composites to reduced hydrogen bonding in CH molecules, leading to an amorphous structure for the polyelectrolyte complex.

Film preparation techniques also may lead to development of different matrix structures. For films based on thermal gelatinization of starch, X-ray diffraction patterns of yam starch films exhibited a B-type crystalline packing arrangement, characteristic of starch from tubers (Roos 1995). This pattern remained virtually unchanged over the course of storage. Starch and glycerol contents did not markedly

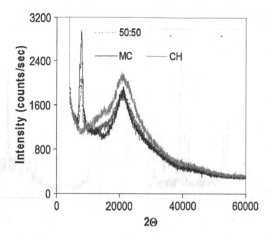

Fig. 6.7 X-ray diffractograms of methylcellulose (MC), chitosan (CH) and composite films

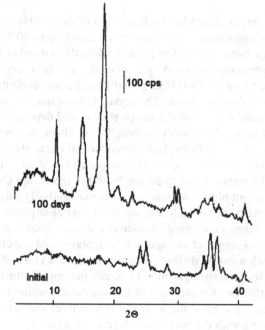

Fig. 6.8 X-ray diffractograms of amylomaize films stored at 20°C and 63.8%relative humidity for 100 days

influence the X-ray pattern of these films (Mali et al. 2002). In contrast, when films were obtained from starch solubilized by alkaline treatment, X-ray diffraction patterns showed that starch films exhibited a crystalline structure of higher stability, namely an A-type starch pattern (Fig. 6.8). This shift in the crystalline pattern was

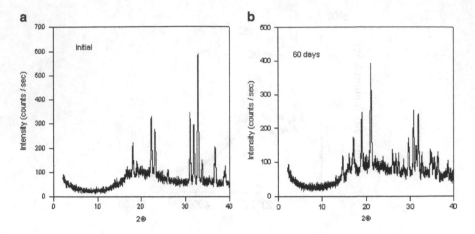

Fig. 6.9 X-ray diffractograms of amylomaize films plasticized with glycerol and sunflower oil, (**a**) at initial time and (**b**) stored at 20°C and 64% relative humidity for 60 days

also observed for amylomaize films by Bader and Göritz (1994a, c) and was attributed to film drying conditions. For drying temperatures up to 60°C, diffractograms of films exhibited a B-type crystalline pattern, while those dried at 100°C exhibited diffractograms corresponding to A-type crystals. As film drying temperature increased over the range of 60–100°C, film diffractograms exhibited a continuous transition from B- to A-type crystals. The crystallinity of starch films was primarily related to the amylose content of the starch source, and depended on both drying and storage conditions. High water content favors chain interactions and, thus, crystallization within starch films. In hydrocolloid films, as described for starch, water has an additional role, since it is incorporated in a structural active state through inter- and intramolecular hydrogen bondings (Even and Carr 1978).

Starch films without plasticizer showed higher crystallinity (higher peaks) than those containing plasticizer, which resulted in a larger amorphous zone and lower peaks (Fig. 6.9). Films containing plasticizers also retained a stable diffraction pattern during initial stages of storage, while unplasticized starch films required more time to reach a stable diffraction pattern (García et al. 2000a). Addition of sunflower oil did not strongly modify X-ray pattern of films during storage (Fig. 6.9b). Nevertheless, the presence of oil facilitated polymeric chain mobility, allowing rapid development of the most stable structure, and also reduced crystal growth by interfering with the polymeric chain arrangement.

6.4.1.9 Differential Scanning Calorimetry of Films

Evolution of the film matrix crystalline structure during storage can also be evaluated by DSC. As described in the results of X-ray diffraction, film preparation and storage conditions impact the thermal properties of composite films. Films obtained by starch thermal treatment, such as those formulated with yam starch, did not show any peaks in DSC thermograms at the beginning of the storage time, indicating that starch gelatinization during film production was complete. With storage (90 days at 20°C and 64% relative humidity), these films did not show any peaks in DSC thermograms (between 40° and 120°C), indicating stability of the matrix (Mali et al. 2002).

However, starch films obtained by alkaline treatment showed an endothermic transition with a peak temperature around 50°C during storage (90 days). This peak became narrower and its melting temperature and the corresponding enthalpy (ΔH) increased with increasing storage time (Fig. 6.10). This, DSC transition could be associated with several processes, such as the crystal growth of short chains (products of hydrolysis) and recrystallization of amylose or other long lateral amylopectin chains.

The presence of plasticizers in films limited crystal growth and recrystallization similar to that previously described for X-ray analysis, leading to lower enthalpy values (ΔH) and peak melting temperatures (Fig. 6.10). Sorbitol is a good plasticizer for starch, since its molecular structure is similar to glucose units of starch chains, increasing the chance to interact with polymeric chains (García et al. 2000a). Additives may interfere with polymeric chain association due to steric hindrances, and in these cases film crystallinity will likely decrease.

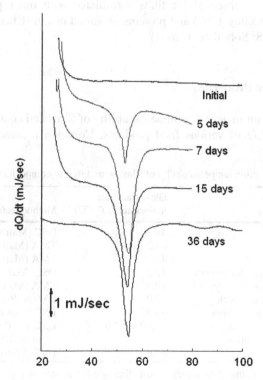

Fig. 6.10 DSC thermograms of amylomaize films stored at 20°C and 63.8% relative humidity

6.4.1.10   Glass Transition Temperature

The glass transition temperature (T_g) is strongly dependent on both the film composition and moisture content, and can denote stability of a film. As water content of an amorphous material increases, T_g decreases. Thus, for a given storage temperature, an amorphous matrix at low water content (in the glassy state) remains stable to molecular change, while the same matrix at higher moisture content, corresponding to the rubbery state, can undergo molecular reorganization. In the rubbery state, polymer chains possess sufficient mobility for crystal growth and perfection. Thus, knowledge of T_g is important, since it impacts mechanical and barrier properties of a film under specific conditions of application and storage.

Both DSC and TMA techniques are commonly used to estimate T_g, though T_g values obtained by DSC are slightly higher than those determined by TMA. Chang and Randall (1992) stressed that the sensitivity of TMA is superior to that of DSC for defining glass transition temperatures. Dynamic mechanical thermal analysis (DMTA) may also be used for determination of T_g. Table 6.7 provides some literature-reported T_g values for various film formulations. Plasticization decreases intermolecular forces between polymer chains and, consequently, reduces T_g (Guilbert and Gontard 1995). For composite starch matrices, T_g values for films without plasticizer were higher than those of films containing glycerol (Table 6.7). In this regard, a similar trend was observed for films formulated with other polysaccharides (Debeaufort and Voilley 1997) and proteins of varied origin (Cherian et al. 1995; Galietta et al. 1998; Sobral et al. 2001).

6.4.1.11   Barrier Properties

Barrier properties ultimately determine the ability of films and coatings to improve storage or shelf life of various food products. Common measurements include

Table 6.7 Glass transition temperature (T_g) of films with different compositions

Film formulation	Glass transition temperature, T_g (°C)	Method (references)
Yam starch	48.0	DSC (Mali et al. 2002)
Yam starch	44.2	TMA (Mali et al. 2002)
Yam starch	44.2	TMA (Mali et al. 2002)
Yam starch plasticized with glycerol	12.2	DSC (Mali et al. 2002)
Yam starch plasticized with glycerol	10.9	TMA (Mali et al. 2002)
Gluten protein plasticized with glycerol	68.0	DMTA (Pommet et al. 2005)
Gelatin plasticized with sorbitol	56.5	DSC (Sobral et al. 2001)
Chitosan (depending on water content and degree of acetylation)	−23.0 to 67.0	DSC and DMTA (Neto et al. 2005)
Pectin	35.0	DSC (Iijimaa et al. 2000)

DSC differential scanning calorimetry; *TMA* thermomechanical analysis; *DMTA* dynamic mechanical thermal analysis

water vapour and gas permeabilities, which are strongly related to film structure, since permeants generally, move through a film or coating via its amorphous zones. Film components, especially plasticizers, affect both barrier and mechanical properties, because they modify film structure, chain mobility and diffusion coefficients of permeants.

Water Vapour Permeability

Determination of water vapour permeability (WVP) is strongly dependent on measurement conditions, such as temperature and the gradient of water vapour pressure. Coated biomaterials, such as fruits and vegetables, are characterized by irregular shapes and high water contents that make WVP measurements difficult. To overcome this challenge, García et al. (1998a, b) used a biological model (coated sliced carrots) that allowed surface area calculations to be accounted for WVP determinations of starch-based coatings (García et al. 1998a, b). WVP testing of films may be conducted according to method E96 (ASTM 1996a) that includes some modifications introduced by Gennadios et al. (1994). Since barrier thickness strongly affects water vapour transport through the film matrix, it is necessary to consider film thickness in WVP calculations. In addition, WVP determinations can be performed using the PERMATRAN, a specific instrument developed by MOCON for analysis of synthetic film materials. It is interesting to note that WVP results for coatings (applied on fruits) and films have been shown to be similar, even though the measuring techniques and driving forces for both systems are different. These findings validate the use of films to characterize barrier properties of coatings (García et al. 1998b, 2000b, 2004c).

Coatings without plasticizer often yield significantly higher WVP values than those with plasticizer (Table 6.8). This difference is attributed to the presence of pores and cracks in unplasticized coatings, as observed by light microscopy and SEM (García et al. 1999). In the absence of cracks or pores, a plasticized film generally would be expected to exhibit higher WVP than an unplasticized film (Banker 1966). These results could be related to structural modifications to the starch network produced by the plasticizer and to the hydrophilic character of glycerol, which favors the absorption and desorption of water molecules to promote permeability. Gontard and coworkers (1993) working on wheat gluten films found similar results on the effect of glycerol. Sorbitol generally resulted in lower WVP values compared to glycerol in plasticized starch-based coatings (Table 6.8e). McHugh and Krochta (1994a, b, c) reported similar results for alginate and pectin films using the same plasticizers at the same concentrations.

Amylomaize films and coatings, which possessed higher amylose content, denser matrix structure and higher degree of crystallinity than those of standard corn starch, also exhibited relatively lower WVP values (Figs. 6.8, 6.10). A similar trend has been observed by others (Noel et al. 1992; Miles et al. 1985a, b). However, the effect of amylose content on the WVP of the starch film matrix was minimized by incorporation of sunflower oil to the coating formulation (Table 6.8).

Table 6.8 Effect of formulation on water vapour permeability (WVP) of composite polysaccharide films

Formulation	Water vapour permeability $(g\ m^{-1}\ s^{-1}\ Pa^{-1}) \times 10^{11}$
Methylcellulose (MC)[a]	[i]7.55 ± 0.60
Composite MC:CH (75:25)[c]	7.24 ± 0.35
Composite MC:CH (50:50)[c]	6.67 ± 0.74
Composite MC:CH (25:75)[c]	6.77 ± 0.92
Chitosan (CH)[b]	5.04 ± 0.81
Corn starch thermally gelatinized (CS)[d]	17.66 ± 2.97
CS + glycerol (G)	8.68 ± 0.20
Composite CH–CS (CH + CS)[e]	8.76 ± 0.87
Composite CH–CS (CH + CS + G)[e]	4.46 ± 0.37
Corn starch[f] (Without additives)	36.80 ± 22.40
Corn starch[f] (Glycerol[g], G)	25.70 ± 10.40
Corn starch[f] (G + Sunflower oil[h], SO)	19.20 ± 4.70
Corn starch[f] (Sorbitol[g], S)	17.50 ± 1.40
Corn starch[f] (S + SO)	12.20 ± 1.10
Amylomaize[f] (Without additives)	26.20 ± 13.90
Amylomaize[f] (G)	21.40 ± 7.50
Amylomaize[f] (G + SO)	17.60 ± 3.70
Amylomaize[f] (S)	12.10 ± 1.50
Amylomaize[f] (S + SO)	9.70 ± 0.80

[a]C concentration = 1%
[b]CH concentration = 1%
[c]CH:MC (w/w) ratio
[d]CS (corn starch) = 5% corn starch
[e]CH–CS (w/w)
[f]Corn starch or amylomaize concentration = 20 g/L
[g]Glycerol or sorbitol concentration = 20 g/L
[h]SO = sunflower oil (2 g/L)
[i]value ± standard deviation

Addition of sunflower oil significantly decreased WVP of starch-based films and coatings. Baldwin and coworkers (1997) reviewed lipid effects on other non-starch composite films. As water vapour transfer generally occurs through the hydrophilic portion of a film, WVP is dependent on the hydrophilic-hydrophobic ratio of film components (Hernandez 1994). In general, WVP increases as polarity and degree of unsaturation/branching of the incorporated lipid increases, though water absorption properties of polar components of the film must also be considered (Gontard et al. 1994).

Yam starch film WVP is lower than those of many edible, biodegradable films such as wheat gluten plasticized with glycerol and amylose and hydroxypropyl methylcellulose containing plasticizer and oil, (Gennadios et al. 1994). In general, starch films exhibit a lower order of magnitude of WVP compared both to protein films and other polysaccharide-based films reported in literature (García et al. 1999, 2001, 2004c; McHugh and Krochta 1994a, b, c; McHugh et al. 1994; Parra et al. 2004; Parris et al. 1995). In comparison to synthetic polymers, starch films have WVP values similar to those of cellophane, but higher than low density polyethylene, the most common synthetic film.

García et al. (2004b) studied the effect of composition on CH–MC composite biopolymer films. Permeabilities of these composite biopolymer films did not differ significantly from those of single component films (Table 6.8). This result, together with X-ray analysis, provided evidence that matrices of composite films were very similar to those based on the single polymers. Table 6.8 also shows WVP values of composite films formulated with chitosan and corn starch. Addition of glycerol decreased WVP of CS films as described previously. McHugh and Krochta (1994c) found similar results for alginate and pectin films. Blending CS with CH decreased WVP of composite films. Plasticized composite CS–CH films have WVP values that were similar to those of CH films, thus blending CH with CS is a good alternative to maintain barrier properties and reduce costs (García et al. 2006).

A summary of WVP values for several hydrophilic edible films and synthetic polymer films is presented in Table 6.9. In comparison to other commercial biopolymers, chitosan films exhibit relatively low WVP. Composite films of CH, MC and

Table 6.9 Comparison of WVP values of biodegradable and synthetic films

Film formulation	WVP (g m^{-1} s^{-1} Pa^{-1})	References
Corn zein	5.35×10^{-10}	Ghanbarzadeh et al. (2007)
Corn zein plasticized with glycerol	8.90×10^{-10}	Gennadios et al. (1994)
Fish skin gelatin	2.59×10^{-10}	Avena-Bustillos et al. (2006)
Whey protein plasticized With sorbitol	7.17×10^{-10}	McHugh et al. (1994)
Wheat gluten plasticized With glycerol	7.00×10^{-10}	Gennadios et al. (1994)
Gelatin (obtained from pigskin) plasticized with sorbitol	1.6×10^{-10}	Sobral et al. (2001)
Amylose	3.8×10^{-10}	Gennadios et al. (1994)
Corn starch plasticized with glycerol	2.57×10^{-10}	García et al. (2001)
Corn starch plasticized with sorbitol	1.75×10^{-10}	García et al. (2001)
Amylomaize starch plasticized with sorbitol	1.21×10^{-10}	García et al. (2001)
Hydroxypropylmethylcellulose with plasticizer and oil	1.90×10^{-10}	Gennadios et al. (1994)
Amylomaize starch with sorbitol and sunflower oil	9.7×10^{-11}	García et al. (2001)
Methylcellulose	8.70×10^{-11}	Donhowe and Fennema (1993b)
Methylcellulose 3%	$8.4–12.1 \times 10^{-11}$	Park et al. (1993)
Chitosan 2% (unknown source)	$3.66–4.80 \times 10^{-11}$	Wong et al. (1992)
Chitosan 3%	$6.19–15.27 \times 10^{-11}$	Caner et al. (1998)
Cellophane	8.4×10^{-11}	Shellhammer and Krochta (1997)
PVDC	2.22×10^{-13}	Shellhammer and Krochta (1997)
LPDE-low density polyethylene	9.14×10^{-13}	Smith (1986)
HDPE-high density polyethylene	2.31×10^{-13}	Smith (1986)

CS attain WVP values close to those of cellophane (Table 6.8), as might be expected, due to similar chemical structures of these polymers (Shellhammer and Krochta 1997). However, WVP values of composite films are not comparable with those of synthetic films, such as low density polyethylene (LDPE), (Smith 1986), the primary polymer used in the food packaging industry (Table 6.9).

Gas Permeability

Measurement of film or coating permeability to CO_2 and O_2 is essential to understanding quality and physiological aspects of coated fruit products during storage. In the case of coatings, gas permeability is commonly measured using isolated films rather than coated products. This simplification is based on the finding that WVP results for coatings and films are similar (García et al. 2000b).

Different methods are used to determine gas permeability of films. Oxygen permeability can be measured using commercial instruments, such as the OXTRAN developed by Mocon Company. In addition, CO_2 and O_2 permeabilities of films can be assessed by using a specially designed cell (García et al. 2000a, b, 2001). This quasi-static method is based on measurement of gas diffusing through a film, which is quantified by a gas chromatograph. As was previously described for WVP determinations, calculated values for gaseous permeability must account for film thickness. Film and coating integrity is a must for attainment of good gas barrier properties. Starch films without plasticizers exhibit much higher CO_2 and O_2 permeabilities and also higher standard deviation coefficients than the same formulations containing plasticizers (Table 6.10). This result was attributed to the presence of pores within films without plasticizers, resulting in a lack of film integrity, as determined by SEM (García et al. 1999, 2000a).

Starch films with sorbitol exhibit lower oxygen permeability values than those formulated with glycerol (Table 6.10). Sorbitol combined with an amylomaize film matrix provided the lowest gas permeability values for all materials tested

Table 6.10 Effect of film formulation on gas permeability of starch-based films

Starch base	Additives	CO_2 permeability (cm^3 m^{-1} s^{-1}) $\times 10^{10}$ LSD$_{0.05}$ = 1.63	O_2 permeability (cm^3 m^{-1} s^{-1}) $\times 10^{10}$ LSD$_{0.05}$ = 1.03
Corn	–	292.1 ± 138.9	15.92 ± 2.99
Corn	Glycerol	56.9 ± 9.7	4.61 ± 0.51
Corn	Glycerol + SO	58.7 ± 5.8	3.83 ± 0.76
Corn	Sorbitol	41.9 ± 8.1	2.48 ± 0.32
Corn	Sorbitol + SO	47.2 ± 6.5	3.77 ± 0.03
Amylomaize	–	280.5 ± 73.7	26.45 ± 2.48
Amylomaize	Glycerol	38.5 ± 12.8	3.21 ± 0.19
Amylomaize	Glycerol + SO	43.9 ± 9.0	2.36 ± 0.04
Amylomaize	Sorbitol	29.6 ± 4.6	2.28 ± 0.26
Amylomaize	Sorbitol + SO	34.3 ± 2.1	2.18 ± 0.04

Glycerol and sorbitol concentration = 20 g/L; SO = sunflower oil (2 g/L)

(Table 6.10). McHugh and Krochta (1994a, b) found similar results working on milk and whey protein films.

Table 6.10 shows that O_2 permeabilities for starch-based films were much lower than those of CO_2, indicating variable permeability of these films to select gases. This effect can be attributed to a higher solubility of CO_2 than O_2 in the film matrix (Young 1984; McHugh and Krochta 1994a). Similar results were obtained by Arvanitoyannis et al. (1994) while working with potato and rice starch films. Further, the addition of lipid, which is necessary to reduce water vapour permeability, maintained the selective gas permeability properties of the starch-based films (Table 6.10).

Synthetic film materials, such as LDPE, show lower gas permeabilities (2.16×10^{-11} and 9.45×10^{-11} cm³ m⁻¹ s⁻¹ Pa⁻¹ for O_2 and CO_2, respectively) than starch-based films. However, LDPE has a relatively low CO_2 to O_2 permeability ratio (≈ 4), compared to an average ratio of 13 for starch-based films (Cuq et al. 1998; García et al. 1998a, b). Cuq et al. (1995) compared CO_2 to O_2 permeability ratios of several synthetic and edible films, and reported that edible films show higher selectivities than synthetic films, with ratios for edible films ranging from 8 to 30. Development of composite edible films and coatings with selective gas permeabilities could be very promising for controlling respiratory exchange and improving conservation of fresh or minimally-processed vegetables (Cuq et al. 1998; García et al. 1998a, b).

Table 6.11 shows that the CO_2 permeability of starch-based films decreased with increasing storage time (García et al. 2000a). Gas and vapour permeabilities of films and coatings depend on several factors such as: (1) ratio of crystalline to amorphous zones, (2) polymeric chain mobility, and (3) specific interactions between the functional groups of the polymers and gases within the amorphous zones. Gaseous permeability measurements of starch stored films correlate with the increase in the crystallinity reported by DSC and X-ray diffraction (Fig. 6.8–6.10). According to Donhowe and Fennema (1993b), film permeability increases with a decreasing crystalline to amorphous zone ratio, since permeation occurs through amorphous zones of the film.

Table 6.11 Effect of plasticizer and starch type on CO_2 permeabilities of starch-based films

Starch base	Plasticizer (20 g/L)	CO_2 permeability [cm³/m sec Pa][a] × 10⁹	
		Initial time	Stored samples[b]
Corn	–	29.21 ± 13.89	8.90 ± 1.28
	Glycerol	5.69 ± 0.97	2.76 ± 0.55
	Sorbitol	4.19 ± 0.81	2.42 ± 0.46
Amylomaize	–	28.05 ± 7.37	7.71 ± 1.40
	Glycerol	3.85 ± 1.28	2.28 ± 0.33
	Sorbitol	2.96 ± 0.46	1.82 ± 0.15

[a]At stadard pressure and temperature conditions
[b]Stored 20 days at 20°C and 63.8% relative humidity

6.4.1.12   Mechanical Properties

Mechanical properties of films and coatings are dependent on additive-matrix interactions, and are also strongly affected by physical, chemical and temperature conditions, which influence film stability and flexibility. Mechanical performance of a film is usually characterized by deformation at break (extension at the moment of rupture, mm), percent elongation at break (deformation divided by initial probe length and multiplying by 100%), tensile strength (force at rupture divided by film cross section, MPa) and elastic modulus (slope of force-deformation curve, N/mm) according to the ASTM D882-91 method (1996b). These can be determined using a texturometer or an Instron machine. These parameters are related to the stretching ability of the film. It is well known that the environmental conditions during production, storage and usage of these materials affect their mechanical properties. Ageing phenomena occurring within the useful lifetime of films also cause great losses to mechanical properties, particularly with regards to film elongation.

Plasticizers interfere with polymeric chain association facilitating their slipping and, thus, decreasing rigidity of the network, producing a less ordered film structure. In the case of amylomaize films, high values of elastic modulus were obtained (Fig. 6.11). These results were attributed to their high amylose content. According to García and coworkers (2001), the plasticizer-matrix interaction showed similar trends for both corn starch and amylomaize films, though plasticized corn starch films were more flexible than those of amylomaize. The combination of lipid and plasticizer addition increased film flexibility, and reduced the effect of starch type on film mechanical properties.

The stress–strain curves for composite films formulated with chitosan (CH) and corn starch (CS) without plasticizer showed a pattern typical of brittle materials (Sarantopoulos et al. 2002; Mali et al. 2005a, b). This characteristic is more evident

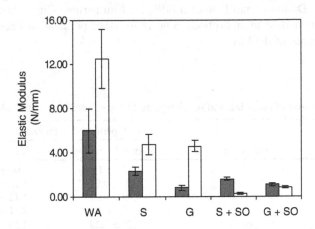

Fig. 6.11 Tensile properties: elastic modulus of corn (*open rectangle*) and amylomaize (*filled dot*) films (WA = without additives; S = with sorbitol; G = with glycerol; S + SO = with sorbitol and sunflower oil and G + SO = with glycerol and sunflower oil)

in CH films than in CS films (Fig. 6.12). Blends of chitosan and corn starch did not show significant differences in film deformation from single component films, although film stress of composite films decreased. These results would indicate that interactions between polymeric chains are weaker, reducing rigidity of the matrix, although both hydrocolloids are compatible (García et al. 2006).

Plasticized films exhibit the stress-strain behavior of ductile polymers compared with unplasticized films (Fig. 6.12). Corn starch films plasticized with glycerol (CS + G) exhibited the most flexible behavior (García et al. 2006). Similar results were obtained by Laohakunjit and Noomhorm (2004) and Mali and coworkers (2005b) for cassava starch films plasticized with glycerol or sorbitol.

For composite films (CS + CH + G), a lower plasticizing effect of glycerol was observed when compared with starch matrices (CS + G). Composite MC–CH films also exhibited different behavior patterns under tensile tests. CH films showed high resistance at break, while MC was more flexible, exhibiting composite films with intermediate percent elongation values (Fig. 6.13) (Pinotti et al. 2007). Similar mechanical properties have been reported in the literature for MC films (Donhowe and Fennema 1993a, b, c; Park et al. 1993, 1994; Debeaufort and Voilley 1997). For CH films, most literature reported values for tensile tests were higher than those obtained by Pinotti et al. (2007). These differences may be due to variations in CH composition and/or suppliers and film preparation (Chen and Lin 1994; Butler et al. 1996; Caner et al. 1998). Caner et al. (1998) reported a wide range of elongation values (14–70%) for CH films, depending on the acid used for chitosan solubilization, plasticizer type, and storage time.

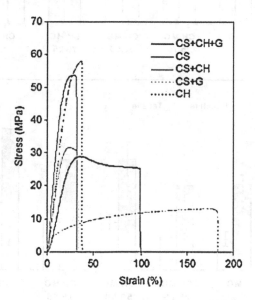

Fig. 6.12 Mechanical properties of composite films formulated with 5% corn starch (CS), chitosan (CH), corn starch plasticized with glycerol (CS + G), corn starch and chitosan (CS + CH) and corn starch and chitosan plasticized with glycerol (CS + G + CH)

For CH–MC composite films, tensile strength and elastic modulus increased with an increasing CH concentration (Fig. 6.13b); accordingly, relative deformation decreased (Fig. 6.13a). The blending of CH with MC enhanced film deformation and increased film flexibility, avoiding the need for plasticizer addition. With regard to puncture tests, the elastic modulus showed the same trend as in tensile tests (Fig. 6.13b). However, all composite films have a similar values of deformation at break, regardless of CH:MC ratio (Fig. 6.13a). Thus, CH may serve as a partial replacement of MC for generation of composite films with good puncture resistance and economic benefit, since MC is itself an expensive commodity.

Table 6.12 compares mechanical properties determined for both biodegradable and synthetic films under identical test conditions. Tensile strength values for

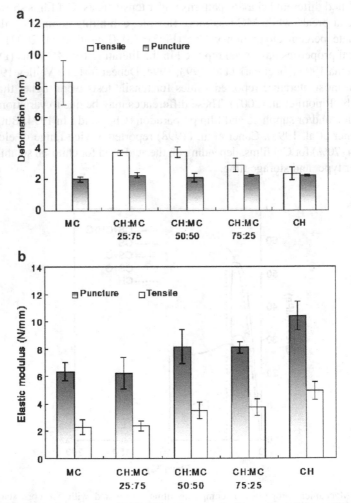

Fig. 6.13 Mechanical properties of MC, CH and composite films with a different CH:MC ratios (Tensile and puncture tests: (a) elongation at break and (b) elastic modulus)

Table 6.12 Comparison of the mechanical properties of corn starch, chitosan and composite films with commercial synthetic films

Film type	Components	Mechanical properties	
		Tensile strength (MPa)	Elongation at break (%)
Biodegradable	CS[a]	47.4 ± 1.5	3.6 ± 0.5
	CS + G	7.1 ± 0.4	22.5 ± 4.2
	CS + CH	24.7 ± 4.0	3.0 ± 0.2
	CS + CH + G	28.7 ± 6.8	11.7 ± 4.0
	CH	60.7 ± 5.8	3.3 ± 0.4
Synthetic	Cellophane[b]	85.8 ± 8.9	14.4 ± 2.4
	LDPE[c]	16.2 ± 4.0	68.7 ± 14.9
	HDPE[c]	27.8 ± 3.2	150.0 ± 18.5

[a]CS: corn starch at a concentration of 5%
[b]Shellhammer and Krochta (1997)
[c]Smith (1986)

composite films were in the range of those for LDPE and HDPE, but were lower than that obtained for cellophane. However, synthetic polymers like LDPE and HDPE exhibited much greater values for elongation at break. Similar results were reported for synthetic films by Cunningham et al. (2000).

6.5 Film and Coating Applications

Various coating applications reported in literature have utilized vegetable, meat, seafood and bread dough products among others (Olorunda and Aworth 1984; Drake et al. 1987, 1991; Dhalla and Hanson 1988; Avena-Bustillos et al. 1993, 1994; Baldwin et al. 1997; García et al. 1998a, b). Since most coating applications are reported for fruits and vegetables, several specific examples will be highlighted here. The diverse array of fruits and vegetables offers many challenges to coating applications that include adhesion of coating to slippery (e.g., tomatoes, mushrooms), irregular or rough (e.g., strawberries) surfaces. These will require addition of ingredients to formulations that will reduce coating solution surface tension to ensure coating uniformity and absence of void holes. In the case of starch-based coatings, those containing plasticizer have been shown to be homogeneous, and covered the whole surface of strawberries, including the achenes (García et al. 1998a). On the other hand, coatings without plasticizer were brittle and possessed undesirable cracks, as revealed by iodine staining of the film surface (García et al. 1998a).

Strategies to extend post-harvest quality of fruits and vegetables need to address several key challenges, like extending maturation and senescence periods, minimizing dehydration and reducing the onset and rate of microbial growth. Edible films and coatings can offer simultaneous solutions to vegetables, such as Brussels sprouts, tomatoes, cucumbers, and red peppers (El Gaouth et al. 1991a, b; Viña et al. 2007), and for fruits like bananas (Banks 1984), apples (Drake et al. 1987), and mangoes (Dhalla and Hanson 1988). Mali and Grossmann (2003) proposed the use of yam starch films as packaging for strawberries contained within plastic trays, and compared performance of these films with that of PVC. Although the PVC film provided better weight and firmness retention for strawberries, yam starch films could be considered a viable packaging alternative.

6.5.1 Application of Coatings to Highly Perishable Fruits

A good example is application of composite active coatings to extend storage life of strawberries, which are highly perishable fruits with only a short period of annual production and high susceptibility to fungal infection (El Gaouth et al. 1991a; García et al. 1998a, b, 2001). Cold gelatinization of starch was the chosen alternative to comply with both requirements: (1) the dipping procedure selected for coating application, and (2) the heat sensitive characteristics of strawberries. Formulations of the tested starch-based coatings are shown in Table 6.2.

6.5.2 Microbiological Analysis

Surface microbial growth is the main cause of spoilage for many food products. The microorganisms that grew on strawberries were mainly yeasts, molds and sugar-fermenting bacteria. Figure 6.14 demonstrates need of a composite coating to decrease microbial counts, as the unplasticized coating did not differ significantly from the control in this regard. Coatings containing potassium sorbate, a well-known, effective antifungal agent, significantly decreased ($P<0.05$) yeast and mold counts on coated strawberries. Since the undissociated form of sorbic acid is the active antimicrobial agent, citric acid was added to the coating formulation to increase potassium sorbate effectivity. Fig. 6.14 shows that 0.2 g/L potassium sorbate used in conjunction with citric acid was the most effective formulation for decreasing microbial counts. Use of active coatings lowers the amount of preservative required to achieve the same antimicrobial efficacy as traditional techniques (Guilbert 1986; Guilbert et al. 1997; Vojdani and Torres 1989a, b).

Storage life, defined as the time necessary to reach 10^6 CFU/g fruit, was 14 days for uncoated fruit stored at 0°C. Coatings containing sorbitol extended storage life to 21 days. At the maximum storage time evaluated (28 days), coating formulations containing both sorbitol and potassium sorbate still exhibited microbial counts below the storage life limit of 10^6 CFU/g of fruit. While addition of potassium sorbate improved antimicrobial characteristics of the starch coatings, addition of both citric acid and potassium sorbate further enhanced antimicrobial effectiveness, yielding a composite coating with a storage life of more than 28 days (Fig. 6.14).

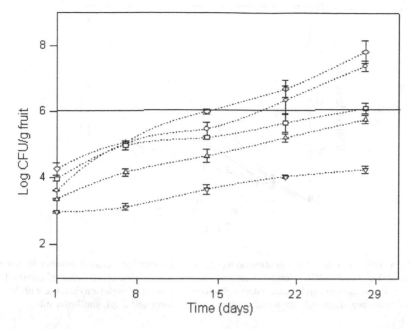

Fig. 6.14 Effect of coating formulation on growth of yeasts and molds (Log CFU/g fruit) in strawberries stored at 0°C and 84%RH (Control: uncoated fruits, (*open diamond*); composition of coating formulation: (*open circle*) amylomaize without plasticizer; (*open square*) amylomaize and 20 g/L sorbitol; (*open up triangle*) amylomaize, 20 g/L sorbitol and 0.2 g/L potassium sorbate and (*open down triangle*) amylomaize, 20 g/L sorbitol, 0.2 g/L potassium sorbate and citric acid; bars indicate standard error)

6.5.3 Quality Attributes of Strawberries

All plasticized coatings effectively reduced strawberry weight loss during storage (Fig. 6.15). Significant differences ($P < 0.05$) were observed among the weight loss values for uncoated strawberries and those coated with common corn or high-amylose corn starch, with lowest weight loss values obtained with amylomaize. Weight losses of fruits coated with starch formulations without plasticizer were similar to those of controlled fruit, due to presence of pores and cracks. Coatings with sorbitol led to significantly lower fruit weight losses than glycerol, regardless of starch type. However, weight losses were unacceptable after 3 weeks of storage for coated fruits without lipid. Addition of sunflower oil was necessary to reduce weight loss and increase storage life, provided that microbial counts could be maintained below the established limit (10^6 CFU/g fruit) even at 28 days of storage. The maximum weight loss reduction (63.2%) after 28 days of storage was obtained for coating formulations that included sunflower oil (Fig. 6.15). Avena-Bustillos et al. (1997), working on apples and celery treated with caseinate

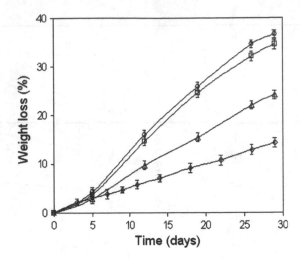

Fig. 6.15 Effect of coating formulation on weight loss of control and coated strawberries during storage at 0°C and 84%RH. (Control: uncoated fruits, (*open circle*); composition of coating formulation: (*open square*) amylomaize without plasticizer; (*open up triangle*) amylomaize with 20 g/L sorbitol and (*open diamond*) amylomaize with 20 g/L sorbitol and 2 g/L sunflower oil)

and acetylated monoglyceride coating, found a 75% increase in resistance to water vapour transfer and consequently, a decrease in weight loss of coated vegetables compared to uncoated controls. Ultimately, changes in texture and appearance are the most important factors that determine post-harvest storage life of fruits and vegetables. The rate and extent of firmness loss during ripening of soft fruits like strawberries is also another factor of primary importance. According to Manning (1993), fruit softening is attributed to degradation of components of the cell wall, primarily pectic substances, due to activity of specific enzymes such as polygalacturonase. For both control and coated fruits, breaking force decreased as a function of storage time. Coating formulations that minimized weight loss also maintained better firmness, since the firmness attribute is highly influenced by water content.

The effect of coatings on surface color modifications of strawberries was analyzed, because this quality attribute may also determine consumer acceptability of fruit. Glycerol or sorbitol significantly ($P < 0.05$) delayed development of surface color, which resulted in a slight change in color lightness with time. Coating formulations containing glycerol gave better surface color results compared to those with sorbitol. Lipid addition did not significantly modify ($P < 0.05$) surface color results, regardless of the plasticizer used in formulations.

Active starch-based coatings containing antimicrobial agents, plasticizer and lipid improved barrier properties to water vapour (Table 6.8), reduced microbial growth and showed selective gas permeability (Table 6.10), and thus, extended storage life of strawberries. The best coating formulation in our case contained high-amylose starch, sorbitol, sunflower oil and potassium sorbate plus citric acid as antimicrobial agents.

6.5.4 Physiological Modifications

Physiological parameters, including titratable acidity, pH, anthocyanin level and sugar content, are good indicators of fruit maturation and senescence. Coatings may be used to alter the natural physiological behavior, modifying organoleptic characteristics of fruits such as color, taste or flavor.

In coated strawberries, these physiological parameters were slowed down, but did reach commercially acceptable values (García et al. 1998b). These results indicate that starch-based coatings retard metabolic reactions, and thus, senescence of coated fruits is delayed. The oxygen and carbon dioxide barriers lead to a reduction in respiration rate by limiting exposure to ambient oxygen, increasing internal carbon dioxide, delaying ripening and senescence, and extending the storage life of treated fruits (Baldwin 1994; Avena-Bustillos et al. 1993, 1994; Drake et al. 1991; García et al. 1998a, b).

Another important parameter in fruits is the degree of maturity reached at the time of application of the coating. For example, when composite starch-based coatings were applied on strawberries at 25, 50, 75 and 100% color maturity ripening stages, 75% was selected as the optimum degree of maturity for prolonging storage life. In this case, development of appropriate anthocyanin content, reducing and non-reducing sugar levels and titratable acidity of strawberries were slowed down during storage, reaching commercially acceptable values at the end of the storage period. The delay in development of physiological parameters observed in coated fruits, regardless of the degree of maturity, was attributed to the differential gaseous permeability of films and influence on respiratory activity.

6.6 Coating Application on Fried Foods

As mentioned previously, another novel application of coatings is to reduce oil uptake of products during deep-fat frying. Cellulose derivatives, such as methyl cellulose (Table 6.2), can be used to decrease oil content of food products during frying. Hydrocolloids with thermal gelling or thickening properties, like proteins and carbohydrates, have been investigated for this purpose. Williams and Mittal (1999a, b) found that methylcellulose (MC) films provided the best oil barrier properties, and reduced fat uptake more than hydroxypropylcellulose and gellan gum films in pastry mix applications. Mallikarjunan et al. (1997) stressed that products coated with cellulose derivatives form a protective layer at the surface of food products during initial stages of frying, due to thermal gelation of the hydrocolloid at temperatures above 60°C. This protective layer limits transfer of moisture and fat between the sample and the frying medium. MC formulations that included sorbitol as plasticizer were applied to dough discs and potato strips to reduce oil uptake during deep fat frying (García et al. 2002, 2004a). Coating application did not modify textural characteristics or sensory properties of the fried samples. MC coating formulations were the most effective, reducing oil uptake between 35 and 40% depending on the product. To obtain a coating with adequate adherence and integrity, it is necessary to incorporate a plasticizer, sorbitol being the most effective in the case of MC coatings. SEM micrographs showed good adhesion between the coating and the dough, and no significant differences between coated and uncoated samples were detected by the sensory panel.

6.7 Future Trends

The use of coatings and packaging films by the food industry has become a topic of great interest, because of their potential for increasing shelf life of many food products (Ahvenainen 2003; Coles et al. 2003; Giles and Bain 2001; Hernandez et al. 2000). By selecting the right materials and packaging technologies, it is possible to maintain product quality and freshness of food products (Brown 1992; Stewart et al. 2002). Nowadays, a large part of materials used in packaging industries is produced from fossil fuels, which are practically not degradable. Preservation of natural resources and recycling has led to a renewed interest in biomaterials and renewable raw materials. A concerted effort to extend shelf life and to enhance food quality, while reducing packaging waste, has encouraged exploration of new bio-based packaging materials, such as edible and biodegradable films from renewable resources (Albertsson and Karlsson 1995; Trznadel 1995; Tharanathan and Kittur 2003). However, like conventional packaging, bio-based packaging has to supply a number of important functions with regards to food applications, including containment and protection, maintenance of sensory quality and safety, and labeling (Robertson 1993).

Polar biopolymers, such as polysaccharides and proteins, are being studied as prospective replacements for synthetic polymers in the film and plastic industries. Their potential benefits are both environmental and cost-related. Traditional techniques for processing thermoplastic synthetic polymers have been adapted to hydrophilic polymers like starch and gelatin.

To date, use of biodegradable films for food packaging has been strongly limited because of the poor barrier properties and weak mechanical properties shown by natural polymers. For this reason, natural polymers are frequently blended with other synthetic polymers or, less frequently, chemically modified with the aim of extending their applications (Guilbert et al. 1997; Petersen et al. 1999). Recent advances in genetic engineering and composite science offer significant opportunities for improving materials from renewable resources with enhanced support of global sustainability, although use of materials obtained by genetic modification is still controversial.

For scientists, the real challenge lies in finding applications that would consume sufficiently large quantities of these materials to reduce their cost, allowing biodegradable polymers to compete economically in the market. Starch and its derivates are promising raw materials, because they are renewable, widely available and relatively low cost (Gonera and Cornillon 2002; Smits et al. 1998). As a packaging material, starch alone does not form films with appropriate mechanical properties, unless it is first plasticized or chemically modified. When starch is treated in an extruder by application of both thermal and mechanical energy, it can be converted to a thermoplastic material. In production of thermoplastic starches, plasticizers are expected to efficiently reduce intra-molecular hydrogen bonds and to provide stability to product properties. There are many opportunities for using starch as packaging materials (Kim and Pometto 1994). As described previously, starch can be employed as packaging for fruits, vegetables, snacks, or other dry products. For example, Novamont, an Italian company, commercializes EverCorn™, a starch-based material available as films or bags. Similarly, Bioenvelope, a Japanese company commercializes Bio-P™, another starch-based film. In these applications, however, efficient mechanical, oxygen and moisture protection is needed, as thermoplastic starch alone cannot meet all these requirements. Because of the hydrophilicity of the starch, performance changes during and after processing, due to water content changes. To overcome this drawback, many different routes have been reported. Some biodegradable materials like Mater-Bi™, produced by Novamont, are available in the market, which is based on 60% starch and 40% alcohol. This is mainly commercialized in Europe and America.

Today's excellent performance of traditional plastics is an outcome of continued R&D efforts of the last several years. However, existing biodegradable polymers have come to public view only recently. Prices of biodegradable biopolymers can be realized through continued research and development efforts focused on optimizing composition and functionality to improve performance of edible films and coatings.

Acknowledgments The financial support provided by CONICET (Consejo Nacional de Investigaciones Científicas y Técnicas), Agencia Nacional de Promoción Científica y Tecnológica and Universidad Nacional de La Plata is gratefully acknowledged.

References

Ahvenainen R (2003) Novel food packaging techniques. Boca Raton, FL: CRC Press LLC
Albertsson AC, Karlsson S (1995) Degradable polymers for the future. Acta Polymers, 46, 114–123
Arvanitoyannis I, Kalichevsky M, Blanshard JMV, Psomiadou E (1994) Study of diffusion and permeation of gases in undrawn and uniaxially drawn films made from potato and rice starch conditioned at different relative humidities. Carbohydr. Polym. 24, 1–15
ASTM (1996a) Standard test methods for water vapour transmission of material, E96 – 95. Annual book of ASTM. Philadelphia, PA: American Society for Testing and Materials

ASTM (1996b) Standard test methods for tensile properties of thin plastic sheeting, D882-91. Annual Book of ASTM. Philadelphia, PA: American Society for Testing and Materials

Avena-Bustillos RJ, Cisneros-Zeballos LA, Krochta JM, Saltveit ME (1993) Optimization of edible coatings on minimally processed carrots using response surface methodology. Am. Soc. Agric. Eng. , 36 3, 801–805

Avena-Bustillos RJ, Krochta JM, Saltveit ME, Rojas-Villegas RJ, Sauceda-Pérez JA (1994) Optimization of edible coating formulations on Zucchini to reduce water loss. J. Food Eng. 21, 197–214

Avena-Bustillos RJ, Krochta JM, Saltveit ME (1997) Water vapour resistance of red delicious apples and celery sticks coated with edible caseinate-acetylated mono-glyceride films. J. Food Sci. 62, 2, 351–354

Avena-Bustillos RJ, Olsen CW, Olson DA, Chiou B, Yee E, Bechtel PJ, McHugh TH (2006) Water vapour permeability of mammalian and fish gelatin films. J. Food Sci. 71, 4, 202–207

Bader HG, Göritz D (1994a) Investigations on high amylose corn starch films. Part 1: wide angle X-ray scattering (WAXS). Starch/Stärke 46, 6, 229–232

Bader HG, Göritz D (1994b) Investigations on high amylose corn starch films. Part 2: water vapour sorption. Starch/Stärke 46, 7, 249–252

Bader HG, Göritz D (1994c) Investigations on high amylose corn starch films. Part 3: stress strain behaviour. Starch/Stärke 46, 11, 435–439

Baker RA, Baldwin EA, Nisperos-Carriedo MO (1994) Edible coatings for processed foods. In: Krochta JM, Baldwin EA, Nisperos-Carriedo M (Eds.) Edible Coatings and Films to Improve Food Quality. Technomic Pub. Co., Lancaster, PA, pp. 89–104

Baldwin EA (1994) Edible coatings for fruits and vegetables, past, present and future. In: Krochta JM, Baldwin EA, Nisperos-Carriedo MO (Eds.) Edible Coatings and Films to Improve Food Quality. Technomic Pub. Co., Lancaster, PA, pp. 25–64

Baldwin EA, Nisperos-Carriedo MO, Hagenmaier RD, Baker RA (1997) Use of lipids in coatings for food products. Food Technol. 51, 6, 56–64

Banker GS (1966) Film coating, theory and practice. J. Pharm. Sci. 55, 81

Banks NH (1984) Some effects of TAL Pro-Long coatings on ripening bananas. J. Exp. Bot. 35, 127

Brown WE (1992) Plastics in Food Packaging. Properties, Design, and Fabrication. New York, NY: Marcel Dekker

Butler B, Vergano R, Testin R, Bunn J, Wiles J (1996) Mechanical and barrier properties of edible chitosan films as affected by composition and storage. J. Food Sci. 61, 5, 953–961

Caner C, Vergano P, Wiles J (1998) Chitosan film mechanical and permeation properties as affected by acid, plasticizer, and storage. J. Food Sci. 63, 6, 1049–1053

Chang BS, Randall CS (1992) Use of subambient thermal analysis of optimize protein lyophilization. Cryobiology 29, 632–656

Chang YP, Cheah PB, Seow CC (2000) Plasticizing – antiplasticizing effects of water on physical properties of tapioca starch films in the glassy state. J. Food Sci. 65, 3, 445–451

Chen RH, Lin JH (1994) Relationships between the chain flexibilities of chitosan molecules and the physical properties of their casted films. Carbohydr. Polym. 24, 41–46

Chen M, Yeh H, Chiang B (1996) Antimicrobial and physicochemical properties of methylcellulose and chitosan films containing a preservative. J. Food Process Preserv. 20, 379–390

Chen M, Deng J, Yang F, Gong Y, Zhao N, Zhang X (2003) Study on physical properties and nerve cell affinity of composite films from chitosan and gelatin solutions. Biomaterials 24, 2871–2880

Cherian G, Gennadios A, Weller C, Chinachoti P (1995) Thermomechanical behavior of gluten films; effect of sucrose, glycerin and sorbitol. Cereal Chem. 72, 1, 1–6

Coles R, McDowell D, Kirwan MJ (2003) Food packaging technology. Oxford, UK: Blackwell Publishing Ltd

Cunningham P, Ogale A, Dawson P, Acton J (2000) Tensile properties of soy protein isolate films produced by a thermal compaction technique. J. Food Sci. 65, 4, 668–671

Cuppet SL (1994) Edible coatings as carriers of food additives, fungicides and naturals antagonists. *In*: Krochta JM, Baldwin EA, Nisperos-Carriedo M (Eds.) Edible Coatings and Films to Improve Food Quality. Technomic Pub. Co., Lancaster, PA, pp. 121–137

Cuq B, Gontard N, Guilbert S (1995) Edible films and coatings as active layers. *In*: Rooney ML (Ed.) Active Food Packaging. Chapman & Hall, UK, pp. 111–142

Cuq B, Gontard N, Guilbert S (1998) Proteins as agricultural polymers for packaging production. Cereal Chem. 75, 1, 1–9

Dhalla R, Hanson SW (1988) Effect of permeability coatings on the storage life of fruits. II. Pro-long treatments of mangoes (*Mangifera indica L.* cv. Julie). Int. J. Food Sci. Technol. 23, 107–112

Debeaufort F, Voilley A (1997) Methylcellulose-based edible films and coatings: 2. Mechanical and thermal properties as a function of plasticizer content. J. Agric. Food Chem. 45, 685–689

Donhowe IG, Fennema OR (1993a) Water vapour and oxygen permeability of wax films. J. Am. Oil Chem. Soc. 70 (9), 867–873

Donhowe IG, Fennema O (1993b) The effects of solution composition and drying temperature on crystallinity, permeability and mechanical properties of methylcellulose films. J. Food Process Preserv. 17, 231–246

Donhowe IG, Fennema O (1993c) The effects of plasticizers on crystallinity, permeability and mechanical properties of methylcellulose films. J. Food Process Preserv. 17, 247–258

Donhowe IG, Fennema OR (1994) Edible films and coatings: characteristics, formation, definitions and testing methods. *In*: Krochta JM, Baldwin EA, Nisperos-Carriedo M (Eds.) Edible Coatings and Films to Improve Food Quality, Technomic Pub. Co., Lancaster, PA, pp. 1–21

Drake SR, Fellman JK, Nelson JW (1987) Post-harvest use of sucrose polyesters for extending the shelf-life of stored Golden Delicious apples. J. Food Sci. 52, 5, 1283–1285

Drake SR, Nelson JW (1990) Storage quality of waxed and nonwaxed "Delicious" and "Golden Delicious" apples. J. Food Qual. 13, 331–334

Drake SR, Cavalieri R, Kupferman EM (1991) Quality attributes of "D'Anjou" pears after different wax drying and refrigerated storage. J. Food Qual. 14, 455–465

El Gaouth A, Arul J, Ponnampalam R, Boulet M (1991a) Chitosan coating effect on storability and quality of fresh strawberry. J. Food Sci. 56, 6, 1618–1620

El Gaouth A, Arul J, Ponnampalam R, Boulet M (1991b) Use of chitosan coating to reduce water loss and maintain quality of cucumber and bell pepper fruits. J. Food Process Preserv. 15, 339–368

Even WR, Carr SH (1978) Micromechanical fracture analysis of amylose. Polymer 19, 583–588

Galietta G, Giola DD, Guilbert S, Cuq B (1998) Mechanical and thermomechanical properties of films based on whey protein as affect by plasticizer and crosslinking agents. J. Dairy Sci. 81, 3123–3130

García MA, Martino MN, Zaritzky NE (1998a) Starch-based coatings: effect on refrigerated strawberry (*Fragaria × Ananassa*) quality. J. Sci. Food Agric. 76, 411–420

García MA, Martino MN, Zaritzky NE (1998b) Plasticizer effect on starch-based coatings applied to strawberries (*Fragaria × Ananassa*). J. Agric. Food Chem. 46, 3758–3767

García MA, Martino MN, Zaritzky NE (1999) Edible starch films and coatings characterization: sem, water vapour and gas permeabilities. J. Scanning Microsc. 21, 5, 348–353

García MA, Martino MN, Zaritzky NE (2000a) Microstructural characterization of plasticized starch-based films. Starch/ Stärke 52, 4, 118–124

García MA, Martino MN, Zaritzky NE (2000b) Lipid addition to improve barrier properties of edible starch-based films and coatings. J. Food Sci. 65, 6, 941–947

García MA , Martino MN, Zaritzky NE (2001) Composite starch-based coatings applied to strawberries (Fragaria × ananassa). Nahrung/Food 45, 4, 267–272

García MA, Ferrero C, Bértola N, Martino MN, Zaritzky NE (2002) Edible coatings from cellulose derivatives to reduce oil uptake in fried products. Innov. Food Sci. Emerg. Technol. 3–4, 391–397

García MA, Ferrero C, Campana A, Bértola N, Martino M, Zaritzky N (2004a). Methylcellulose coatings reduce oil uptake in fried products. Food Sci. Technol. Int. 10, 5, 339–346

García MA, Martino MN, Zaritzky NE (2004b) Research advances in edible coatings and films from starch. *In*: Mohan RM (Ed.) Research Advances In Food Science. Global Research Network. Research Advances in Food Science 4, 107–128

García MA, Pinotti A, Martino MN, Zaritzky NE (2004c) Characterization of composite hydrocolloid films. Carbohydr. Polym. 56, 3, 339–345

García MA, Pinotti A, Zaritzky NE (2006) Physicochemical, water vapour barrier and mechanical properties of corn starch and chitosan composite films. Starch/Starke 58, 9, 453–463

Gennadios A, Weller CL, Gooding CH (1994) Measurement errors in water vapour permeability of highly permeable, hydrophilic edible films. J. Food Eng. 21, 395–409

Ghanbarzadeh B, Musavi M, Oromiehie AR, Rezayi K, Razmi E, Milani J (2007) Effect of plasticizing sugars on water vapour permeability, surface energy and microstructure properties of zein films. LWT 40, 1191–1197

Giles GA, Bain DA (2001) Technology of Plastics Packaging for the Consumer Market. Boca Raton, FL: CRC Press LLC

Gonera A, Cornillon P (2002) Gelatinization of starch/gum/sugar system studied by using DSC, NMR and CSLM. Starch/Sta¨rke, 54, 508–516

Gontard N, Guilbert S, Cuq B (1992) Edible wheat gluten films: influence of the main process variables on film properties using response surface methodology. J. Food Sci. 57, 1, 190–199

Gontard N, Guilbert S, Cuq JL (1993) Water and glycerol as plasticizers affect mechanical and water vapour barrier properties of an edible wheat gluten film. J. Food Sci. 58, 1, 206–211

Gontard N, Duchez C, Cuq JL, Guilbert S (1994) Edible composite films of wheat gluten and lipids: water vapour permeability and other physical properties. Int. J. Food Sci. Technol. 19: 39–50

Greener IK, Fennema OR (1989) Evaluation of edible, bilayer films for use as moisture barriers for foods. J. Food Sci. 54, 1400–1406

Guilbert S (1986) Technology and application of edible protective films. *In*: Mathlouthi M (Ed.) Food Packaging and Preservation. Theory and Practice. Elsevier Applied Science Publishing Co. London, UK, pp. 371–394

Guilbert S, Gontard N (1995) Technology and applications of edible protective films. *In*: VII Biotechnology and Food Research – "New Shelf-Life Technologies and Safety Assessments". Helsink, Finland, pp. 49–60

Guilbert S, Cuq B, Gontard N (1997) Recent innovations in edible and/or biodegradable packaging materials. Food Addit. Contam. 14, 6, 741–751

Hagenmaier RD, Baker RA (1994) Wax microemulsions and emulsions as citrus coatings. J. Agric. Food Chem. 42, 899–902.

Hardenburg RE (1967) Wax and related coatings for horticultural products – a bibliography. Agric. Res. Serv. Bull. 965. Cornell University, Ithaca, NY

Hernandez E (1994) Edible coatings for lipids and resins. *In*: Krochta JM, Baldwin EA, Nisperos-Carriedo M (Eds.) Edible Coatings and Films to Improve Food Quality. Technomic Pub. Co., Lancaster, PA, pp. 279–304

Hernandez R, Selke SEM, Cultler J (2000) Plastics Packaging: Properties, Processing, Applications, Regulations. Munich, Germany: Hanser Gardner Publications

Hershko V, Nussinovitch A (1998) Relationship between hydrocolloid coating and mushroom structure. J. Agric. Food Chem. 46, 8, 2988–2997

Iijimaa M, Nakamura K, Hatakeyama T, Hatakeyamab H (2000) Phase transition of pectin with sorbed water. Carbohydr. Polym. 41, 101–106

Kester JJ, Fennema OR (1986) Edible films and coatings: a review. Food Technol. 12, 47–59

Kim M, Pometto III OR (1994) Food packaging potential of some novel degradable starchepolyethylene plastics. J. Food Prot., 57, 1007–1012

Klug HP, Alexander LE (1974) Crystallite size and lattice strains from line broadening. In: X-ray Diffraction Procedures for Polycrystalline and Amorphous Materials. New York, USA: Wiley-Intersciences, pp. 618–708

Köksel H, ahbaz F Özboy Ö (1983) Influence of wheat-drying temperatures on the birefringence and X-ray diffraction patterns of wet-harvested wheat starch. Cereal Chem. 70, 4, 481–483

Krochta JM, De Mulder-Johnson C (1997) Edible and biodegradable polymer films: challenges and opportunities. Food Technol. 51, 2, 61–77

Kumar MNVR (2000) A review of chitin and chitosan applications. React Funct. Polym. 46, 1–27

Laohakunjit N, Noomhorm A (2004) Effect of plasticizers on mechanical properties of rice starch film. Starch/Stärke 5, 8, 348–356

Mali S, Grossmann MVE (2003) Effects of yam starch films on storability and quality of fresh strawberries (*Fragaria ananassa*). J. Agric. Food Chem. 51, 24, 7005–7011

Mali S, Grossmann MV, García MA, Martino MN, Zaritzky NE (2002) Microstructural characterization of yam starch films. Carbohydr. Polym. 50, 4, 379–386

Mali S, Grossmann MV, García MA, Martino M, Zaritzky N (2005a) Mechanical and thermal properties of yam starch films. Food Hydrocol. 19, 157–164

Mali S, Sakanaka LS, Yamashita F, Grossmann MVE (2005b) Water sorption and mechanical properties of cassava starch films and their relation to plasticizing effect. Carbohydr. Polym. 60, 283–289

Mallikarjunan P, Chinnan MS, Balasubramaniam VM, Phillips RD (1997) Edible coatings for deep-fat frying of starchy products. LWT 30, 709–714

Manning K (1993) Soft fruits. In: Seymour GB, Taylor JE, Tucker GA (Eds.) Biochemistry of Fruit Ripenning. Chapman & Hall, London, pp. 347–373

McHugh TH, Krochta JM (1994a) Milk-protein-based edible films and coatings. Food Technol. 48, 1, 97–103

McHugh TH, Krochta JM (1994b) Sorbitol- vs glycerol- plasticized whey protein edible films: integrated oxygen permeability and tensile property evaluation. J. Agric. Food Chem. 42, 4, 841–845.

McHugh TH, Krochta JM (1994c) Permeability properties of edible films. In: Krochta JM, Baldwin EA, Nisperos-Carriedo M (Eds.) Edible Coatings and Films to Improve Food Quality. Technomic Pub. Co., Lancaster, PA, pp 139–187

McHugh TH, Aujard JF, Krochta JM (1994) Plasticized whey protein edible films: water vapour permeability properties. J. Food Sci. 59, 2, 416–419

Miles MJ, Morris VJ, Ring SG (1985a) Gelation of amylose. Carbohydr. Res. 135, 257–269

Miles MJ, Morris VJ, Orford PD, Ring SG (1985b) The roles of amylose and amylopectin in the gelation and retrogradation of starch. Carbohydr. Res. 135, 271–281

Muzzarelli R, Baldassarre V, Conti F, Ferrara P, Biagini G, Gazzanelli G, Vasi V (1988) Biological activity of chitosan: ultrastructural study. Biomaterials 9, 3, 247–252

Neto CGT, Giacomettib JA, Jobb AE, Ferreirab FC, Fonsecaa JLC, Pereiraa MR (2005) Thermal analysis of chitosan based networks. Carbohydr. Polym. 62, 97–103

Nisperos-Carriedo MO (1994) Edible coatings and films based on polysaccharides. In: Krochta JM, Baldwin EA, Nisperos-Carriedo MO (Eds.) Edible Coatings and Films to Improve Food Quality. Technomic Pub. Co., Lancaster, PA, pp. 305–330

Noel TR, Ring SG, Whittman MA (1992) The structure and gelatinization of starch [Review]. Food Sci. Technol. Today, 6, 159

Oh JH, Wang B, Field PD, Aglan HA (2004) Characteristics of edible films made from dairy proteins and zein hydrolysate cross-linked with transglutaminase. Int. J. Food Sci. Technol. 39, 1–8

Olorunda AO, Aworh OC (1984) Effects of Pro-long, a surface coating agent, on the shelf-life and quality attributes of plantain. J. Sci. Food Agric. 35, 573–578

Park H, Weller C, Vergano P, Testin R (1993) Permeability and mechanical properties of cellulose-based edible films. J. Food Sci. 58, 6, 1361–1364, 1370

Park HJ, Bunn JM, Weller CL, Vergano PJ, Testin RF (1994) Water vapour permeability and mechanical properties of grain protein-based films as affected by mixtures of polyethylene glycol and glycerin plasticizers. Trans ASAE 37, 4, 1281–1285

Parra DF, Tadini CC, Ponce P, Lugão AB (2004) Mechanical properties and water vapour transmission in some blends of cassava starch edible films. Carbohydr. Polym. 58, 475–481

Parris N, Coffin DR, Joubran RF, Pessen H (1995) Composition factors affecting the water vapour permeability and tensile properties of hydrophilic films. J. Agric. Food Chem. 43, 1432–1435

Paull RE, Chen NJ (1989) Waxing and plastic wraps influence water loss from papaya fruit during storage and ripening. J. Am. Soc. Hortic. Sci. 114, 937–942

Pawlak A, Mucha M (2003) Thermogravimetric and FTIR studies of chitosan blends. Thermochimica Acta, 396, 153–166

Petersen K, Nielsen PV, Bertelsen G, Lawther M, Olsen MB, Nilssonk NH (1999) Potential of biobased materials for food packaging. Trends Food Sci. Technol. 10, 52–68

Pinotti A, García MA, Martino MN, Zaritzky NE (2007) Study on microstructure and physical properties of composite films based on chitosan and methylcellulose. Food Hydrocol. 21, 1, 66–72

Pommet M, Redl A, Guilbert S, Morel MH (2005) Intrinsic influence of various plasticizers on functional properties and reactivity of wheat gluten thermoplastic materials. J. Cereal Sci. 42, 81–91

Robertson GL (1993) Food Packaging. Principles and Practice. New York, NY: Marcel Dekker

Romero-Bastida CA, Bello-Pérez LA, García MA, Martino MN, Solorza-Feria J, Zaritzky NE (2005) Mechanical and microstructural characterization of films prepared by thermal or cold gelatinization of non-conventional starches. Carbohydr. Polym. 60, 2, 235–244

Roos YH (1995) Phase Transitions in Foods. Taylor, S.L. Academic Press San Diego, CA, USA

Sarantopoulos C, de Oliveira M, Padula Coltro L, Vercelino Alves RM, Correa García E (2002) Embalagens Plásticas Flexíveis. Centro de Tecnología de Embalagem. Campinas, Brasil

Shellhammer TH, Krochta JM (1997) Whey protein emulsion film performance as affected by lipid type amount. J. Food Sci. 62, 2, 390–394

Sionkowska A, Wisniewski J, Skopinska J, Kennedy CJ, Wess TJ (2004) Molecular interactions in collagen and chitosan blends. Biomaterials, 25, 795–801

Smith SA (1986) Polyethylene, low density. In: Bakker M. (Ed.) The Wiley Encyclopedia of Packaging Technology. John Wiley & Sons, New York, pp. 514–523

Smith JP, Hoshino J, Abe Y (1995) Interactive packaging involving sachet technology. In: Rooney ML (Ed.) Active Food Packaging. Blackie Academic and Professional, Glasgow, pp. 143–173

Smits ALM, Ruhnau FC, Vliegenthart JFG (1998) Ageing of starch based systems as observed by FT-IR and solid state NMR spectroscopy. Starch/Starke, 50, 11–12, 478–483

Snyder RL, Bish DL (1989) Quantitative analysis. In: Bish DL, Post JE (Eds.) Modern Powder Diffraction. Mineralogical Society of America, Washington, DC, pp. 101–144

Sobral PJA, Menegalli FC, Hubinger MD, Roques MA (2001) Mechanical, water vapour barrier and thermal properties of gelatin based edible films. Food Hydrocol. 15, 423–432

Sohail SS, Wang B, Biswas MAS, Oh JH (2006) Physical, morphological, and barrier properties of edible casein films with wax applications. J. Food Sci. 71, 4, 255–259

Spiess WEL, Wolf WR (1993) The results of the COST 90 project on water activity. In: Jowitt R, Escher F, Hallstrom FB, Meffert MF, Spiess WEL, Vos G (Eds.) Physical Properties of Foods. Applied Science Publishers, London, pp. 65–91

Stepto RFT, Tomka I (1987) Injection moulding of natural hydrophilic polymers in the presence of water. Chimia 41, 3, 76–81

Stewart CM, Tompkin RB, Cole MB (2002) Food safety: new concepts for the new millennium. Innov. Food Sci. Emerg. Technol. 3, 105–112

Tharanathan NR, Kittur SF (2003) Chitin-the undisputed biomolecule of great potential. Crit. Rev. Food Sci. 43, 61–83

Trezza TA Krochta JM (2000a) The gloss of edible coatings as affected by surfactants, lipids, relative humidity and time. J. Food Sci. 65, 4, 658–662

Trezza TA, Krochta JM (200b) Color stability of edible coatings during prolonged storage. J. Food Sci. 65, 7, 1166–1169

Trznadel M (1995) Biodegradable polymer materials. Int. Polym. Sci. Technol. 22, 12, 58–65

Van Soest JJG, Vliegenthart JFG (1997) Crystallinity in starch plastics: consequences for material properties. Trends Biotechnol. 15, 208–213

Van Soest JJG, Hulleman SHD, de Wit D, Vliegenthart JFG (1996) Crystallinity in starch bioplastics. Ind. Crops Prod. 5, 11–22

Vermeiren L, Devlieghere F, van Beest M, de Kruijf N, Debevere J (1999) Developments in the active packaging of foods. Trends Food Sci. Technol. 10, 77–86

Viña SZ, Mugridge A, García MA, Ferreyra RM, Martino MN, Chaves AR, Zaritzky NE (2007) Quality of refrigerated Brussels sprouts. Effect of plastic packaging film and edible starch coatings. Food Chem. 103, 701–709

Vojdani F, Torres JA (1989a) Potassium sorbate permeability of methylcellulose and hydroxypro-pylmethylcellulose multi-layer films. J. Food Process. Preserv. 13, 417–430

Vojdani F, Torres JA (1989b) Potassium sorbate permeability of polysaccharide films: chitosan, methylcellulose and hydroxypro-pylmethylcellulose. J. Food Process. Eng. 12, 33–48

Vojdani F, Torres JA (1990) Potassium sorbate permeability of methylcellulose and hydroxypro-pylmethylcellulose coatings: effect of fatty acids. J. Food Sci. 55, 3, 841–846

Wanchoo RK, Sharma PK (2003) Viscometric study on the compatibility of some water-soluble polymer-polymer mixtures. Eur. Polym. J. 39, 1481–1490

Were L, Hettiarachachy NS, Coleman M (1999) Properties of cysteine-added soy protein-wheat gluten films. J. Food Sci. 64, 3, 514–518

Williams R, Mittal GS (1999a) Water and fat transfer properties of polysaccharide films on fried pastry mix. LWT, 32, 440–445

Williams R, Mittal GS (1999b) Low-fat fried foods with edible coatings: modeling and simulation. J. Food Sci. 64(2), 317–322

Wong DWS, Gastineau FA, Gregorski KS, Tillin SJ, Pavlath AE (1992) Chitosan-lipids films: microstructure and surface energy. J. Agric. Food Chem. 40, 540–544

Wu T, Zivanovic S, Draughon FA Conway WS, Sams CE (2005) Physicochemical properties and bioactivity of fungal chitin and chitosan. J. Agric. Food Chem. 53, 3888–3894

Young AH (1984) Fractionation of Starch. In: Whistler RL, Be Miller JN, Paschall EF (Eds.) Starch, Chemistry and Technology, 2nd edn. Academic Press, Orlando, FL, USA, pp. 249–283

Zaccaron C, Oliveira R, Guiotoku M, Pires A, Soldi V (2005). Blends of hydroxypropyl methyl-cellulose and poly(1-vinylpyrrolidone-co-vinyl acetate): miscibility and thermal stability. Polym. Degrad. Stab. 90, 1, 21–27

Chapter 7
Edible Films and Coatings for Fruits and Vegetables

Guadalupe I. Olivas and Gustavo Barbosa-Cánovas

7.1 Introduction

There is a growing trend toward increased consumption of fresh fruits and vegetables. According to the USDA, fresh fruit consumption in the United States in 2000 was 28% above average annual fruit consumption of the 1970s, and fresh vegetable consumption was 26% above average annual vegetable consumption for the same period (USDA 2001–2002).

Higher consumption of fruits and vegetables has been associated with growing interest in a healthier diet, and is expected to increase over time. International organizations (World Health Organization/WHO 2002; Food and Agriculture Organization/ FAO 2003) have been urging nations everywhere to promote consumption of fruits and vegetables, as a diet high in such foods has been found to be associated with decreased incidences of birth defects, mental and physical retardation, weakened immune systems, blindness, cardio-vascular diseases, and some forms of cancer and diabetes (Ford and Mookdad 2001; Genkinger et al. 2004; Hung et al. 2004; FAO 2003). WHO (2002) has estimated that low fruit and vegetable intake is among the top ten risk factors contributing to mortality (2.7 million deaths each year). They also reported that 19% of gastrointestinal cancer, 31% of ischemic heart disease, and 11% of stroke occurrences are caused by low intake of fruits and vegetables.

Consumers today have higher expectations than ever before, insisting their food be more nutritious and safer to eat, with wider variety and longer shelf life. In the case of produce, maintaining high quality food for long periods of time is difficult, since fruits and vegetables are composed of living tissue which undergoes major

G.I. Olivas (✉)
Centro de Investigación en Alimentación y Desarrollo (CIAD), Av. Río Conchos
S/N, Cuauhtémoc, Chihuahua 31570, Mexico
e-mail: golivas@ciad.mx

G. Barbosa-Cánovas
Center for Nonthermal Processing of Food, Washington State University,
Pullman, WA 99164-6120, USA
Barbosa@wsu.edu

M.E. Embuscado and K.C. Huber (eds.), *Edible Films and Coatings for Food Applications*, 211
DOI 10.1007/978-0-387-92824-1_7, © Springer Science+Business Media, LLC 2009

changes in chemical and physical states due to synthetic and degradative biochemical processes. Nutritional, flavor, textural, and pigmentation changes can all affect quality of produce (Kays and Paull 2004).

Maintaining quality of fruits and vegetables becomes even more difficult when produce has been minimally processed. Thus, fresh-cut or minimally-processed fruits and vegetables have become very important issues among food scientists and technologists in recent years. Initially, fresh cut produce was only offered to restaurants or supplied to the food service industry, where commodities were then sold to consumers within a short time; thus, there was no need to maintain quality of food for long periods. Now fresh cut fruits and vegetables have expanded to supermarkets in response to consumer demand.

As with intact fresh produce, consumers likewise expect minimally processed food commodities to be nutritious, attractive, and ready-to-eat, as well as to exhibit high quality and long shelf life, with no differences in flavor and texture from the original counterpart. These expectations are difficult to meet, since minimally processed commodities undergo rapid deterioration, wherein increased respiration rate and ethylene production, and consumption of sugars, lipids and organic acids accelerate ripening process. These changes cause senescence and, thus loss of texture and water, as well as undesirable changes in flavor and color (Kays 1991).

Use of edible films and coatings has been studied as a good alternative for preservation of intact and fresh-cut fruits and vegetables, since such films can create semipermeable barriers to gases and water vapor, maintaining quality of the product. Edible films and coatings have also been studied as potential carriers of additives to help preserve, or even improve quality of produce. Application of edible coatings consists of applying a thin layer of edible material to the food's surface, while generation of edible films involves casting a film of edible material to cover the commodity. However, in studies on edible films and coatings, the two terms are generally considered the same, and results are often reported interchangeably (Gordon 2005).

Overall, the purpose of using edible films and coatings for fruits and vegetables is to retard transfer of gas, vapor and volatiles, thus providing food with a modified atmosphere that decreases respiration and senescence, reduces aroma loss, retains moisture and delays color changes throughout storage.

7.2 Postharvest Physiology

Since fruits and vegetables consist of living tissue, subsequent physiological and biochemical changes which cause detrimental changes in quality and shelf life of produce are common after harvesting. Respiration, transpiration and ethylene production are main factors contributing to deterioration of fruits and vegetables (DeEll et al. 2003). After harvest fresh, produce can also experience biotic and abiotic stresses, which can modify their characteristics prior to consumption. Extent to which quality and shelf life of produce is actually affected will also depend on factors, such as cultivar, variety and stage of maturity.

7.2.1 Postharvest Physiology of Intact Fruits and Vegetables

7.2.1.1 Ethylene Production

Ethylene is a hormone produced when vegetable or fruit undergoes stress. Ethylene triggers ripening and senescence, and is partially responsible for changes in flavor, color and texture of fruits and vegetables (Kays and Paull 2004). On the other hand, removal of ethylene will slow ripening and senescence. Controlled atmosphere (CA) or modified atmosphere (MA) storage of produce can reduce ethylene production/action, preserving quality of fruits and vegetables for longer periods.

7.2.1.2 Water Loss

One important factor, which is necessary to preserve quality of fruits and vegetables, is avoidance of water loss. Moisture is lost during transpiration when water is converted from liquid to gas. Before harvest, the fruit or vegetable preserves the amount of water stored inside by replacing or taking back water lost (due to transpiration) through roots. When produce is harvested, it loses its source of water, so recuperation from water loss is not possible. When water is lost, turgor of produce decreases, as does firmness. Water stress also causes metabolic alterations changes in enzyme activation causing accelerated senescence, reduced flavor and aroma, decline in nutritional value, and increased susceptibility to chilling injury and pathogen invasion. Maximum acceptable water loss allowed before the commodity is considered unmarketable varies; apples and oranges can tolerate 5% water loss, while celery and cabbage can tolerate 10% (Kays and Paull 2004).

7.2.1.3 Respiration

Enzymatic reactions occur in fruits and vegetables during respiration where oxygen is consumed, carbon dioxide is produced, and heat and energy are released. Respiration is required to keep produce alive and to support any developmental changes. Respiration rate itself is influenced by temperature, humidity and gas composition. When oxygen composition is lowered, respiration decreases and hence, senescence does as well. However when oxygen concentration falls below threshold level anaerobic respiration occurs, which in turn causes ethanol production, off-flavor formation and loss of harvested produce (Kays and Paull 2004).

7.2.2 Postharvest Physiology of Minimally Processed Fruits and Vegetables

Problems mentioned above increase with partial or total loss of skin in minimally processed fruits and vegetables. Skin protects produce against water loss and pathogen invasion and provides partial barrier to gases. Slicing, chopping, and peeling, etc., of fruits and vegetables can cause injury not only to cells immediately exposed by the action, but also to unexposed cells deep within the tissue, increasing extent of damage (Saltveit 1997).

7.2.2.1 Ethylene Production

Generally, there is noticeable increase in ethylene production when produce tissue is wounded, inducing deep physiological changes such as ripening in climacteric fruits (Abeles et al. 1992). For example, in green tissues, ethylene triggers chlorophyll degradation, whereas in carrots, high concentration of ethylene triggers increase in phenolics and isocumarin, giving a bitter taste to the vegetable (Poenicke et al. 1977; Sarkar and Phan 1979). Rate of ethylene production experienced in minimally-processed fruits and vegetables will depend on produce type and maturity stage at which it is cut (Toivonen and DeEll 2002). For instance, in fresh kiwifruit, tomatoes and strawberries, all three undergo increase in ethylene production when cut (Watada et al. 1990), whereas pears do not (Gorny et al. 2000; Rosen and Kader 1989).

7.2.2.2 Water Loss

Cutting, slicing, and peeling, etc., of fruits and vegetables, will increase water transpiration rate due to exposure of tissue following removal of natural epidermal layer, and will also increase surface area of exposure (Toivonen and DeEll 2002). Water loss through transpiration cannot be replaced and is problematic, since loss of small amounts of water will severely impact produce quality. With water loss, turgor of produce drops, as well as turgidity and firmness. As mentioned, water stress also causes metabolic alterations, changes in enzyme activation, causing accelerated rate of senescence, reduced flavor and aroma, decreased nutritional value, and increased susceptibility to chilling injury and pathogen invasion (Sams 1999).

7.2.2.3 Respiration

Wounding of fruits and vegetables also causes increase in respiration, thereby increasing production of carbon dioxide and consumption of oxygen, further causing decrease in stored reserves. This increase will also depend on commodity type, storage temperature and degree of wounding (Ahvenainen 1996). For instance, slicing a mature green tomato can increase respiration up to 40% at 8°C, compared with intact fruit (Mencarelli et al. 1989). However minimally-processed apples do not show increased respiration (Lakakul et al. 1999). When minimally-processed produce is packaged under modified atmosphere with very low oxygen concentration, anaerobic respiration will occur, with consequent formation of off-flavors, ethanol, ketoncs and aldehydes. Increase in respiration depends on the commodity, as some fruits and vegetables do not show increase in these compounds (e.g., bananas), while others such as tomatoes do (Toivonen and DeEll 2002; Mencarelli et al. 1989).

7.2.2.4 Browning

Enzymatic browning is one of the main problems encountered in maintaining quality of fresh-cut fruits and vegetables. When a fruit or vegetable is cut, reaction occurs that is catalyzed by the enzyme polyphenol oxidase, wherein phenolic compounds react with atmosphere oxygen in the presence of copper, causing formation of dark compounds (Ahvenainen 1996). Enzymatic browning is the main problem occurring in fruits such as apples, pears, bananas and peaches, as well as in vegetables such as potatoes and lettuce (Sapers 1993).

7.2.2.5 Microbial Contamination

When fruit or vegetable tissue is wounded, surface moisture and cytoplasmic remains of ruptured cells with high concentrations of sugars and proteins will be present, which are ideal conditions for microbial growth (Saltveit 1997). This microbial growth could cause food borne illnesses and food spoilage (Nguyen and Carlin 1994). Minimally-processed fruits and vegetables are susceptible to invasion by microorganisms, such as *Pseudomonas spp., Erwinia herbicola, Flavobacterium, Xanthomona, Eterobacter Agglomerans, Leuconostoc mesenteroides, Lactobacillus spp.*, molds, and yeasts (Zagory 1999). It is important to be aware that a modified atmosphere can cause growth of and toxin production by *C. Botulinum* on fruits with pH higher than 4.8, so provision should be made to coat tropical fruits in particular to avoid *C. Botulinum* growth (Olivas and Barbosa-Cánovas 2005).

7.3 History of Edible Films and Coatings for Fruits and Vegetables

Edible films and coatings have been extensively studied for the last 20 years, and much has been published on their use for fresh and minimally-processed fruits and vegetables (Tables 7.1 and 7.2). Reviews found in literature on the topic of edible coatings for whole fruits and vegetables (Platenius 1939; Claypool 1940; Hardenburg 1967; Curtis 1988; Park and Chinnan 1990; Hagenmaier and Shaw 1992; Banks et al. 1993; Baldwin 1994; Park 1999) and minimally-processed fruits and vegetables (Wong et al. 1994; Baldwin et al. 1995a, b; Nisperos and Baldwin 1996; Olivas and Barbosa-Cánovas 2005; Lin and Zhao 2007) have addressed the issue indicating different perspectives.

Table 7.1 Edible coatings on minimally processed fruits and vegetables

Fruits	Coatings and concentrations	Additives and concentrations	Variables evaluated	Results	References
Apple slices	Apple puree–alginate	CaCl$_2$, N-acetylcysteine, oregano oil, lemon grass oil, vanillin	C, EtOH, Et, F, ST, MG	Coatings extended the shelf life of fresh-cut apple	Rojas-Graü et al. (2007)
Apple and papaya cylinders	Alginate or gellan	Glycerol, CaCl$_2$, B. lactis, sunflower oil	WVP, WS, B. lactis	Gellan coatings avoided WL better than alginate coatings. Alginate- and gellan-based edible coatings carried successfully viable probiotics on papaya and apple cylinders	Tapia et al. (2007)
Apple slices (Gala)	CaCl$_2$ + Alginate or alginate/acetylated monoglyceride–linoleic acid or alginate butter–linoleic acid	Potassium sorbate, alpha-monostearate	WL, C, F, MG, TA, SS, FV	All coatings preserved F, and C. WL was delayed, mainly with Alg–monoglyceride–linoleic acid coating. Higher amounts of Butanol, hexanal, 1-hexanol, trans-2-hexenal on coated samples at end of storage.	Olivas et al. (2007)
Mango slices	Chitosan	–	ST, WL, C	The coating retarded WL and drop in sensory quality, increasing the SS content, TA and AA content and inhibiting MG.	Chien et al. (2007b)
Shelled Pecans	CMC	PG, Sorbitol, lecithin, α-tocopherol, BHA, BHT, citric acid	ST, R, C	Imparted shine, F was not preserved, R was delayed.	Baldwin and Wood (2006)
Apple slices	WPC + Beeswax	AA, cysteine Cys, or 4-hexylresorcinol	WL, C, BI, ST	WPC-AA and WPC-BW coatings reduced B. WL was not prevented by coatings use.	Pérez-Gago et al. (2006)
Apple slices	WPI, WPC or HMPC + Beeswax or carnauba	Glycerol, stearic acid	WL, BI	WL was not avoided. WPI and WPC were more efficient at preserving C compared with HMPC.	Pérez-Gago et al. (2005a)

Persimmon pieces	WPC–Beeswax or WPC–Beeswax + AA	Glycerol	WL, F, C,	WPC–BW–Ascorbate coatings reduced B, WL, and F loss in fresh-cut persimmons	Pérez-Gago et al. (2005a)
Apple slices	WPI-beeswax	Glycerol, stearic acid	WL, C, ST	Lower BI than uncoated apples. WL was not reduced by use of coating	Pérez-Gago et al. (2003b)
Pear wedges	Methylcellulose (MC) and MC–stearic acid (SA)	AA, potassium sorbate, $CaCl_2$	WL, F, MG, C, TA, SS, FV	Coating retarded B. WL was delayed on pears coated with MC–SA. Pear wedges coated with MC–SA formulation contained higher amounts of hexyl acetate	Olivas et al. (2003)
Baby carrots	Xanthan gum	Calcium lactate calcium gluconate, vitamin E	C, ST, Vitamin E, β-carotene, Calcium	Calcium and vitamin E increased on coated carrots. Edible coatings improved C	Mei et al. (2002)

WPI whey protein isolate; *F* firmness; *WL* weight loss; *TA* titratable acidity; *SS* soluble solids; *RR* respiration rate; *AA* ascorbic acid; *C* color; *MG* microbial growth; *IGC* internal gas composition; *FV* flavor volatiles; *DS* decay symptoms; *ST* sensory test; *EtOH* ethanol; *G* gloss; *Wh* whitening; *S* sugars; *Et* ethylene; *BI* browning index; *B* browning; *PG* propylene glycol; *BHA* butylated hydroxyanisole; *BHT* butylated hydroxytoluene; *R* rancidity; *WS* water solubility

Table 7.2 Edible coatings on intact fruits and vegetables

Fruits	Coatings	Additives	Variables evaluated	Results	References
Mandarin	Low molecular weight chitosan (LMWC) or High molecular weight chitosan (HMWC)	–	WL, F, TA, AA, Fungal decay	LMWC coating improved firmness, titratable acidity, ascorbic acidity, and water content. Mandarin coated with LMWC exhibited greater antifungal resistance than TBZ	Chien et al. (2007a)
Strawberry	Starch	Sorbitol	C, F, WL, SS, MG C, F, WL, SS, MG C, F, WL, SS, MG	Opaque coating	Ribeiro et al. (2007)
	Carrageenan	Glycerol + CaCl$_2$ Tween 80		Preserved better firmness	
	Chitosan	Tween 80 + CaCl$_2$		Delayed WL Delayed water loss	
Cherry	Aloe vera gel and water 1:3	–	RR, WL, C, F, MG	Delay in WL, F loss	Martinez-Romero et al., (2006)
Strawberry	Chitosan	Tween 80, oleic acid	SSD, WVR, TA, pH, SS, anthocyanin, Fungal decay, RR, MP, ST	Anthocyanin content decreased throughout storage in coated samples. Coatings had no significant effects on TA, pH and SS content. Coated fruits had decrease in aroma and flavor, especially when the ratio oleic acid:chitosan was high	Vargas et al. (2006)
Strawberry	Calcium gluconate or chitosan or chitosan + Calcium gluconate		Fungal decay, WL, F, C, pH, TA, SS	Coatings decreased surface damage and delayed fungal decay. Chitosan coatings retained F and delayed changes in external C	Hernandez-Munoz et al. (2006)
Cherry	LBG + Beeswax or shellac	Polysorbate 80, glycerol, NH3	F, C, WL, ST, EtOH	C decay was not prevented by the coating. WL and F loss was delayed	Rojas-Argudo et al. 2005
Strawberry	Cactus–mucilage	Glycerol	F, C, ST	Coating increased shelf life of strawberry	Del-Valle et al. (2005)

Guava	Dextrans	Glycerol	FV, C	Coatings showed poor water retention on fruits stored at room temperature. Best efficiency of dextrans was at 4°C, by increasing guava quality (size, color, aroma, water content)	Quezada-Gallo et al. (2005)
Grapes	Aloe vera gel	—	WL, C, F, SS, TA, ST, RR, MG, Et	A. vera gel delayed postharvest quality losses, extending storability from 7 days at 1°C to 35 days. Coating reduced MG	Valverde et al. (2005)
Apricots and green peppers	MC + Stearic acid	PEG, PEG + CA, PEG + AA	WL, Vitamin C	Coatings delayed WL of fresh apricots and green peppers. Coatings containing AA or CA lowered vitamin C loss	Ayrancy and Tunc (2004)
Cherry	Sodium caseinate or milk protein concentrate	Glycerol and Bees wax or stearic–palmitic acid blend	SS, TA, pH, WL	Coatings reduced WL of the fruits. Edible coatings had a beneficial effect on sensory quality of cherries	Certel et al. (2004)
Cherry	Semperfresh™ 10 and 20 g/L	—	F, WL, TA, SS, SC, AA, C	Effective in reducing WL, and preserving F, AA, TA and C. Shelf life increased by 21% at 30°C and 26% at 0°	Yaman and Bayo (2002)
Apple	Shellac, candelilla, shellac–carnauba			Shellac caused anaerobic respiration on Braeburn and Granny Smith apples. This research recommends shellac for "Delicious," carnauba–shellac for "Braeburn" and "Fuji" apples	Bai et al. (2003a)
Apple	Zein	Propylene glycol	F, G, Wh, TA, FV, S, IGC, Et, ST	Good overall apple quality comparable to a commercial shellac coating	Bai et al. (2003b)
Plum	HPMC + Beeswax or shellac	Stearic acid, glycerol	WL, F, Et, EtOH	WL of coated plums decreased as lipid content increased. F was not affected by coating after short-term storage at 20°C. In prolonged storage, coatings significantly reduced F loss and internal breakdown	Pérez-Gago et al. (2003a)

(continued)

Table7.2 (continued)

Fruits	Coatings	Additives	Variables evaluated	Results	References
Grapes	Chitosan	NaOH to modify pH to 5.6.	Inhibition of *Botrytis cinerea*	They observed that inhibitory effect originated from combination of coating and antifungal effects and stimulation of defense responses	Romanazzi et al. (2002)
Pear	Carnauba wax	–		Decrease in pear friction discoloration and internal O_2 concentration	Amarante et al. (2001)
Pear	Carnauba wax	–		Modification of O_2 instead of CO_2 was the means of delaying ripening	Amarante et al. (2001)
Kiwifruit	Soy protein + pullulan + stearic acid	Glycerol	Peroxidase, pectin, TA, RR, Et	Coating retarded softening of kiwifruit	Xu et al. (2001)
Mango	Cellulose. Nature Seal® 2020 hydroxypropyl-cellulose Carnauba. Tropical Fruit Coating 213 (TFC)	Preservative, acidulant and emulsifier Fatty acid soaps	IGC, WL, FV, SS, TA, DS, ST, F	Both coatings created modified atmospheres, reduced decay, and improved appearance. Polysaccharide coating delayed ripening and increased concentrations of flavor volatiles. Carnauba coating reduced WL	Baldwin et al. (1999)

F firmness; *WL* weight loss; *TA* titratable acidity; *SS* soluble solids; *RR* respiration rate; *CA* citric acid; *AA* ascorbic acid; *C* color; *MG* microbial growth; *IGC* internal gas composition; *FV* flavor volatiles; *DS* decay symptoms; *ST* sensory test; *EtOH* ethanol; *G* gloss; *Wh* whitening; *S* sugars; *Et* ethylene; *LGB* Locust bean gum; *SSD* surface solid density; *WVR* water vapor resistance; *MP* mechanical properties; *TBZ* thiabendazole

Application of edible coatings to preserve quality of fruits and vegetables is not new. Hardenburg (1967) comments that their use dates back to the XII century, where oranges and lemons were coated with wax in China to preserve the fruits longer. In the U.S., the first patent on use of edible coatings dates back to 1916, where a method preserving whole fruits with molten wax was patented by Hoffman (Hoffman 1916). In 1972, Bryan patented a method to preserve grapefruit halves with edible coating composed of low methoxyl pectin and locust bean gum dispersed in grapefruit juice.

7.4   Technical Development of Edible Films and Coatings

The main purpose of employing edible films and coatings is to provide a semi-permeable barrier against gases and vapor; however, there are other purposes (e.g., to carry additives such as texture enhancers, antimicrobials, antioxidants, etc.; Table 7.3). In general, edible films and coatings represent an interesting option for intact and minimally-processed fruits and vegetables, since a barrier can be formed that protects the produce and decreases rate of physiological postharvest deterioration.

Successful application of edible films or coatings as barriers for fruits and vegetables mainly depends on developing film or coating that can provide appropriate internal gas composition for a specific fruit/vegetable (Park 1999). Some factors/questions need to be addressed when developing a film or coating: (1) how properties of coating solution will affect the product, (2) how coating will change with time, (3) how coating will interact with the product that might lead to flavor/off-flavor generation, and (4) how storage conditions will affect coating. Other considerations include importance of thickness, color, and flavor of coating, since these parameters can change final quality of the coated product.

Thus, to develop the appropriate film or coating, full understanding of the physiology of the fruit/vegetable being coated is needed, as well as a clear idea of the function of the film or coating relative to the product. Different purposes of films and coatings developed for fruits and vegetables are more fully detailed in the next section.

Table 7.3 List of additives incorporated into edible films and coatings for fruits and vegetables

Additives	Examples
Antibrowning compounds	Ascorbic acid
Antimicrobial agents	Potassium sorbate
Texture enhancers	Calcium chloride
Nutrients	Vitamin (E)
Aroma precursors	Linoleic acid
Probiotics	*B. lactis*
Flavors and colorants	Apple puree

7.4.1 Edible Films and Coatings for Preservation of Fruits and Vegetables

7.4.1.1 Water Loss

Edible films/coatings decrease water vapor transmission rate by forming a barrier on the fruit or vegetable surface. This barrier prevents texture decay, since water is essential for preservation of cell turgor (Garcia and Barret 2002). Metabolic alterations that can cause accelerated rate of senescence due to water loss can also be avoided with their use. Ability of films and coatings to function as barriers to water vapor relies on external conditions, which include (1) temperature and relative humidity, (2) characteristics of commodity such as type of product, variety, maturity and water activity, and (3) characteristics of film such as composition, concentration of solids, viscosity, chemical structure, polymer morphology, degree of cross-linking, solvents used in casting the film, and type of plasticizer used (Olivas and Barbosa-Cánovas 2005). In case of minimally-processed fruits, there is usually very high water activity present at the surface, making it difficult to develop a coating that delays water loss, since capacity of films to work as barriers to water vapor decreases as water activity increases in the commodity (or relative humidity of the environment) (Hagenmaier and Shaw 1990). However a number of coatings have been developed that can delay water loss in some minimally processed fruits and vegetables (Table 7.1).

7.4.1.2 Texture

When coating or film is applied to a commodity, a modified atmosphere is developed. Reduction of internal oxygen and increase of carbon dioxide in the commodity will in some cases delay softening (Kader 1986). Intact apples retained firmness when stored under modified atmosphere (Knee 1980). Edible coatings can also preserve texture of fruits and vegetables by acting as a partial barrier to water and serving as carriers of texture enhancers. Loss in texture has been delayed with application of edible coating to whole avocados (Maftoonazad et al. 2007). Olivas et al. (2007); likewise preserved texture of cut-apples when coated with alginate films containing calcium chloride.

7.4.1.3 Respiration

Edible films or coatings can reduce respiration and, hence increase shelf life of a commodity. In selection of a coating, several considerations should be addressed to avoid extremely low oxygen concentration inside the commodity. Low oxygen concentration in the product could lead to anaerobic respiration, which can result in deterioration of product due to production of off-flavors and accelerated senescence (Kays and Paull 2004). Some studies (Tables 7.1 and 7.2) show how edible films/coatings have favorably delayed respiration, and in some cases caused anaerobic respiration.

7.4.1.4   Ethylene

A well selected edible coating will produce a modified atmosphere inside the fruit, reducing levels of internal oxygen. If oxygen concentration inside the commodity drops below 8%, there will be a decrease in ethylene production (Kader 1986), and the commodity's quality will be preserved longer.

7.4.1.5   Color

One of the most important attributes of fruits and vegetables is color (Kays 1999). For some minimally-processed fruits and vegetables, browning is a big problem that can be controlled by use of films or coatings as carriers of anti-browning agents. The most common antioxidant used on fruits and vegetables is ascorbic acid; however other compounds have been successfully used such as cysteine, 4-hexylresorcinol, citric acid and calcium chloride (Olivas et al. 2007; Pérez-Gago et al. 2006). Baldwin et al. (1996) coated apple slices and potato cores with Nature Seal™ and found that ascorbic acid delayed browning more effectively when applied in edible coating than in aqueous solution. Depending on coating formulation, appearance of fruit can be affected positively or negatively by selected coating. For instance, candelilla wax gives a natural, noncoated appearance to apples, whereas shellac or carnauba coating (when in contact with water) will give whitish color to apples (Bai et al. 2003a).

7.4.1.6   Flavor

Perhaps the most important attribute of fruits and vegetables is flavor. Consumers may buy the commodity based on its appearance when first purchased, but if the flavor is not acceptable, they will avoid buying the product a second time. Flavor can be preserved or modified with edible films or coatings by two different means: (1) as a barrier to aroma volatiles, and (2) as a carrier of flavors. Baldwin et al. (1999), for instance, found that a polysaccharide coating worked as a barrier to aroma volatiles on whole mango. Edible coatings can also modify internal atmosphere of the commodity, causing low oxygen and high carbon dioxide concentration.This is not beneficial to flavor, since it could lead to a decrease in production of characteristic flavor compounds (Ke et al. 1994; Fellman et al. 2003). Some works have even suggested the possibility that edible coatings on cut fruits can supply fruits with volatile precursors (Olivas et al. 2003, 2007). Pear wedges coated with a methylcellulose–stearic acid formulation contained higher amounts of hexyl acetate throughout storage, probably due to synthesis in wounded tissue from the stearic acid contained in the coating (Olivas et al. 2003). Higher production of hexanol was observed in apples coated with alginate–linoleic acid. According to Paillard (1979), hexanol can be produced from fatty acids such as linoleic acid.

7.4.1.7 Microbial Contamination

In case of minimally-processed fruits and vegetables, where natural protection (skin) has been eliminated, opportunity for microorganisms to invade and grow on the surface of the fruit is present. Incorporating antimicrobial compounds into edible films or coatings will preserve quality of fresh-cut fruits and vegetables. Since antimicrobials are needed just on the surface of the product, their application as part of a coating will help minimize antimicrobial usage (Vojdani and Torres 1990). Retention of antimicrobial on coated produce surface will depend on coating attributes (composition, hydrophilic characteristics and manufacturing procedure) and commodity type (pH and water activity), as well as storage conditions (temperature and duration) (Cagri et al. 2004). Antimicrobials most commonly used include potassium sorbate, sodium benzoate, sorbic acid, benzoic acid and propionic acid; other natural antimicrobials such as lemongrass, oregano oil and vanillin have also been used (Rojas-Graü et al. 2007).

7.4.1.8 Nutritional Quality

Edible films and coatings can affect nutritional quality of fruits and vegetables. On one hand, they can be used as carriers of nutrients. On the other hand, they can produce abiotic stress, which could modify metabolism of the commodity, affecting production of nutrients. Some works have determined effects of coatings on nutritional quality, and on phenolics and other phytochemicals. Viña et al. (2007) coated Brussels sprouts with starch and analyzed effect of coating on amount of ascorbic acid and total flavonoids retained in the vegetable after storage. Han et al. (2004) found higher amounts of vitamin E and calcium on strawberries coated with chitosan, containing calcium and vitamin E in the formulation, due to diffusion of these nutrients into the fruit. Romanazi et al. (2002) observed an increase in phenylalanine ammonia–lyase activity, a key enzyme for synthesis of phenolic compounds, on grapes coated with chitosan. Edible coatings have also been used as carriers of probiotics. *Bacillus lactis* was maintained for 10 days on fresh-cut fruits under refrigeration, when applied on alginate- and gellan-based edible coatings (Tapia et al. 2007).

7.5 Properties of Edible Films and Coatings

7.5.1 Edible Films and Coatings as Partial Barriers to Water Vapor and Gases

Edible coating or film mainly works as a partial barrier to water vapor and gases by decreasing transmission rate of a given partial pressure difference between internal and external atmospheres. This partial barrier is conducive to a modified internal atmosphere low in oxygen and high in carbon dioxide, suppresses respiration rate and reduces transpiration losses. Scientific literature provides vast amounts of information on barrier properties of edible films; however, comparisons between different films are sometimes difficult or impossible due to use of different types of equipment and dissimilar test conditions during measurements (Tuil et al. 2000). Choosing a proper coating is complex, because it depends on specific respiration and transpiration rates of the commodity and outside environmental conditions. When applying a coating to a fruit, whole or cut, permeability to gases is modified due to the barrier formed by coating. This could lead to anaerobic respiration.

Respiration and ripening of fruits and vegetables involve gas exchange with the environment. Carbon dioxide, oxygen, water and other metabolic byproducts, such as ethylene and other volatile compounds, are main substances exchanged during storage. Surface coatings modify gas exchange rate between environment and fresh produce, and thus control transpiration, respiration and other metabolic processes that lead to loss of quality. The mechanics and main factors involved in mass transfer processes through fruit skin and coated fruits follow.

Fruits and vegetables exchange gases with their storage environment through a phenomenon known as gas diffusion. Gas diffusion is a passive transport phenomenon in which Gibbs free energy is minimized through mass transfer from a region with high concentration of a given chemical species to a region with lower concentration of the same chemical species. During this spontaneous process, the interface between the two mentioned regions (natural skin or an artificial coating) may present some opposition to the mass transfer process, affecting mass transfer rate. Unless the interface in a system completely hinders mass transfer, diffusion carries on until equilibrium is reached (i.e., when a concentration gradient ceases to exist). Interphases causing extremely high opposition to a specific compound are referred to as "impermeable". Unidimensional steady state mass transfer, as the one witnessed in fresh produce stored under specific conditions, may be mathematically described by Fick's first law of diffusion, which states that mass flux per unit area of a given chemical species (J) is proportional to its concentration gradient ($\partial \phi / \partial x$).

$$J = D \frac{\partial \phi}{\partial x} \qquad (7.1)$$

Magnitude of the proportionality constant (D), known as diffusivity, theoretically depends on temperature (T), viscosity (η) of medium in which diffusion is taking place, and on size of diffusing particles (r), according to the Stokes–Einstein relation,

$$D = \frac{k_B T}{6\pi \eta r} \tag{7.2}$$

where k_B is the Boltzmann constant $(1.380 \times 10^{-23} \text{J/K})$. In practice, however D values for specific situations may depart from this theoretical value due to structural peculiarities or chemical affinity effects, making empirical correction necessary.

In order to have a mathematical expression to model and describe gas diffusion (normal to the surface) through biological membranes, Eq. (7.1) may be solved by separating the variables and integrating from point x_1, where concentration c_1 of a given chemical species is found to be point x_2, and where concentration c_2 of the same chemical species exists, yielding:

$$J = D \frac{(c_1 - c_2)}{(x_2 - x_1)} \tag{7.3}$$

Typically $(x_2 - x_1)$ represents thickness of the membrane, and since the chemical species being diffused through the membrane is a gas, Eq. (7.3) may be substituted by Eq. (7.4):

$$J = DS \frac{\rho_1 - \rho_2}{\text{thickness}} \tag{7.4}$$

where gas concentrations and concentration gradient are expressed as a function of the partial pressure of diffusing gas on both sides of the membrane $(\rho_1$ and $\rho_2)$, using Henry's law:

$$c = S\rho \tag{7.5}$$

where concentration (c) of a gas is proportional to its partial pressure (ρ), depending on a proportionality constant known as solubility (S). Diffusivity multiplied by solubility of a specific gas through a specific membrane is known as permeability coefficient P, changing Eq. (7.4) to its final form as:

$$J = P \frac{\rho_1 - \rho_2}{\text{thickness}} \tag{7.6}$$

7.5.2 Edible Films and Coatings for Active Packaging

Besides working as partial barriers to vapor and gases, edible films and coatings can serve as carriers of ingredients to help preserve quality and improve nutritional value of fruits and vegetables. When coating is used for more than providing a barrier to external conditions, it is called active coating (Rooney 1995). Antimicrobial agents, texture enhancers and anti-browning compounds can be added to preserve quality of a commodity longer. Incorporation of some nutrients to the coating can increase nutritional value of the commodity. Aroma precursors can also be incorporated into a coating to increase flavor production (Olivas et al. 2003).

7.5.3 Factors Affecting the Properties of Films and Coatings

Properties of films (or coatings) are affected by characteristics of the film, commodity, and environment, such as coating solution composition, viscosity of coating solution, coating thickness, type of product, plant variety, fruit/vegetable maturity, previous fruit treatment, fruit surface coverage (how coating adheres to surface of fruit) and external atmosphere (Hagenmaier and Shaw 1992; Banks et al. 1993; Cisneros-Zevallos and Krochta 2002). A clear understanding of how these factors affect performance of a film will aid in developing a proper coating for a specific product. Some of these factors are discussed next.

7.5.3.1   Coating Composition

Composition of film or coating affects its permeability. Chemical structure, concentration, chemical nature, physical state, method of preparation, crystallinity, polarity, orientation, presence of additives and use of compound blends will all affect performance of a film (Miller and Krochta 1997; Morillon et al. 2002). A coating containing hydrophilic compounds is not a very good barrier to water vapor, while hydrophobic substances perform well as barriers to moisture migration, since lipid-based edible coatings have low affinity for water due to their apolar nature (Morillon et al. 2002). Length of the hydrocarbon chain affects WVP (water vapor permeability) of lipid compounds. Higher the carbon number, lower the WVP, as the apolar part of the molecule generally increases with increasing number of carbons (McHugh and Krochta 1994). However, if number of carbons is higher than 18, this behavior reverses, presumably due to heterogeneous nature of coating produced (Koelsch and Labuza 1992). Lipids with higher affinity to water when used in coatings will have greater permeability to water vapor. Factors such as lipid concentration, physical state, degree of nonsaturation and chemical structure also affect WVP of lipid coatings. Morillon et al. addressed these important issues in a review published in 2002.

7.5.3.2   Coating Thickness

Cisneros-Zevallos and Krochta (2003b) published a work in which physical principles of dip-plate coatings were applied to fruit coatings. They demonstrated that internal modified atmosphere of a coated commodity depends on coating thickness, while at the same time coating thickness depends on viscosity, concentration, density and drainage time of coating solution. Debeaufort et al. (1993) found that water vapor transfer rate and permeability decreased when thickness of a triglyceride coating increased from 0 to 60 μm.

7.5.3.3   Aging of Films

It is important to bear in mind that a coating will change over time along with its properties. Edible coatings, depending on their formulation, are subject to chemical and physical changes that can cause degradation of polymer chains, migration of low molecular weight additives, water absorption, etc. If there is migration of plasticizer, mechanical properties of film will be affected (Anker et al. 2001).

7.5.3.4   Type of Commodity

Quality of coated fruits can vary greatly, since each fruit, even those of the same variety, is different in skin resistance, gas diffusion, fruit respiration rate, etc. Also, the effect of individual applications of coatings will vary between fruits, affecting coating thickness and proportion of pores blocked by coating material (Banks et al. 1993). Coatings developed for one variety of fruit may not be appropriate for another. Bai et al. (2003a) studied coatings for apples and found that shellac is the best coating for Delicious apples; however, this coating caused anaerobiosis on Braeburn and Granny Smith apples.

7.5.3.5   Fruit Surface Coverage

According to Banks et al. (1993), fruit coatings work as barriers to gases by blocking pores on the fruit's surface. In addition, they found that coating of fruit cuticle can affect water vapor diffusion.

7.5.3.6   Storage Conditions

Film permeability depends on relative humidity (RH) and gas composition of the environment, which can also affect properties of coating. In case of cut fruits, water activity of the coated fruit is also affected, as shown by Cisneros-Zevallos and Krochta (2002). When cut fruit was coated with a hydrophilic film, they found that the surrounding environment's RH greatly affected barrier properties of the film. Temperature also affects properties of edible films or coatings. Kester and Fenema (1989) studied WVP of films at temperatures varying from 15 to 40°C, and observed that WVP decreases when temperature decreases.

7.6   Composition of Edible Films and Coatings

Edible films and coatings are composed of hydrocolloids, which consist of either polysaccharides or proteins, or hydrophobic compounds (e.g., lipids or waxes). Edible films may also be composed of a mixture of hydrocolloids and hydrophobic compounds (composite films or coatings). Figure 7.1 shows the most common compounds used in edible films and coatings.

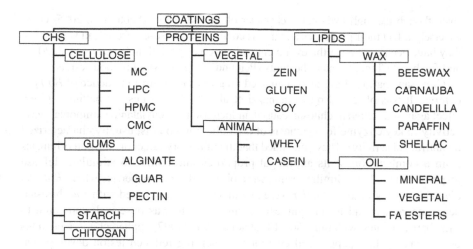

Fig. 7.1 Main components of edible films and coatings for intact and minimally processed fruits and vegetables (CHS carbohydrates; FA fatty acids; MC methylcellulose; HPMC hydroxypropyl-methylcellulose; HPC hydroxypropylcellulose, CMC carboxymethylcellulose)

7.6.1   Carbohydrates

7.6.1.1   Chitosan

The compound chitosan comes from marine invertebrates and is a derivate of chitin, which after cellulose, is the most abundant polysaccharide resource on earth (Tuil et al. 2000). Chitosan is a high-molecular-weight cationic polysaccharide composed of (1 → 4)-linked 2-acetamido-2-deoxy-β-D-glucopyranosyl and 2-amino-2-deoxy-β-D-glucopyranosyl units (Sebti et al. 2005), and produces transparent films. Chitosan is not water-soluble, so a coating solution comprised of weak organic acid (acetic acid) must be used. Chitosan has been shown to be a natural food preservative, though the antimicrobial mechanism involved is not well elucidated. It is believed that positively charged chitosan molecules interact with negatively charged microbial cell membranes, causing change in microbial cell permeability that leads to leakage of cell constituents (No et al. 2007).

Chitosan films or coatings can increase shelf life and preserve quality of fruits and vegetables by decreasing respiration rates, inhibiting microbial development and delaying ripening. They have been used on fruits and vegetables with good results, showing antimicrobial activity against *Bacillus cereus*, *Brochothrix thermosphacta*, *Lactobacillus curvatus*, *Lactobacillus sakey*, *Listeria monocytogenes*, *Pediococcus acidilactici*, *Photobacterium phosphoreum*, *Pseudomona fluorescens*, *Candida lambica*, *Cryptococcus humiculus*, and *Botrytis cinerea* (Devlieghere et al. 2004; Romanazzi et al. 2002). Chitosan is considered ideal coating for fruits and vegetables, mainly because it can form a good film on the commodity's surface and can control microbial growth (Muzzarelli 1986; No et al. 2007).

Even though the high hydrophilic character of chitosan films (Sebti et al. 2005) could adversely affect their properties and use as coatings for fresh-cut fruits and vegetables, they have been used nevertheless on cut fruits such as peeled litchi (Dong et al. 2004) and fresh-cut Chinese water chestnuts (Pen and Jiang 2003) with good results.

Chitosan coatings have also been used on grapes to control incidence of *Botrytis cinerea*. It was observed by Romanazzi et al. (2002) that besides having a direct effect against *B. cinerea*, chitosan caused an increase in phenylalanine ammonia–lyase activity (a key enzyme for synthesis of phenolic compounds, usually characterized by antifungal activity). They concluded that the inhibitory effect of chitosan originates from a combination of its antifungal properties and ability to stimulate defense responses in the host. Similar results were observed for strawberries where *Botrytis cinerea* and *Rhyzopus stolonifer* were inoculated and then coated with 1% chitosan; decay decreased and lower synthesis of anthocyanin was observed, compared to strawberries treated with fungicide. El-Ghaouth et al. (1997) applied chitosan coatings on strawberry, bell pepper and cucumber, observing reduced lesion development due to *Botrytis cinerea* and *Rhizopus stolonifer*, as well as delayed ripening. Others found (El-Ghaouth et al. 1992) that chitosan-coated tomatoes showed reduction in respiration rate and ethylene production, were firmer with less decay, and had less red pigmentation, compared to noncoated tomatoes. Litchi has also been coated with chitosan, where there was delay in reduction of anthocyanin content and increase of PPO activity (Zhang and Quantick 1997; Jiang et al. 2005).

Composite coatings containing chitosan are likewise a good option for fruits and vegetables. Chitosan coatings containing oleic acid had good water retention properties on coated strawberries (Vargas et al. 2006), while apples coated with *N,O*-carboxymethyl chitosan films and placed in cold storage could be maintained in fresh condition for more than 6 months (Davies et al. 1989).

Chitosan has been approved for use as a food additive in Japan and Korea. In the U.S., Generally Recognized as Safe (GRAS) notices were submitted (by a "notifier") to the FDA in 2001 and 2005 to evaluate chitosan (from shrimp) and its suitability for use; however (at notifier's request) the FDA ceased evaluation of both notices (No et al. 2007; US FDA/CFSAN 2006).

7.6.1.2 Starch

Starch is an abundant, inexpensive polysaccharide obtained from cereals, legumes, and tubers, and is extensively utilized in the food industry. Starch is a polysaccharide consisting of α–$(1 \rightarrow 4)$ D-glucopyranosyl units with D-glucopyranosyl chains linked by α–$(1 \rightarrow 6)$ glycosidic bonds. It is composed of two macromolecules, amylose, essentially linear, and amylopectin, highly branched. Ratio of amylose and amylopectin varies with source. Regular or standard wheat, corn, and potato starches generally contain lower amounts of amylose than those of most legumes (Hoover and Sosulsky 1991; BeMiller and Whistler 1996). Films formed with starch are often very brittle and have poor mechanical properties (Peressini et al. 2003). These films are formed by drying of a gelatinized dispersion, as hydrogen bonds form between hydroxyl groups (Lourdin et al. 1995). Since these interactions are weak, mechanical properties of starch-based films are of poor quality. To overcome this problem, it is necessary to blend starch with other compounds or chemically modify the starch (Tuil et al. 2000).

Proportion of amylose and amylopectin within starch will affect film-forming properties. A higher proportion of amylose will improve mechanical properties of the film (Han et al. 2006), as the branch-on-branch structure of amylopectin interferes with intermolecular association and disrupts film formation (Peressini et al. 2003). Rice and pea starch films (30 and 40% amylose, respectively, by weight) containing glycerol as plasticizer (starch:plasticizer ratio of 2:1) were found to have low oxygen permeability and good elastic properties, with pea films showing highest elasticity and WVP (Mehyar and Han 2004). Han et al. (2006) reported higher oxygen permeability and lower elongation (one order of magnitude) for pea starch films (~40% amylose) with glycerol as plasticizer (starch:plasticizer ratio 3:2). Nevertheless, these starch films still showed low oxygen permeability compared to other films, while WVP of starch films was observed to be comparable to that of protein films.

García et al. (1998) compared quality of strawberries coated with starch from different sources (corn starch – 25% amylose; potato starch – 23% amylose; high-amylose corn starch – 50% amylose; high-amylose corn starch – 65% amylose), and found a significant effect of amylose on color, weight loss and firmness of coated strawberries. Strawberries coated with 50% and 65% high-amylose starches best retained their quality attributes compared to other treatments.

Since starch films are hydrophilic, their properties will change with fluctuations in RH; for example, their barrier properties decrease with increasing RH. Starch is therefore not the best option when working with minimally-processed high-water activity commodities. Bai et al. (2002) coated apples with starch solutions and observed high gloss at the beginning of storage. However, they observed a large decrease in gloss during storage, though firmness, internal oxygen and carbon dioxide concentrations of starch-coated apples had values similar to shellac-coated apples (a coating used commercially).

7.6.1.3 Cellulose

Cellulose, the most abundant natural polymer on earth, is composed of linear segments $(1 \rightarrow 4)$-linked β-D-glucopyranosyl units. Although very inexpensive, cellulose is difficult to use as coating, because of its water-insolubility and highly associated crystalline structure. However, some cellulose derivatives produced commercially such as carboxymethylcellulose (CMC), methylcellulose (MC), hydroxypropylcellulose (HPC), and hydroxypropylmethylcellulose (HPMC), can overcome limitations associated with native cellulose. Due to the cellulose linear structure, these derivatives tend to form good films, though film properties will depend on cellulose structure and molecular weight (Park et al. 1993; Ayranci et al. 1997). Ayranci et al. (1997) found that WVP of HPMC films decreased with increasing molecular weight of HPMC. However, Park et al. (1993) observed the opposite result for MC and HPC films, wherein WVP of these films increased with increasing molecular weight of HPC and MC. Comparison between MC, HPC, and CMC coatings, applied to shelled pecan nuts, showed CMC to be the best coating material for this application (Baldwin and Wood 2006). CMC coating imparted sheen and delayed rancidity; however firmness was not preserved. WVP of HPMC and MC films was also observed to decrease with increasing polymer molecular weight (Ayranci et al. 1997). Further, MC has been used to preserve green color and firmness of avocados and lower their respiration rates during storage (Maftoonazad and Ramaswamy 2005). Coating apricots and green peppers with a composite film of MC and stearic acid effectively reduced water loss; when coated with a film containing citric acid or ascorbic acid, vitamin C loss was also reduced (Ayrancy and Tunc 2004).

Carboxymethylcellulose has been widely used as coating for fruits. Some coating formulations containing CMC are already on the market, such as Tal Pro-long™ and Semperfresh™. Tal Pro-long™ (or Pro-long) is composed of sucrose polyesters of fatty acids and the sodium salt of CMC. Semperfresh™ is composed of sucrose esters of fatty acids, sodium salt of CMC, and mono- and diglycerides. These coatings have been shown to retard ripening of fruits (Fig. 7.1). Quality of cherries coated with Semperfresh™ can be preserved for longer periods, reducing weight loss and preserving firmness and skin color. Another cellulose product on the market, Nature Seal™, is composed of cellulose derivatives, but without sucrose fatty acid esters. A combination of Nature Seal™ and soy protein coatings carrying anti-browning agents and preservatives prolonged shelf life of cut apples by 1 week when stored at 4°C (Baldwin et al. 1996).

7.6.1.4 Alginate

Alginate, the sodium salt of alginic acid, is a block copolymer of β-D-mannuronic acid (M) and α-L-guluronic acid (G), and is isolated from brown seaweed (Sime 1990). Alginate forms strong, translucent, glossy films. Compared to CMC, gelatin, whey protein isolate, potato starch and sodium caseinate films, those of sodium alginate have lowest WVP, oxygen permeability and elongation percentage and highest tensile strength (Wang et al. 2007). Sodium alginate films are soluble in water, acid, and alkali (Wang et al. 2007), making them a good choice for coating whole fruits and vegetables. These films are not as effective for coating minimally-processed fruits and vegetables, though incorporation of calcium cations to decrease solubility of polymers comprising such films would make them a viable option (Olivas et al. 2007).

Composition of alginate (M:G ratio) also affects WVP of films. Alginate–Ca^{2+} films with higher concentrations of G have lower WVP than films with higher concentrations of M due to greater ability of G to form intermolecular cross-links via calcium salt-bridges. Although properties of alginate films are influenced by surrounding RH, alginate–Ca^{2+} films retain their strength, even at high RH values (Olivas and Barbosa-Cánovas 2008).

Alginate–Ca^{2+} coatings have been used successfully to prolong shelf life of fresh-cut Gala apples without causing undesirable anaerobic respiration. These coatings minimized weight loss and browning, and preserved firmness during storage.

7.6.2 *Proteins*

Proteins can form edible films due to the ability of their side chains to form intermolecular cross-links. Properties of these films depend on the nature of these linkages. In general, protein films are considered to have good gas barrier properties, though their water barrier properties are generally poor (Gennadios et al. 1994), since the latter depend on the RH of the environment and/or water activity of the food. Proteins generally used as coatings for fruits and vegetables are reviewed in paragraphs that follow.

7.6.2.1 Zein Coatings

Zein is a protein composed of prolamines found within corn endosperm, and is soluble in aqueous alcohol. Zein films have strong yellow color relative to HPMC, whey protein concentrate, shellac or whey protein isolate coatings (Trezza and Krochta 2000). Applied to tomatoes, zein coatings will delay color change, weight loss and softening without ethanol production (Park et al. 1994). Zein coatings containing vegetable oils, citric acid and antioxidants have been used to prevent rancidity of nuts and inhibit moisture transfer from fruit pieces in dry mixes (Andres 1984). Park et al. (1996) found that apples coated with zein solutions experienced decrease in respiration rate, whereas zein-coated pears had an increased respiration rate, compared to uncoated controls. Nevertheless, zein coatings delayed weight loss in both apples and pears during storage. A zein coating formulation, containing 10% zein and 10% propylene glycol dissolved in aqueous alcohol, was developed for Gala apples by Bai et al. (2003b). This coating maintained overall fruit quality comparable to commercial shellac coating. Special attention should be taken when coating fruits and vegetables with zein solutions, since, zein coatings can become white in color on contact with water, depending on the coating's concentration (Bai et al. 2003b).

7.6.2.2 Gluten and Soy Proteins

Gluten is the main storage protein in wheat and corn. Gluten films have good oxygen and carbon dioxide barrier properties, but exhibit relatively high WVP (Gennadios and Weller 1990). Mechanical treatment of gluten leads to disulfide bridge formation created by the amino acid, cysteine, which is relatively abundant in gluten. Wheat gluten has been used to preserve quality of fruits and vegetables. Peanuts were coated with a composite coating of soy protein isolate and wheat gluten with the objective of preventing fat deterioration. Soy protein also contains cysteine residues that can form disulphide bridges. Soy protein films are therefore similar to gluten films in mechanical properties. Baldwin et al. (1996) improved properties of Nature Seal™ coating on sliced apples and potatoes by adding soy protein, which reduced coating permeability to oxygen and water vapor.

7.6.2.3 Whey

Milk proteins contain 20% whey protein of which β-lactoglobulin is the main protein component. Whey protein is water-soluble, but β-lactoglobulin denatures when heated, exposing internal sulfur groups of cysteine, which then cross-link to form an insoluble film (McHugh and Krochta 1994). Whey protein has been studied extensively as a coating for different foods, including intact and minimally-processed fruits and vegetables. Whey protein was shown to produce a translucent and flexible film with excellent oxygen and aroma barrier properties at low RH (McHugh and Krochta 1994; Miller and Krochta 1997). Though some have found that whey protein films provide a poor moisture barrier, others have reported that incorporation of lipids reduces WVP of whey protein films (McHugh and Krochta 1994; Shellhammer and Krochta 1997; Pérez-Gago and Krochta 2001). Another study indicated that WPI (whey protein isolate) films are good gas barriers, but that they are influenced by the RH of the environment, affecting resistance of coating to oxygen and carbon dioxide permeation. As RH decreased, resistance of the coating to gas transfer increased. At low RH, oxygen decreased and carbon dioxide increased in coated fruits. At RH values ranging from 70 to 80%, anaerobic metabolism was induced due to low oxygen levels (Cisneros-Zevallos and Krochta 2003a).

Whey protein-based coatings were more effective in reducing enzymatic browning of Golden Delicious apple slices than HPMC-based coatings - probably due to the antioxidant effect of amino acids, such as cysteine and/or higher oxygen barrier imparted by the protein. No differences in browning were found between WPI- and WPC- (whey protein concentrate) based films. Lipid inclusion also affected degree of browning (measured with a colorimeter), but these differences were less evident at the end of storage, as assessed by a sensory panel. Results suggest that addition of anti-browning agents to whey protein coatings, combined with proper storage conditions, could significantly extend shelf life of fresh-cut apples (Pérez-Gago et al. 2005a).

7.6.3 Lipids

Lipids can be included in formulation of edible coatings or films within a lipid-based film, as a single layer of lipids dispersed in a hydrocolloid network, or as a secondary layer (a lipid layer over a hydrocolloid layer). Coatings containing lipids generally have good moisture barrier properties, since lipids have very low affinity for water (Krochta 1997). The properties lipids confer to films and coatings will depend on the characteristics of the lipid component, such as its physical state, degree of saturation and chain length of fatty acids. Saturated long-chain fatty acids provide coatings with best water vapor barrier properties (among fatty acids), as they produce more densely packed structure and have less mobility than unsaturated short-chain fatty acids (Kamper and Fennema 1984). Lipids which are solid at the desired storage temperature will form coatings with better water vapor barrier properties than lipids which are liquid under same conditions, mainly because solubility of water vapor in lipids is lower in films having more ordered molecular organization (Kester and Fennema 1989).

7.6.3.1   Carnuba and Shellac Wax

Carnauba is a natural plant wax and is GRAS (generally recognized as safe). It is relatively permeable to gases, and in microemulsion form, is quite shiny. The primary problems with carnauba wax are its loss of gloss during storage and its relatively high gas permeability, which does not effectively delay ripening (Baldwin et al. 1999). However, it is an excellent barrier to water vapor and can be combined with shellac to create a coating of moderate permeability to gases and low permeability to water vapor. Natural Shine™ 8000 (carnauba wax) is used to reduce oxygen transfer into and water vapor out of the fruit, while shellac is used to give a shiny appearance. Johnfresh™ is composed of both carnauba wax and shellac.

Most Delicious apples marketed in the U.S. are coated with shellac or shellac combined with carnauba wax (Bai et al. 2002). Shellac and carnauba wax are often used commercially as a coating for apples and citrus fruits to improve their appearance by adding gloss, to prevent water loss leading to shriveling and loss of marketability, and to maintain quality through delayed ripening and senescence. Unfortunately, both materials are also associated with non-food uses; shellac also has problems with low gas permeability, which can lead to delayed ripening in some fruits (Baldwin et al. 1999) and cause anaerobic conditions. The apple industry is especially concerned that consumers may object to shellac, which does not currently have GRAS status. Shellac has further problems with whitening, or "blushing" as it is referred to in the industry, where water condenses on coated fruit surface after removal from cold storage. Nevertheless, shellac is recognized as one of the shiniest coatings available, and was found to improve appearance of apples. In one study, subsequent sales of red and green apple cultivars, such as Delicious and Granny Smith, respectively, increased due to shellac coatings (Bai et al. 2003a).

7.6.3.2   Beeswax

Beeswax can be used as a composite film. Han et al. (2006) added beeswax to pea starch films and found a decrease in WVP and an increase in oxygen permeability with beeswax concentration higher than 30%. Beeswax-pea starch films were homogeneous and translucent. Thickness of pea starch coating significantly increased with addition of beeswax (Han et al. 2006). Beeswax has been used in composite coatings in combination with WPI, WPC, and HMPC to preserve intact fruits such as plums (Pérez-Gago et al. 2003a) and minimally-processed fruits, such as apple slices and persimmon pieces (Pérez-Gago et al. 2005a, b; Pérez-Gago et al. 2006). In most cases, weight loss was not prevented with beeswax, with the exception of coated plums and persimmon pieces (see Tables 7.1 and 7.2).

7.6.4   Other Films and Coatings

Other compounds have been used to form coatings for fruits and vegetables, such as pullulan, gellan, aloe vera, cactus–mucilage and fruit puree. Aloe vera coatings have been studied for their ability to preserve quality of cherries (Martínez-Romero et al. 2006) and grapes (Valverde et al. 2005). Martínez-Romero et al. (2006) found reduction in the microbial population of coated cherries, while control cherries experienced considerable increase in microbial population; aloe vera coatings also preserved texture and color, and decreased water loss in cherries. Valverde et al. (2005) extended storage life of grapes at 1°C from 7 to 35 days, with reduction in microbial load also observed for coated grapes. A cactus–mucilage film was studied to preserve quality of strawberries (Del-Valle et al. 2005). Coatings containing a mixture of fruit puree and hydrocolloid have likewise been used. Alginate–apple puree was used as coating to preserve quality of apple slices (Rojas-Graü et al. 2007).

7.6.5   Additives in Films and Coatings

Additives in coatings and films have been used for different purposes. Those already discussed are shown in Table 7.3. Anti-browning compounds, antimicrobial agents, texture enhancers, nutrients, probiotics and flavors are examples of some additives used in coatings. However, plasticizers are the main additives used in films and coatings. Plasticizer is added to the formulation to improve mechanical properties of films and coatings. Without plasticizers, most films and coatings are brittle, and it is difficult to form a homogenous coating. Plasticizer combines with the main component of the film, moving the component's chains apart, and thus reduces rigidity of the structure (Guilbert and Biquet 1996). Plasticizer also attracts water molecules around it, which reduces intermolecular interactions of the main component (Ke and Sun 2001). Major plasticizers used have been polyols such as glycerol, sorbitol and polyethylene glycol (Sothornvit and Krochta 2005), but lately, disaccharides such as sucrose and monosaccharides (e.g., fructose, glucose, and mannose) have been investigated (Zhang and Han 2006). Monosaccharides have proved to be effective as plasticizers (Olivas and Barbosa-Cánovas 2008; Zhang and Han 2006), exhibiting films with lower WVP compared to those containing polyols as plasticizers. Whey protein films that have been plasticized with sucrose have excellent oxygen barrier properties and display high gloss; however, sucrose tends to crystallize with time and lose its properties (Dangaran and Krochta 2007). How plasticizers affect properties of films will depend on factors such as type of plasticizer (molecular size, total number of hydroxyl groups, configuration), its concentration and type of polymer. Plasticizers normally generate a homogenous mixture without phase-separation. However, some works attribute phase separation observed between plasticizer and polymer to an excess of plasticizer or incompatibility between plasticizer and polymer (Aulton et al. 1981; Donhowe and Fennema 1993; Ayranci et al. 1997; Jagchud and Chinnan 1999).

7.7 Challenges and Future Trends

Not much work has been done to identify the relationship between the internal atmosphere caused by the coating and velocity of physiological ripening processes, such as respiration, tissue softening, metabolic reactions, production of metabolites and secondary compounds generated during storage. Most work in this area has dealt with changes in quality due to application of coating.

To preserve quality of fruits by means of decreasing oxygen in the internal atmosphere, special care should be taken not to minimize oxygen concentration to a point where anaerobic respiration may occur. Thus, for each fruit, it is necessary to know the optimum oxygen concentration at which rate of consumption is minimized without promoting development of anaerobic respiration.

Although extensive research has been done in the area of edible films and coatings for fresh and minimally-processed fruits and vegetables, most work has been empirical in nature. Many factors should be studied to understand their effects on coating performance for specific types of produce and properties such as:

- Properties of coating solution: composition, concentration, viscosity and density
- Properties of film: mechanical, gas and vapor barrier properties
- Properties of coating: thickness, temperature, and atmospheric conditions
- Properties of the produce (fresh or minimally-processed): respiration rate, water activity, composition, etc

Thus it is necessary to go beyond empirical studies to investigate how these factors/properties will affect the final product and how they can be controlled or modified to enhance quality.

The study of edible coatings for fruits and vegetables deals with a wide range of scientific fields including: chemistry, biochemistry, horticulture, food science and engineering. The final objective should be to develop edible films and coatings that can be marketed commercially. Thus, to place edible coatings on the market, future research will need to concentrate on whole interdisciplinary works that provide comprehensive information that addresses all important issues relative to development of a specific coating for a specific product.

Future research should include:

- Characterization of physicochemical properties of coating solutions (composition, concentration, solubility, viscosity, density, surface tension, etc), including evaluation of how variation in chemical and physical conditions (e.g., pH, temperature, time, etc.) could affect these properties.
- Characterization of films intended for use as coatings, under similar conditions as the commodity being treated. This includes gas and vapor permeability, mechanical and sensorial properties, thickness, solubility, digestibility, etc. Focus should be on understanding possible changes to properties of a film when chemical and physical conditions are modified, simulating likely events a commodity could face during actual handling and storage.
- Study of metabolic reactions occurring within commodities and the extent to which they can be modified with coatings. This includes not only respiration,

but also metabolic reactions conducive to production of biochemicals that could be affected by modified atmosphere, or by the commodity itself while in contact and reacting with chemicals contained in a coating.

- Study of internal gas composition of coated commodities and relationships between internal atmosphere and velocity of physiological ripening processes, such as respiration, metabolic reactions and production of secondary compounds during storage.
- Study of impact of coatings on quality and shelf life of commodities, taking into account all possible conditions they could face during handling and storage.
- Determination of optimal methods of applying coatings most conducive to obtaining high quality product at lowest possible cost.
- Study of consumer acceptability of coatings.
- Study of impact of edible coatings on final cost of commodities.

Studies should focus on commercial viability of edible film and coating technology for fruits and vegetables. There is also a need to study effects of coatings on biochemicals and secondary metabolites.

References

Abeles FB, Morgan PW, Saltveit ME (1992) Ethylene in Plan Biology, 2nd edn. Academic Press, New York

Ahvenainen R (1996) New approaches in improving the shelf life of minimally processed fruit and vegetables. Trends in Food Science and Technology 7(6), 179–187

Amarante C, Banks NH, Ganesh S (2001) Effects of coating concentration, ripening stage, water status and fruit temperature on pear susceptibility to friction discoloration. Postharvest Biology and Technology 21, 283–290

Andres C (1984) Natural edible coating has excellent moisture and grease barrier properties. Food Processing 45(13), 48–49

Anker M, Standing M, Hermansson AM (2001) Aging of whey protein films and the effect on mechanical and barrier properties. Journal of Agricultural and Food Chemistry 49, 989–995

Aulton ME, Abdul-Razzak MH, Hogan JE (1981) The mechanical properties of hydroxypropyl-methylcellulose films derived from aqueous systems. Part 1. The influence of plasticizers. Drug Development and Industrial Pharmacy 7, 649–668

Ayrancy E, Tunc S (2004) The effect of edible coatings on water and vitamin C loss of apricots (Armeniaca vulgaris Lam.) and green peppers (Capsicum annuum L.). Food Chemistry 87(3), 339–342

Ayranci E, Buyuktas S, Cetin E (1997) The effect of molecular weight of constituents on properties of cellulose-based edible films. Lebensmittel-Wissenschaft und-Technologie 30, 101–104

Bai J, Baldwin EA, Hagenmaier RD (2002) Alternatives to shellac coatings provide comparable gloss, internal gas modification, and quality for "Delicious" apple fruit. HortScience 37(3), 559–563

Bai J, Hagenmaier RD, Baldwin EA (2003a) Coating selection for "Delicious" and other apples. Postharvest Biology and Technology 28, 381–390

Bai J, Alleyne V, Hagenmaier RD, Mattheis JP, Baldwin EA (2003b) Formulation of zein coatings for apples (Malus domestica). Postharvest Biology and Technology 28, 259–268

Baldwin EA (1994) Edible coatings for fresh fruits and vegetables: past, present, and future. In: Krochta JM, Baldwin EA, Nisperos-Carriedo MO (Eds.) Edible Coatings and Films to Improve Food Quality. Technomic, Lancaster, PA, pp. 25–64

Baldwin EA, Nisperos-Carriedo MO, Baker RA (1995a) Edible coatings for lightly processed fruits and vegetables. HortScience 30(1), 35–40

Baldwin EA, Nisperos-Carriedo MO, Baker RA (1995b) Use of edible coatings to preserve quality of lightly (and slightly) processed products. Critical Reviews in Food Science and Nutrition 35(6), 509–524

Baldwin EA, Nisperos MO, Chen X, Hagenmaier RD (1996) Improving storage life of cut apple and potato with edible coating. Postharvest Biology and Technology 9(2), 151–163

Baldwin EA, Burns JK, Kazokas W, Brecht JK, Hagenmaier RD, Bender RJ, Pesis DE (1999) Effect of two edible coatings with different permeability characteristics on mango (Mangifera indica L.) ripening during storage. Postharvest Biology and Technology 17, 215–226

Baldwin EA, Wood B (2006) Use of edible coating to preserve pecans at room temperature. HortScience 41(1), 188–192

Banks N, Dadzie B, Cleland D (1993) Reducing gas exchange of fruits with surface coatings. Postharvest Biology and Technology 3, 269–284

BeMiller JN, Whistler RL (1996) Carbohydrates. *In*: Fennema OR (Ed.) Food Chemistry, 3rd ed. Marcel Dekker, New York, pp. 157–223

Cagri A, Ustunol Z, Ryser ET (2004) Antimicrobial edible films and coatings. Journal of Food Protection 67(4), 833–848

Certel M, Uslu MK, Ozedemir F (2004) Effects of sodium caseinate- and milk protein concentrate-based edible coatings on the postharvest quality of Bing cherries. Journal of the Science of Food and Agriculture 84(10), 1229–1234

Chien PJ, Sheub F, Ling HR (2007a) Coating citrus (*Murcott tangor*) fruit with low molecular weight chitosan increases postharvest quality and shelf life. Food Chemistry 100, 1160–1164

Chien PJ, Sheu F, Yang FH (2007b) Effects of edible chitosan coating on quality and shelf life of sliced mango fruit. Journal of Food Engineering 78 (1), 225–229

Cisneros-Zevallos L, Krochta JM (2002) Internal modified atmospheres of coated fresh fruits and vegetables: Understanding relative humidity effects. Journal of Food Science 67(6), 1990–1995

Cisneros-Zevallos L, Krochta JM (2003a) Whey protein coatings for fresh fruits and relative humidity effects. Journal of Food Science 68(1), 176–181

Cisneros-Zevallos L, Krochta JM (2003b) Dependence of coating thickness on viscosity of coating solution applied to fruits and vegetables by dipping method. Journal of Food Science 68(2), 503–510

Claypool LL (1940) The waxing of deciduous fruits. American Society of Horticultural Science Proceedings 37, 443–447

Curtis GJ (1988) Some experiments with edible coatings on the long-term storage of citrus fruits. Proceedings of the Sixth International Citrus Congress 3, 1514–1520

Dangaran KL, Krochta JM (2007) Preventing the loss of tensile, barrier and appearance properties caused by plasticizer crystallization in whey protein films. International Journal of Food Science and Technology 42(9), 1094–1100

Davies DH, Elson CM, Hayes ER (1989) *N,O*-carboxymethyl chitosan, a new water soluble chitin derivative. In: Skjåk-Bræk G, Anthonsen T, Sandford P (Eds.) Chitin and Chitosan: Sources, Chemistry, Biochemistry, Physical Properties and Applications. Elsevier, London, UK, pp. 467–72

Debeaufort F, Martín-Polo M, Voilley A (1993) Polarity, homogeneity and structure affect water vapour permeability of model edible films. Journal of Food Science 58, 426–429, 434

DeEll JR, Prange RK, Peppelenbos HW (2003) Postharvest physiology of fresh fruits and vegetables. *In*: Chakraverty A, Jujumdar AS, Raghava GSV, Ramaswamy HS (Eds.) Handbook of Postharvest Technology. Marcel Dekker, New York, pp. 455–484

Del-Valle V, Hernandez-Munoz P, Guarda A, Galotto MJ (2005) Development of a cactus-mucilage edible coating (*Opuntia ficus* indica) and its application to extend strawberry (*Fragaria ananassa*) shelf-life. Food Chemistry 91(4), 751–756

Devlieghere F, Vermeulen A, Debevere J (2004) Chitosan: antimicrobial activity, interactions with food components and applicability as a coating on fruit and vegetables. Food Microbiology 21, 703–714

Dong H, Cheng L, Tan J, Zheng K, Jiang Y (2004) Physicochemical properties and application of pullulan edible films and coatings in fruits preservation. Journal of Food Engineering 64, 355–358

Donhowe IG, Fennema O (1993) The effects of plasticizers on crystallinity, permeability, and mechanical-properties of methylcellulose films. Journal of Food Processing and Preservation 17, 247–257

El-Ghaouth A, Arul J, Grenier J, Asselin A (1992) Chitosan coating to extend the storage life of tomatoes. HortScience 27(9), 1016–1018

El-Ghaouth A, Arul J, Wilson C Benhamou N (1997) Biochemical and cytochemical aspects of the interactions of chitosan and *Botrytis cinerea* in bell pepper fruit. Posthharvest Biology and Technology 12, 183–194

Fellman JK, Rudell DR, Mattinson DS, Mattheis JP (2003) Relationship of harvest maturity to flavor regeneratin after CA storage of 'Delicious' apples. Postharvest Biology and Technology 27, 39–51

Food and Agriculture Organization of the United Nations (2003) Increasing fruit and vegetable consumption becomes a global priority. Taken from: http://www.fao.org/english/newsroom/focus/2003/fruitveg1.htm in 2007

Ford ES, Mokdad AH (2001) Fruit and vegetable consumption and diabetes mellitus incidence among U.S. adults. Preventive Medicine 32, 33–39

Garcia E, Barret DM (2002) Preservative treatments for fresh cut fruits and vegetables. In: Lamikanra O (Ed.) Fresh-Cut Fruits and Vegetables. CRC Press, Boca Raton, FL, pp. 267–304

García MA, Martino MN, Zaritky NE (1998) Starch-based coatings: effect on refrigerated strawberry (Fragaria ananassa) quality. Journal of the Science of Food and Agriculture 76(3), 411–420

Genkinger JM, Platz EA, Hoffman SC, Comstock GW, Helzlsouer KJ (2004) Fruit, vegetable, and antioxidant intake and all-cause, cancer, and cardiovascular disease mortality in a community-dwelling population in Washington County, Maryland. American Journal of Epidemiology 60, 1223–1233

Gennadios A, Weller CL (1990) Edible films and coatings from wheat and corn proteins. Food Technology 44, 63–69

Gennadios A, McHugh TH, Weller CL, Krochta JM (1994) Edible coatings and films based on proteins. *In*: Krochta JM, Baldwin EA, Nisperos-Carriedo MO (Eds.) Edible Coatings and Films to Improve Food Quality. CRC Press, Boca Raton, FL, pp. 201–278

Gordon LR (2005) Edible and biobased food packaging materials. *In*: Food Packaging Principles and Practice. CRC Press, Boca Raton, FL, pp. 43–54

Gorny R, Cifuentes RA, Hess-Pierce B, Kader AA (2000) Quality changes in fresh-cut pear slices as affected by cultivar, ripeness stage, fruit size, and storage regime. Journal of Food Science 65, 541–544

Guilbert S, Biquet B (1996) Edible films and coatings. *In*: Bureau G, Multon JL (Eds.) Food Packaging Technology. VCH Publishers, New York

Hagenmaier RD, Shaw PE (1990) Moisture permeability of edible films made with fatty acid and (hydroxypropyl) methylcellulose. Journal of Agriculture and Food Chemistry 38, 1799–1803

Hagenmaier RD, Shaw PE (1992) Gas-permeability of fruit coating waxes. Journal of the American Society for Horticultural Science 117(1), 105–109

Han C, Zhao Y, Leonard SW, Traber MG (2004) Edible coatings to improve storability and enhance nutritional value of fresh and frozen strawberries *Fragaria ananassa*) and raspberries (*Rubus ideaus*). Postharvest Biology and Technology 33, 67–78

Han JH, Seo GH, Park IM, Kim GN, Lee DS (2006) Physical and mechanical properties of pea starch edible films containing beeswax emulsions. Journal of Food Science 71(6), E290–E296

Hardenburg RE (1967) Wax and related coatings for horticultural products. A bibliography. Agricultural Research Service Bulletin 51–15, USDA, Washington, DC

Hernandez-Munoz P, Almenar E, Ocio MJ, Gavara R (2006) Effect of calcium dips and chitosan coatings on postharvest life of strawberries (Fragaria × Ananassa). Postharvest Biology and Technology 39 (3), 247–253

Hoffman AF (1916) Preserving Fruit. US patent 19,160,104

Hoover R, Sosulski FW (1991) Composition, structure, functionality and chemical modification of legume starches. Canadian Journal of Physiology and Pharmacology 69, 79–92

Hung HC, Joshipura KJ, Jiang R (2004) Fruit and vegetable intake and risk of major chronic disease. Journal of National Cancer Institute 96, 1577–1584

Jangchud, A. and Chinnan, M.S. (1999) Properties of peanut protein film: Sorption isotherm and plasticizer effect. Lebensmittel-Wissenschaft und-Technologie 32, 89–94.

Jiang Y, Li J, Jiang W (2005) Effects of chitosan coating on shelf life of cold-stored litchi fruit at ambient temperature. Lebensmittel-Wissenschaft und-Technologie 38, 757–761

Kader AA (1986) Biochemical and physiological basis for effects of controlled and modified atmospheres on fruits and vegetables. Food Technology 40(5), 99–104

Kamper SL, Fennema O (1984) Water vapor permeability of an edible, fatty acid, bilayer film. Journal of Food Science 49(6): 1482–1485.

Kays SJ (1991) Metabolic processes in harvested products. In: Postharvest. Physiology of Perishable Plant Products. Van Nostrand Reinhold, New York, pp. 75–142

Kays SJ (1999) Preharvest factors affecting appearance. Postharvest Biology and Technology 15(3), 233–247

Kays SJ, Paull RE (2004) Stress in harvested products. In: Postharvest Biology. Exon Press, Athens, GA, pp. 355–414

Ke T, Sun X (2001) Thermal and mechanical properties of poly (lactic acid) and starch blends with various plasticizers. Transactions of the ASAE 44(4), 945–953

Ke D, Yahia E, Mateos M, Kader A (1994) Ethanolic fementation of Barlett pears as influenced by ripening stage and atmospheric composition. Journal of the American Society for Horticultural Science. 32(6), 593–600

Kester JJ, Fennema O (1989) An edible film of lipids and cellulose ethers: barrier properties to moisture vapor transmission and structural evaluation. Journal of Food Science 54, 1383–1389

Knee M (1980) Physiological responses of apple fruits to oxygen concentrations. Annals of Applied Biology 96, 243–253

Koelsch CM, Labuza TP (1992) Functional, physical and morphological properties of methyl cellulose and fatty acid-based edible barriers. Lebensmittel-Wissenschaft und-Technologie 25, 404–411

Krochta JM (1997) Edible composite moisture-barrier films. In: Blakistone B (Ed.) 1996 Packaging Yearbook. National Food Processing Association, Washington, DC, pp. 38–54

Lakakul R, Beaudry RM, Hernandez RJ (1999) Modeling respiration of apple slices n modified-atmosphere packages. Journal of Food Science 64, 105–110

Lin D, Zhao Y (2007) Innovations in the development and application of edible coatings for fresh and minimally processed fruits and vegetables. Comprehensive Reviews in Food Science and Food Safety 6, 60–75

Lourdin D, Valle GD, Colonna P (1995) Influence of amylose content on starch films and foams. Carbohydrate Polymers 27(4), 261–270

Maftoonazad N, Ramaswamy HS (2005) Postharvest shelf life extension of avocados using methyl cellulose-based coatings. Lebensmittel-Wissenschaft und-Technologie 38, 617–624

Maftoonazad N, Ramaswamy HS, Moalemiyan M, Kushalappa AC (2007) Effect of pectin-based edible emulsion coating on changes in quality of avocado exposed to *Lasiodiplodia theobromae* infection. Carbohydrate Polymers 68, 341–349

Martínez-Romero D, Alburquerque N, Valverde JM, Guillén F, Castillo S, Valero D, Serrano M (2006) Postharvest sweet cherry quality and safety maintenance by *Aloe vera* treatment: a new edible coating. Postharvest Biology and Technology 39, 93–100

McHugh TH, Krochta JM (1994) Milk-protein-based edible films and coatings. Food Technology 48(1), 97–103

Mehyar GF, Han JH (2004) Physical and mechanical properties of high-amylose rice and pea starch films as affected by relative humidity and plasticizer. Journal of Food Science 69(9), E449–E454

Mei Y, Zhao Y, Yang J, Furr HC (2002) Using edible coating to enhance nutritional and sensory qualities of baby carrots. Journal of Food Science 67, 1964–1968

Mencarelli F, Saltveit ME, Massantini R (1989) Lightly processed foods: ripening of tomato slices. Acta Horticulturae 244, 193–200

Miller S, Krochta JM (1997) Oxygen and aroma barrier properties of edible films: a review. Trends in Food Science Technology 8, 228–37

Morillon V, Debeaufort F, Blond G, Cappelle M, Voilley A (2002) Factors affecting the moisture permeability of lipid-based edible films: a review. Critical Reviews in Food Science and Nutrition 42(1), 67–89

Muzzarelli RAA (1986) Filmogenic properties of chitin/chitosan. In: Muzzarelli RAA, Jeuniaux C, Gooday GW (Eds.) Chitin in Nature and Technology. New York, Plenum Press, pp. 389–396

Nguyen C, Carlin F (1994) The microbiology of minimally processed fresh fruits and vegetables. Critical Reviews in Food Science and Nutrition 34(4), 371–401

Nisperos MO, Baldwin EA (1996) Edible coatings for whole and minimally processed fruits and vegetables. Food Australia 48(1), 27–31

No HK, Meyers SP, Prinyawiwatkul W, Xu Z (2007) Applications of chitosan for improvement of quality and shelf life of Foods: a Review. Journal of Food Science 72(5), R87–R100

Olivas GI, Barbosa-Cánovas GV (2005) Edible coatings for fresh cut fruits. Critical Reviews in Food Science and Nutrition 45, 657–670

Olivas GI, Barbosa-Cánovas GV (2008) Alginate-calcium films: water vapor permeability and mechanical properties as affected by plasticizer and relative humidity. LWT-Food Science and Technology 41, 359–366

Olivas GI, Rodriguez JJ, Barbosa-Cánovas GV (2003) Edible coatings composed of methylcellulose stearic acid, and additives to preserve quality of pear wedges. Journal of Food Processing and Preservation 27, 299–320

Olivas GI, Mattinson DS, Barbosa-Cánovas GV (2007) Alginate coatings for preservation of minimally processed "Gala" apples. Postharvest Biology and Technology 45, 89–96

Paillard NMM (1979) Biosynthese des produits volatils de la pomme: formation des alcools et des esters a partir des acides gras. Phytochemistry 18, 1165–1171.

Park HJ, Chinnan MS (1990) Properties of edible coatings for fruits and vegetables. American Society of Agricultural Engineers. Paper no. 90–6510. p. 19

Park HJ, Weller CL, Vergano PJ, Testin RF (1993) Permeability and mechanical properties of cellulose based edible films. Journal of Food Science 58, 1361–1364, 1370

Park HJ, Chinnan MS, Shewfelt RL (1994) Edible coating effects on storage life and quality of tomatoes. Journal of Food Science 59(3), 568–570

Park HJ, Rhim JW, Lee HY (1996) Edible coating effects on respiration rate and storage life of Fuji apples and Shingo pears. Foods and Biotechnology 5(1), 59–63

Park HJ (1999) Development of advanced edible coatings for fruits. Trends in Food Science and Technology 10, 254–260

Pen LT, Jiang YM (2003) Effects of chitosan coating on shelf life and quality of fresh-cut Chinese water chestnut. Lebensmitte–Wissenschaft und Technologie 36, 359–364

Peressini D, Bravin B, Lapasin R, Rizzotti C, Sensidoni A (2003) Starch-methylcellulose based edible films: rheological properties of film-forming dispersions. Journal of Food Engineering 59, 25–32

Pérez-Gago MB, Krochta JM (2001) Lipid particle size effect on water vapor permeability and mechanical properties of whey protein/beeswax emulsion films. Journal of Agricultural and Food Chemistry 49, 996–1002

Pérez-Gago MB, Rojas C, Del Río MA (2003a) Effect of hydroxypropyl methylcellulose–lipid edible composite coatings on plum (cv. Autumn giant) quality during storage. Journal of Food Science 68(3), 879–883

Pérez-Gago MB, Serra M, Alonso M, Mateos M, Del Río MA (2003b) Effect of solid content and lipid content of whey protein isolate-beeswax edible coatings on color change of fresh-cut apples. Journal of Food Science 68(7), 2186–2191

Pérez-Gago MB, Serra M, Alonso M, Mateos M, Del Río MA (2005a) Effect of whey protein- and hydroxypropyl methylcellulose-based edible composite coatings on color change of fresh-cut apples. Postharvest Biology and Technology 36, 77–85

Pérez-Gago MB, Del Río MA, Serra M (2005b) Effect of whey protein-beeswax edible composite coating on color change of fresh-cut persimmons cv. "Rojo Brillante." Acta Horticulturae 682 (3), 1917–1923

Pérez-Gago MB, Serra M, Del Río MA (2006) Color change of fresh-cut apples coated with whey protein concentrate-based edible coatings. Postharvest Biology and Technology 39, 84–92

Platenius, H (1939) Wax Emulsions for Vegetables. Cornell University Agricultural Experimental Station Bulletin No.723. p 43

Poenicke EF, Kays SJ, Smittle DA, Williamson RE (1977) Ethylene in relation to postharvest quality deterioration in processing cucumbers. Journal of the American Society for Horticultural Science 102, 303–306

Quezada-Gallo J, Gramin A, Pattyn C, Diaz Amaro M, Debeaufort F, Voilley A (2005) Biopolymers used as edible coating to limit water transfer, colour degradation and aroma compound 2-pentanone lost in Mexican fruits. Acta Horticulturae 682(3), 1709–1716

Ribeiro C, Vicente AA, Teixeira JA, Miranda C (2007) Optimization of edible coating composition to retard strawberry fruit senescence. Postharvest Biology and Technology 44, 63–70

Rojas-Argudo C, Pérez-Gago MB, Del Río MA (2005) Postharvest quality of coated cherries cv. "Burlat" as affected by coating composition and solids content. Food Science and Technology International 11(6), 417–424

Rojas-Graü M, Raybaudi-Massilia RM, Soliva-Fortuny RC, Avena-Bustillos RJ, McHugh TH, Martín-Belloso O (2007) Apple puree-alginate edible coating as carrier of antimicrobial agents to prolong shelf-life of fresh-cut apples. Postharvest Biology and Technology 45, 254–264

Romanazzi G, Nigro F, Ippolito A, Di Venere D, Salerno M (2002) Effects of pre- and postharvest chitosan treatments to control storage grey mold of table grapes. Journal of Food Science 67(5), 1862–1867

Rooney ML (1995) Overview of active food packaging. In: Rooney ML (Ed.) Active Food Packaging. Blackie Academic and Professional, Glasgow, pp. 1–37

Rosen JC, Kader AA (1989) Postharvest physiology and quality maintenance of slices pear and strawberry fruits. Journal of Food Science 54, 656–659

Saltveit ME (1997) Physical and physiological changes in minimally processed fruits and vegetables. In: Tomas-Barberan FA, Robins RJ (Eds.) Phytochemistry of Fruit and Vegetables. Clarendon Press, Oxford, New York, pp. 205–220

Sams CE (1999) Preharvest factors affecting postharvest texture. Postharvest Biology and Technology 15(3), 249–254

Sapers GM (1993) Browning of foods: control by sulfites, antioxidants, and other means. Food Technology 47(10), 75–84

Sarkar SK, Phan CT (1979) Naturally-occurring and ethylene-induced compounds in the carrot root. Journal of Food Protection 42, 526–534

Sebti I, Martial-Gros A, Carnet-Pantiez A, Grelier S, Coma V (2005) Chitosan polymer as bioactive coating and film against *Aspergillus niger* contamination. Journal of Food Science 70(2), M100–M104

Sime WJ (1990) Alginates. In: P. Harris (Ed.), Food Gels, Elsevier Applied Science, London pp. 53–58.

Sothornvit R, Krochta JM (2005) Plasticizers in edible films and coatings. In: Han JH (Ed.) Innovations in Food Packaging. Elsevier Academic Press, London, pp. 403–433

Tapia M, Rojas-Graü M, Rodríguez F, Ramírez J, Carmona A, Martin-Belloso O (2007) Alginate- and gellan-based edible films for probiotic coatings on fresh-cut fruits. Journal of Food Science 72(4), E190–E196

Toivonen PMA, DeEll JR (2002) Physiology of fresh-cut fruits and vegetables. In: Olusola Lamikanra (Ed.) Fresh-cut fruits and vegetables. CRC Press, Boca Raton, FL, pp. 91–123

Trezza TA, Krochta JM (2000) Color stability of edible coatings during prolonged storage. Journal of Food Science 65(7), 1166–1169

Tuil R, Fowler P, Lawther M, Weber CJ (2000) Properties of biobased packaging materials. *In*: Weber CJ (Ed.) Biobased Packaging Materials for the Food Industry: Status and Perspectives. The Royal Veterinary and Agricultural University, Denmark, pp. 13–45

United States Department of Agriculture, USDA (2001–2002) Agriculture Fact Book

US FDA/CFSAN (2006) Inventory of GRAS notices: summary of all GRAS notices. Available from: http://www.cfsah.fda.gov/~rdb/opa-gras.html. Accessed July 2007

Valverde JM, Valero D, Martinez-Romero D, Guillen F, Castillo S, Serrano M (2005) Novel edible coating based on Aloe vera gel to maintain table grape quality and safety. Journal of Agricultural and Food Chemistry 53(20), 7807–7813

Vargas M, Albors A, Chiralt A, González-Martínez C (2006) Quality of cold-stored strawberries as affected by chitosan-oleic acid edible coatings. PostHarvest Biology and Technology 41 (2), 164–171

Viña S, Mugridge A, García MA, Ferreyra RM, Martino MN, Chaves AR, Zaritzky NE (2007) Effects of polyvinylchloride films and edible starch coatings on quality aspects of refrigerated Brussels sprouts. Food Chemistry 103(3), 701–709

Vojdani F, Torres JA (1990) Potassium sorbate permeability of methylcellulose and hydroxypropyl methylcellulose coatings – effect of fatty-acids. Journal of Food Science 55(3), 841–846

Wang LZ, Liu L, Holmes J, Kerry JF, Kerry JP (2007) Assessment of film-forming potential and properties of protein and polysaccharide-based biopolymer films. International Journal of Food Science and Technology 42(9), 1128–1138

Watada AE, Abe K, Yamauchi N (1990) Physiological activities of partially processed fruits and vegetables. Food Technology 20, 116, 118, 120–122

Wong DWS, Camirand WM, Pavlath AE (1994) Development of edible coatings for minimally processed fruits and vegetables. In: Krochta JM, Baldwin EA, Nisperos-Carriedo MO (Eds.) Edible Coatings and Films to Improve Food Quality. Technomic, Lancaster, PA, pp. 65–88

World Health Organisation. (2002) The world health report: reducing risks, promoting healthy life. Geneva: World Health Organisation

Xu S, Chen X, Sun D (2001) Preservation of kiwifruit coated with an edible film at ambient temperature. Journal of Food Engineering 50(4), 211–216

Yaman Ö, Bayoındırlı L (2002) Effects of an edible coating and cold storage on shelf-life and quality of cherries. Lebensmittel-Wissenschaft und-Technologie 35, 146–150

Zagory D (1999) Effects of post-processing handling and packaging on microbial populations. PostHarvest Biology and Technology 15: 313–321

Zhang D, Quantick PC (1997) Effects of chitosan coating on enzymatic browning and decay during postharvest storage of litchi (Litchichinensis Sonn.) fruit. Postharvest Biology and Technology 12, 195–202

Zhang Y, Han JH (2006) Mechanical and thermal characteristics of pea starch films plasticized with monosaccharides and polyols. Journal of Food Science 71(2), 109–E118

Chapter 8
Edible Films and Coatings for Meat and Poultry

Zey Ustunol

8.1 Introduction

Edible films and coatings are defined as continuous matrices that can be prepared from proteins, polysaccharides and/or lipids to alter the surface characteristics of a food. Although the terms films and coatings are used interchangeably, films in general are preformed and are freestanding, whereas, coatings are formed directly on the food product. Proteins used in edible films include wheat gluten, collagen, corn zein, casein and whey protein. Alginate, dextrin, pectin chitosan, starch and cellulose derivatives are commonly used in polysaccharide films. Suitable lipids for use in films and coatings include waxes, acylglycerol, and fatty acids (Kester and Fennema 1986). Composite films containing both lipid and hydrocolloid components have also been developed.

Plasticizers are often added to film-forming solutions to enhance the properties of the final film. These film additives are typically small molecules of low molecular weight and high boiling point which are highly compatible with the polymer. Common food-grade plasticizers such as sorbitol, glycerol, mannitol, sucrose and polyethylene glycol decrease brittleness and increase flexibility of the film, which are important attributes in packaging applications. Plasticizers used for protein-based edible films decrease protein interactions and increase both polymer chain mobility and intermolecular spacing (Lieberman and Guilbert 1973). The type and concentration of plasticizer influence properties of protein films (Cuq et al. 1997); mechanical strength, barrier properties, and elasticity decrease when high levels of plasticizers are used (Cherian et al.1995; Galietta et al. 1998; Gontard et al. 1993). Water is another important plasticizer for protein films (Krochta 2002). Similar to other plasticizers, water content impacts film properties.

Z. Ustunol (✉)
Department of Food Science & Human Nutrition, Michigan State University,
2105 South Anthony Hall, Lansing, MI 48824-1225, USA
e-mail: ustunol@anr.msu.edu

M.E. Embuscado and K.C. Huber (eds.), *Edible Films and Coatings for Food Applications*, 245
DOI 10.1007/978-0-387-92824-1_8, © Springer Science+Business Media, LLC 2009

Common covalent cross-linking agents such as gluteraldehyde, calcium chloride, tannic acid, and lactic acid are used to improve water resistance, cohesiveness, rigidity, mechanical strength and barrier properties of edible films and coatings (Guilbert 1986; Marquie et al. 1995; Ustunol and Mert 2004). Exposure to UV light increases cohesiveness of protein films and coatings by promoting cross-linking (Brault et al. 1997). Alternatively, enzymatic cross-linking treatments with transglutaminases or peroxidases can be used to stabilize films.

Renewed interest in edible films and coatings in recent years has been due to concerns about the environment and a need to reduce the amount of disposable packaging, as well as consumer demand for higher quality food products and extended shelf life. Edible films and coatings are not intended to replace nonedible synthetic packaging materials or to be their biodegradable counterpart. Edible films are secondary packaging materials; they may offer protection to a food product after the primary packaging has been opened. They also provide food processors with a number of new unique opportunities for product processing, handling and development.

Edible films and coatings can provide protection to a food product by serving as barriers to moisture migration, and preventing diffusion of gases important in food deterioration, such as O_2 or CO_2. They can also enhance quality and appearance of a food product by preventing flavor and aroma migration and by providing structural integrity. Edible films and coatings can also serve as carriers for antimicrobials, antioxidants, nutrients, color, herbs and spices, and provide for localized or delayed activity if needed. Currently, edible films and coatings are used in a variety of food applications. Collagen films and sausage casings are probably the most successful commercial application of edible films in meat products. This chapter reviews application of edible films and coatings for meat and poultry products.

8.2   Historical Background

Edible films and protective coatings have been used for centuries to prevent quality loss such as shrinkage, oxidative off-flavors, microbial contamination, and discoloration in meat and poultry products. Yuba, the first freestanding edible film, was developed in Japan from soymilk during the fifteenth century, and was used for food preservation purposes. In sixteenth century England, cut meats were coated in fats to reduce moisture loss and, thus, shrinkage in a process called "larding" (Kester and Fennema 1986). Since then, a number of lipid coating formulations have been used to enhance quality of meat and meat products. Letney (1958) proposed coating meat with melted fat and letting it solidify to form a film to extend storage life of meat products during refrigerated storage and maintain "bloom" (a term used in the industry to signify the conversion from the purple state to the red state in the presence of oxygen). Carnauba wax, beeswax, and candelilla have been used to coat frozen meat to increase its shelf life (Daniels 1973). Use of acylated acylglycerol containing chlortetracycline (Ayers 1959), mixtures of mono-, di- and triacylglycerols in alcohol (Anderson 1960, 1961a, b), and acetylated mono- and diacylglycerol coatings have been suggested to reduce off-flavors and moisture loss, as well to maintain color and prevent freezer-burn in meat products. Applications of paraffinic acid mono-, di- and triacylglycerol with or without carboxylic acid have also been reported to improve meat quality and storage life (Schneide 1972). Other researchers have reported application of acetylated monoacylglycerol in vacuum-packed meats (Griffin et al. 1987; Leu et al. 1987). Emulsion coatings containing lipids have also been demonstrated to be useful in enhancing meat quality (Zabik and Dawson 1963; Bauer et al. 1968; Bauer and Neuser 1969; Kroger and Igoe 1971; Hernandez 1994; Baldwin et al. 1997). Films and coatings made from lipids alone, however, lack structural integrity and are brittle. Hirasa (1991) reported that highly saturated acetylated monoacyglycerol coatings tend to flake and crack during cold storage and have bitter aftertaste (Morgan 1971). Also, unsaturated acylglycerols are susceptible to oxidation. More recently, lipids have been incorporated in formation of composite protein or polysaccharide films to improve moisture barrier characteristics and provide flexibility to these films (Baldwin et al. 1995).

Various edible polysaccharide films and coatings such as starch and its derivatives, alginates, carrageenans, cellulose ethers and pectin also have been used to improve quality of meat and poultry products (Kester and Fennema 1986). Polysaccharide films are nongreasy and have visual appeal, which make them desirable for application as wraps in meat products (Labell 1991). They are good barriers to gases; however, because of their hydrophilic nature, they are poor barriers to moisture. Alginate coatings require a gelling agent, such as calcium chloride, which may impart bitterness (Earle 1968; Lazarus et al. 1976; Williams et al. 1978). Earle (1968) reported extending shelf life of meat products by coating them with an aqueous solution of algin and dextrose, followed by application of calcium solution. Carboxymethylcellulose (CMC) has been reported to improve adhesion of breading after baking when used in commercial breading mix (Suderman et al. 1981). Other derivatives of cellulose, methylcellulose (MC) or hydroxypropyl methylcellulose (HPMC) form reversible thermal gels, and have been used to produce glazed sauces for poultry and seafood. This treatment minimizes runoff during cooking, thereby reducing moisture loss (Baker et al. 1994).

Protein films and coatings have also been investigated throughout history to enhance quality of meat and poultry products. In the nineteenth century, use of gelatin films to preserve meat was proposed by Harvard and Harmony (1869) and Morris and Parker (1896). Patents as early as 1869 disclose gelatin-based films (Gennadios et al. 1997). Additional patents have been granted for acidic, aqueous solutions of gelatin and a metaphosphate polymer (Keil et al. 1960), and aqueous solutions of metal gelatinates (Keil 1961) as coatings on processed meats such as sausages, Canadian bacon, and boned hams to inhibit mold growth and lipid oxidation, and reduce handling damage. Incorporation of polyhydric alcohols (i.e., propylene glycol, ethylene glycol, glycerol, or sorbitol) into gelatin-film-forming solutions produce quick-setting, flexible films that exhibit good barrier properties (Whitman and Rosenthal 1971). Gelatin films have been applied to poultry products to prevent microbial growth and moisture loss and retard oil absorption during frying. Gelatin films also have been used as carriers of antioxidants for poultry products (Klose et al. 1952; Childs 1957; Gennadios et al. 1997). Villegas et al. (1999) reported on improved oxidative stability and color retention in frozen meat pieces used as pizza topping when they were gelatin-coated.

However, gelatin films lack the strength of those from collagen. Today, reconstituted or regenerated collagen films and sausage casings are probably the most successful commercial applications of edible films in the meat industry. Collagen films are used as wraps in hams and netted roast (Gennadios et al. 1997). Sausage casings are used in sausage production to hold meat batter together until it is heat-set to obtain the desired shape, form, and size (Wang 1986).

Natural casings have been the traditional sausage casings used throughout history. The collagen casing came about as a more sanitary and uniform alternative to gut casings. Compared to natural casings, they also have improved strength and flexibility and, therefore, better machinability. Collagen casings provide for more uniform products and better control of net product weight at high processing speeds. In addition, because of rapid uptake of smoke and smoke color by collagen, the smoking cycle may be shortened, thereby reducing product shrinkage due to smoking. Historical development of the collagen casing has been reviewed by Hood (1987). The coating of meat with gelatin obtained from partially hydrolyzed collagen is believed to be the origin of modern casing technology. However, a lack of collagen-like fibrillar structure within gelatin films has limited commercial development of gelatin-based casings, owing to their inelasticity and brittleness (Hood 1987).

The work of German scientist Oscar Becker during the 1920s and 1930s on synthetic collagen casings is believed to have led to manufacture of regenerated collagen casings, particularly those based on the dry extrusion principle (Becker 1936, 1938, 1939). Historically, industrial manufacture of collagen casings is divided into two distinct processes. Collagen casings have been produced from corium (the flesh side) of cattle hides using a "dry" process ("dry spinning technology") originally developed in Europe during the 1930s or the "wet" process ("wet spinning technology") developed in the 1960s in the United States.

The dry process involves alkaline treatment of hide coriums followed by acidification to pH 3. Acid-swollen coriums are then shredded to preserve maximum fiber structure of collagen, and are mixed to produce a dough high in solids (>12% solids). Plasticizers and cross-linking agents are added, and dough is pumped and extruded to form tubular casings followed by drying, conditioning, neutralizing, and/or providing for additional cross-linking. This process produces a tough casing, which was not acceptable to American sausage makers and therefore was not adopted by US casing manufacturers. In contrast, the wet process starts with acid- or alkaline-dehairing of cattle hides. The hide corium is decalcified, ground and mixed with acid to produce a swollen slurry of 4–5% solids using high shear homogenization. Cellulose and carboxymethylcellulose may be added to improve mechanical properties of the casing. The slurry is then homogenized and extruded, after which it passes through a coagulation bath of brine and shaped into a tubular casing. Casings are washed free of salts, and treated with plasticizing and cross-linking agents to provide for improved flexibility and strength, respectively. They are then shirred (or collapsed like an accordion) to fit over various-sized sausage stuffing horns, dried to 13–18% moisture, sealed in plastic bags and packaged (Hood 1987). Their shelf life depends upon maintaining favorable temperature and relative humidity conditions during storage.

Proper alignment of the collagen fibers is important in obtaining desirable properties for collagen casings. Orientation of the fibers in the direction of extrusion is undesirable because the casing can be readily split or torn lengthwise. Extrusion should realign collagen fibers as they exit the orifice of the extruder to produce a woven, crosshatched, fibrous collagen casing structure. The design of the extruder rotating disc and its action has a critical role in arrangement of collagen fibers, influencing strength and flexibility of the final collagen casings. The wet extrusion process typically produces shorter collagen fibers and provides for faster processing and larger production volumes. Dry extrusion technology produces thicker, longer collagen fibers and more expensive casings (Hood 1987).

A number of patents have been issued on improvements to collagen casings. Burke (1976) described a process for preparing collagen casings from limed bovine hide collagen that was soaked in an edible acid at pH < 5.5. Hides were then neutralized and washed, ground and formed into collagen slurry, and processed into collagen casings. Wilson and Burke (1977) reported that water and a mixture of partial fatty acid esters of glycerin and sorbitol could be used to coat the inside of unshirred collagen casings to improve the shirring process. Ziolko (1977, 1979) produced tubular collagen casing by extruding two gel "ropes" from collagen, with each individual rope consisting of several smaller ropes. The casing extruder was equipped with inner and outer orifices and extruded the ropes, orienting fibers in the direction of extrusion. Directing them outward in a radial and helical direction formed the first tubular layer; the second tubular layer was formed by directing inward in a radial and helical direction opposite from the first layer that was concentric to the first. The two layers were then united by a hardening step.

It is important that a sausage casing be able to stretch and shrink to withstand contraction and expansion of meat batter during various steps of processing (Schmidt 1986). Collagen casings must be strong enough to withstand the rigors of high-speed filling and linking, as well as to withstand high temperature and humidity conditions encountered during sausage manufacturing. Collagen fiber content, drying conditions, and extent of cross-linking influence the strength of collagen casings. Cross-linking agents, such as gluteraldehyde, provide for stronger and more coherent films of collagen fibrils that possess improved longitudinal and transverse strength, regardless of whether the casing is re-wetted or dry (Rose 1968). Glyceraldehyde (which lacks toxicity associated with other aldehydes used for cross-linking) also promotes cross-linking of fibrils and increases film strength and temperature resistance (Gennadios et al. 1994). Carboxymethylcellulose and salt treatments prevent disintegration of collagen casings during high-temperature cooking or frying (McKnight 1964a, b). UV irradiation of 180–420 nm increased collagen casing strength (Miller and Marder 1998).

Tsuzuki and Lieberman (1972) reported that enzymatic treatments employed in manufacture resulted in collagen casings with a more uniform diameter, greater thickness, and increased tenderness. Addition of protease from *Aspergillus niger* var. *macroporous* to collagen batter partially solubilized collagen prior to the extrusion step. Similarly, Miller (1983) reported that the wet process could be improved by incorporating proteases such as papain, bromolain, ficin, fungal protease, trypsin, chymotrypsin, or pepsin into the collagen batter prior to extrusion. Boni (1988) reported that mechanical properties of collagen casings at low temperatures were improved by incorporating an alkyl diol into the collagen batter to reduce intermolecular hydrogen bonding and increase molecular spacing. More recently, coating of sausage batter with a thin layer of collagen material has been accomplished by co-extruding sausage batter and collagen (Deacon and Kindleysides 1973; Hood 1987; Smits 1985; Waldman 1985; Kobussen et al. 1999).

In addition to collagen, other proteins such as casein, whey proteins, soy protein, wheat gluten, corn zein and egg albumin have also been investigated in the production of edible films and coatings for meat applications (Ben and Kurth 1995). Such films are good barriers to gases such as oxygen and carbon dioxide, and adhere well to hydrophilic surfaces. But being hydrophilic in nature, these coatings are typically poor barriers to moisture, limiting their application in meat products (Gennadios and Weller 1990; Baldwin et al. 1995). Also, there has been some concern regarding their potential allergenicity, especially in the use of milk, egg, peanut, soybean and rice proteins in films and coatings.

8.3 Technical Developments

8.3.1 Protein-Based Films and Coatings

Co-extrusion technology is an alternative to the conventional method of stuffing sausage batter into a preformed collagen casing for manufacture of sausage (Deacon and Kindleysides 1973). Smits (1985) further refined the process. This process involves simultaneous co-extrusion of collagen and sausage emulsion to create a thin layer of collagen coating at the meat batter surface. The end of the extruder barrel is equipped with a nozzle through which sausage batter flows out. A thin layer of collagen suspension is extruded directly onto the surface of the sausage as it is pumped through the extruder. Counter-rotating concentric cones are employed to orient the collagen fibers in a woven structure. MC is incorporated to provide additional strength to the casing and reduce its disintegration during cooking (Hood 1987). The collagen-coated sausage is then immersed in a brine bath to further set the collagen casing. Smoking and drying of sausages in subsequent steps allow further interactions between the collagen fibers and meat proteins. Co-extrusion produces a casing that is more tender than preformed or reconstituted collagen casings (Waldman 1985). Smits (1985) has described a continuous co-extrusion process that is capable of large production volumes. Over the years, various modifications to the co-extrusion process have been reported. Kobussen et al. (1999) developed a modified extrusion attachment for sausage stuffing machines. This attachment consists of three passageways, where meat emulsion is extruded from two passageways and is coated with a thin layer of extruded collagen pumped through the third passageway. The collagen coating is set with liquid smoke and a dehydrating agent (i.e., alkali or salt solution) to produce a casing that is strong enough that it can be twisted to produce sausage links.

The "hybrid technology" for manufacture of collagen casings was described by Osburn (2002). This technology combines fundamentals of dry and wet processing technologies described previously for collagen-casing manufacture. In the hybrid process, collagen slurry (no more than 5% collagen) is prepared similar to wet processing (dehairing of cattle hides by acid or alkaline, decalcification, grinding and mixing with acid to produce swollen slurry, high shear homogenization). The tubular collagen casing is extruded, and inflated with an ammonia–air mixture to prevent the collapse of the casing. In the next steps, the casing tube is flattened, washed and plasticized in sequence in a manner similar to the wet extrusion process. Then, the casings are re-inflated, dried, and shirred. The hybrid process produces collagen casings with improved tensile strength and smaller diameter. They are usually straight and suitable for manufacture of small sausage casings. But this process is limited in the manufacturing of larger diameter and longer sausage casings and where curved casings may be needed.

Collagen is also used for production of sheet films (cut to the dimensions required by the customer) and in rolls (continuous film reeled on a cylindrical core; typical film length: 50 m or 100 m; typical film widths between 380 and 620 mm), which are used as wraps for netted meats, such as hams or roasts, to prevent the elastic net from sticking to the meat product (Gennadios et al. 1997). In the course of the cooking and smoking process, the collagen film becomes an integral component of the meat, allowing the elastic net to be easily removed from finished product without doing harm to the meat surface, providing for an attractive appearance. Collagen films are also reported to reduce shrink loss and increase juiciness in these netted products. Farouk et al. (1990) studied Coffi®, an edible collagen film (Breechteen Co.) widely used in specialty smoked meats and roasted meat products, which are heat-processed in elastic stretch netting or coarse stockinette. They observed significant reduction in exudation of meat products, including beef round steaks, upon wrapping in Coffi® compared to unwrapped controls. However, wraps had no significant effect on color or lipid oxidation of the meats.

Over the years, the growing US processed-meat industry has relied heavily on imported collagen to meet the needs of production. Since the late 1990s, the safety of imported collagen has become a concern owing to the link between bovine spongiform encephalopathy (BSE or "mad cow" disease) and Creutzfeldt–Jacob disease, a highly fatal disease in humans. On May 20, 2003, the USDA placed an import restriction on ruminant material coming from Canada because of possible risk of BSE. Whey protein-based films and casings have been proposed as alternatives to collagen films and sausage casings (Kim and Ustunol 2001a–c; Simelane and

Ustunol 2005; Amin and Ustunol 2007). Kim and Ustunol (2001a) developed whey protein/lipid emulsion edible films that were heat-sealable, and could be formed into tubular casing. They reported increased hydrogen and covalent bonding (involving C–O–H and N–C) as the main forces responsible for the sealed joint formation of the films. These films were further modified through heat-curing to make them more closely resemble collagen films that would be amenable to hot dog manufacture (Amin and Ustunol 2007). Heat-cured films were further tested under various combinations of time, temperature and relative humidity typically encountered in meat processing. Manufacture of whey protein-based sausage casings is viable because of the high film degradation temperatures (>130°C) and heat sealability of these whey-based films. Furthermore, the ability of these films to withstand temperature/time/relative humidity conditions that would be typically encountered during sausage manufacturing has confirmed their suitability (Simelane and Ustunol 2005).

The same research group subsequently developed heat-sealable whey protein–based casings containing sorbic acid, p-aminobenzoic acid, or sorbic acid–p-aminobenzoic acid (1:1) for use in hot dog manufacture, and compared these casings to commercial collagen and natural casings (Cagri et al. 2002, 2003). Antimicrobial properties of these casings are further discussed in Sect. 8.3.5.

Limited information exists on use cereal and oilseed proteins (i.e., soy protein, wheat gluten, corn zein) as edible films in meat and poultry applications. Corn zein and corn zein–tocopherol coatings were reported to reduce lipid oxidation in precooked pork chops, but did not retard shrinkage (Hargens-Madsen et al. 1995). Herald et al. (1996) reported that addition of an antioxidant, emulsifier and plasticizer to corn zein films reduced rancidity in cooked turkey breast slices; however, they also reported production of off-flavors. Kunte (1996) reported that a 7S soy protein fraction-based coating was not effective in preventing warmed-over flavor (WOF) formation in pre-cooked chicken breast, compared to phosphates. Soy protein and wheat gluten coatings were shown to be as effective as polyvinyl chloride films in reducing moisture loss on coated precooked beef patties during 3 days of refrigerated storage (Wu et al. 2000). Furthermore, both soy protein and wheat gluten coatings reduced lipid oxidation, as indicated by decreased thiobarbituric acid and hexanal values, for coated meat samples relative to those that were not coated. Rhim et al. (2000) reported that wheat gluten and soy protein coatings were as effective as polyvinyl chloride films in reducing moisture loss. Wheat gluten coatings were also effective in reducing lipid oxidation.

8.3.2   Lipid Films and Coatings

Lipids suitable for use in edible films and coatings include waxes (i.e, candelilla, carnauba, beeswax), acylglycerol (vegetable or animal origin), as well as fatty acids and derivates (Kester and Fennema 1986). Lipids, because of their hydrophobicity and tightly packed crystalline structure, are good barriers to moisture and gas migration. Their effectiveness, however, is dependent on lipid type, chemical structure, chemical arrangement, polarity/hydrophobicity and physical state (solid or liquid), and on its interactions with other components (i.e., proteins and polysaccharides). Lipid films and coatings lack structural integrity, and do not adhere well to hydrophilic surfaces (Ben and Kurth 1995). Lipids are also used with proteins and polysaccharides in production of composite films and coatings to improve moisture barrier characteristics and provide flexibility (Baldwin et al. 1995). They can also serve as "release agents" or lubricants to prevent coated foods from sticking together or from sticking to other surfaces such as packaging material (Baldwin et al. 1995).

"Larding" is a process of coating meats with fat to extend shelf life, and was practiced in sixteenth century England. Fats are still used today to coat a variety of foods, including fresh and frozen meat and poultry (Hernandez 1994). McGarth (1955) reported that a coating of wax eliminated downgrading of fresh meats due to discoloration. The wax provided a transparent, attractive coating that also afforded ease in handling product at the supermarket by minimizing mechanical damage. Since then, carnauba wax, beeswax, and candelilla, have been used successfully to coat frozen meat pieces to reduce dehydration during frozen storage (Daniels 1973). Letney (1958) used molten fats as coatings by allowing them to solidify over fresh meat surfaces to form a film/coating to lengthen storage life at refrigerated temperatures, lessen surface hydration and maintain "bloom." Ayers (1959) reported that an acetylated acylglycerol coating with incorporated chlortetracycline was effective in reducing off-flavors and retaining moisture; however, it produced an unappealing meat color. Three patents for lipid-based coatings were issued to Anderson (1960, 1961a, b). Coating solutions, containing 40% mono-, di- and tri-acylglycerol in alcohol, were heated to 50°C, and sprayed directly on the meat surface or applied to meat products in the molten state by dipping. Although initial studies with fresh lamb were not successful, application of coatings to other fresh meats prior to freezing improved color and texture. Schneide (1972) coated beef, veal, pork steaks and fish fillets with different compositions of mono-, di- and triacylglycerols of paraffinic and/or carboxylic acids to improve keeping quality of meat and maintain its desirable sensory properties. Stemmler and Stemmler (1976) described coating formulations containing cellulose proprionate and acetylated monoacylglycerol (obtained from lard) used to prolong freshness, color, aroma, tenderness and microbiological stability of fresh beef and pork cuts. Heine et al. (1979) reported that an acetylated mixture of fatty acid mono-, di-, and triacylglycerols applied to fresh beef and pork pieces helped retain desirable color and reduced moisture loss during 14 days of refrigerated storage. A product called Myvacet® (Eastman Chemical Products, Inc., Kingsport, TN), a distilled acetylated monoacylglycerol, has been marketed as a protective coating for frozen poultry. It is applied as a dip or spray prior to the freezing of meat; the coating is then left on product during cooking (Hernandez 1994). Griffin et al. (1987) and Leu et al. (1987) coated vacuum-packaged strips of loins and top rounds with an acetylated monoacylglycerol coating. They observed no differences in microbial counts, color, or odor of vacuum-packaged steaks or roasts after 4 and 7 weeks of refrigerated storage, respectively. Acetostearin films have also been reported to provide oxidative stability when used on meat surfaces, whereas acetoolein films were less resistant to oxidation (Feuge 1955; Baldwin 1994).

Water-in-oil emulsions from animal fats or vegetable oils have been used as protective coatings and as flavor carriers for meat and poultry products (Bauer et al. 1968; Hernandez 1994; Baldwin et al. 1997). Emulsion coatings were effective in protecting frozen chicken pieces and pork chops from dehydration. Increased meat yield, decreased moisture loss and flavor, and improved tenderness were observed when these meats were coated (Zabik and Dawson 1963; Kroger and Igoe 1971; Baldwin et al. 1997). Addition of MC to water-in-oil emulsion coatings provided for more stable coatings at lower temperatures and prevented excessive moisture loss during cooking for pork, beef, chicken, comminuted meat, sausages, fish and seafood (Bauer and Neuser 1969).

8.3.3 Polysaccharide Films and Coatings

Starch, alginate, dextrins, pectin, chitosan and carrageenans are used in polysaccharide films and coatings. Water-soluble polysaccharides are long-chain polymers that are typically used in foods for thickening and/or gelling abilities (Glicksman 1983; Whistler and Daniel 1990; Nisperos-Carriedo 1994). Polysaccharide films are good barriers to gases, and possess resistance to fats and oils; however, their hydrophilic nature makes them poor water vapor barriers (Ben and Kurth 1995). They have been used to extend shelf life of meat and poultry products by retarding dehydration, oxidative rancidity and surface browning. The ability of some polysaccharides (e.g., methylcellulose) to form thermally induced gelatinous coatings has also made them desirable for reducing oil absorption during frying (Nisperos-Carriedo 1994).

Starch, being one of the largest component biomasses produced on earth, is abundant and readily available for use in edible films and coatings. Starch is available from different botanical sources, including wheat, corn, rice, potato, cassava, yam, and barley, among others; corn represents the major commercial source (Riaz 1999). There are also several genetic mutant varieties of corn created to alter starch content and the ratio of amylose to amylopectin, the two main polymers of starch. Starch-based films are similar to plastic films in their properties; they are odorless, tasteless and colorless. They are nontoxic, biologically absorbable and semipermeable to carbon dioxide and are good barriers to oxygen (Nisperos-Carriedo 1994). Ediflex®, an extruded hydroxypropylated high-amylose starch film, was developed and used as a wrap for frozen meats and poultry in the late 1960s (Anonymous 1967; Kroger and Igoe 1971; Morgan 1971; Sacharow 1972). Ediflex® was flexible, a good oxygen barrier, oil-resistant and heat-sealable, and therefore was effective in protecting meat products during frozen storage. The coating also dissolved during thawing and cooking (Kroger and Igoe 1971; Morgan 1971; Sacharow 1972).

Alginates are derived from brown seaweed and, in the presence of divalent cations, produce films that are particularly useful for applications that enhance quality of meat and poultry products (Kester and Fennema 1986; Nisperos-Carriedo 1994). Calcium is the most common and effective divalent cation in gelling alginates, though magnesium, manganese, aluminum, ferrous, or ferric ions are also used (Kester and Fennema 1986). At levels of 5 M or greater, calcium chloride may provide a bitter taste to foods, so other calcium salts may be substituted or lower levels used (Baker et al. 1994). Alginate film strength may further be improved in the presence of modified starches, oligosaccharides or simple sugars (Gennadios et al. 1997).

Allen et al. (1963) coated beefsteaks, pork chops and chicken drumsticks in alginate or cornstarch by dipping them first in coating solution and then in calcium chloride solution. Coatings were effective in reducing moisture loss and, therefore, shrinkage during 1 week of refrigerated storage. They also provided improved juiciness, texture, color and odor within the meats. However, calcium chloride contributed a bitter taste. Calcium gluconate, nitrate, or proprionate provides more acceptable flavor; however, due to their weak ionizing properties, the coatings are not as strong as those formed with calcium chloride (Allen et al. 1963; Hartal 1966). Lazarus et al. (1976) reported that alginate coatings gelled by calcium chloride reduced weight loss of lamb carcasses during refrigerated storage. Although the alginate coatings were not as effective as plastic films in preventing weight loss, bacterial counts were significantly reduced in the alginate-coated carcasses. In other studies, alginate coatings were not reported to affect cooking loss, flavor, odor, or overall acceptability of coated beef (Mountney and Winter 1961), pork (Nisperos-Carriedo 1994) and poultry products (Williams et al. 1978). Alginate coatings also did not decrease weight loss, off-odor and drip of beef cuts compared to noncoated controls (Williams et al. 1978). The coated meats were acceptable in flavor, tenderness, appearance and cooking loss. However, after 4.5 months of refrigerated storage, drying of the coating was noted.

As consumer demand for more convenient foods has increased, considerable research effort has been focused on improving quality of precooked meat products. Precooked meat products are susceptible to lipid oxidation, which results in a rancid or warmed-over flavor (WOF) during refrigerated storage (Tims and Watts 1958; St. Angelo et al. 1987; Love 1988). Moisture loss is also a critical factor affecting quality and shelf life of precooked meats. Among the many techniques used, edible films and coatings have also been studied as a means for controlling quality of precooked meats. Wanstedt et al. (1981) observed that alginate-coated precooked, frozen and stored pork patties had reduced oxidative off-flavors (thus, WOF), better sensory properties and greater desirability than the uncoated controls. Coatings of starch–alginate, starch–alginate–tocopherol and starch–alginate–rosemary have been reported to reduce WOF in precooked, refrigerated pork chops and beef patties (Hargens-Madsen et al. 1995; Ma-Edmonds et al. 1995; Handley et al. 1996). Starch–alginate–stearic acid composite films were effective in controlling lipid oxidation, WOF and moisture loss in precooked ground beef patties (Wu et al. 2001). Tocopherol-treated films were more effective than the non-tocopherol controls;

however, none of the films was as effective as polyester vacuum bags in retarding moisture loss and lipid oxidation.

Cellulose and its derivatives also produce edible films and coatings that are water-soluble, tough, flexible and resistant to fats and oils (Krumel and Lindsay 1976; Baldwin et al. 1995). However, their application to meat and poultry products has been limited. Ayers (1959) reported that application of methylcellulose (MC)–lipid coatings prevented desiccation and extended shelf life of beefsteaks. These coatings were transparent and peeled easily. MC and HPMC coatings on meats and poultry are used during deep fat frying to reduce oil uptake and moisture loss (Balasubramaniam et al. 1997; Meyers 1990), and extend shelf life of the frying oil (Balasubramaniam et al. 1997; Holownia et al. 2000). They can also help maintain integrity of the breading or batter during frying. Specifically, CMC has been shown to improve breading adhesion (Suderman et al. 1981).

Carrageenan is a complex mixture of polysaccharides extracted from red seaweed (Nisperos-Carriedo 1994). Carrageenan-based coatings have been reported to extend shelf life of poultry (Pearce and Lavers 1949; Meyers et al. 1959). Carrageenan and agar (another seaweed-derived polysaccharide) coatings containing antibiotics were shown to extend shelf life of coated poultry (Meyers et al. 1959) and beef (Ayers 1959), though coatings did not reduce moisture loss. More recently, incorporation of nisin (a bacteriocin) into agar coatings together with food-grade chelators (EDTA, citric acid, or polyoxyethylene sorbitan monolaureate) effectively reduced levels of *Salmonella typhimurium* on poultry products (Natrajan and Sheldon 1995).

Limited reports are available on application of chitosan and pectin films and coatings to meat and poultry products. Stubbs and Cornforth (1980) showed that calcium pectinate gel coatings significantly reduced shrinkage and bacterial growth for beef patties.

8.3.4 Composite Films

A single-component film generally has either good barrier or good mechanical properties, but typically not both. Thus, multiple components are often combined to form composite films with desired properties. Polysaccharides and proteins, owing to their intermolecular interactions, form films and coatings that have good structural integrity and mechanical properties; but they are generally not effective barriers to moisture because of their hydrophilic nature. Lipids, on the contrary, are good barriers to moisture as a result of their hydrophobic character, though their films lack structural integrity, are brittle, and do not adhere well to hydrophilic surfaces. In formation of composite films and coatings, two or more components are combined to improve mechanical properties, gas exchange, adherence to surfaces and/or moisture barrier properties (Baldwin et al. 1995). They may be applied as emulsions or bi-layers. Plasticizers, such as glycerol, polyethylene glycol, or sorbitol, may be added to modify film mechanical properties and provide increased flexibility (Ben and Kurth 1995). Cross-linking of proteins with enzymes (e.g., transglutaminase) or polysaccharides with polyvalent ions (e.g., calcium cations) may be employed during the film-forming process, and various additives to improve specific film attributes may also be incorporated. Ben and Kurth (1995) reported that clarity of caseinate–lipid composite films could be improved by addition of three enzymes to cross-link sodium caseinate molecules. Resulting films had improved appearance, and meat was juicier because of reduced drip loss. Packaging waste and handling was also reduced owing to the elimination of an absorbent pad (Ben and Kurth 1995).

8.3.5   Antimicrobial Films and Coatings

Edible films and coatings can serve as carriers for a wide range of food additives, including antimicrobials, which can reduce microbial growth at meat and poultry surfaces to improve product safety and extend product shelf life. The primary advantage of antimicrobial edible films and coatings is that inhibitory agents in these films can be specifically targeted to postprocessing contaminants on the food surface, with diffusion rates of antimicrobials into the food product being partially controlled by properties of the antimicrobial agent itself, as well as the properties of the film.

As consumer demand for convenience continues to increase, demand for ready-to-eat (RTE) foods will also continue to grow. Quality, safety and shelf life of RTE foods is often dictated by the type and numbers of pathogenic and spoilage bacteria present on the food surface. Approximately two-thirds of all microbiologically related class I recalls in the United States result from postprocessing contamination during subsequent handling and packaging, as opposed to processing itself. Most of these recalls are prompted by contamination with *Listeria monocytogenes,* for which the United States has maintained a policy of zero tolerance since 1985. From January 1998 to February 2003, over 130 *Listeria*-related class I recalls involving more than 80 million pounds of cooked RTE meats have been issued (CDC 2000). More than 35 million pounds of hot dogs and luncheon meats were voluntarily recalled in 1998 by one Michigan manufacturer in response to an outbreak that resulted in 101 cases of listeriosis (including 21 fatalities) spread across 22 states. Two years later, another listeriosis outbreak involving 29 cases and ten states (including seven fatalities) prompted recall of approximately 14.5 million pounds of turkey and chicken delicatessen meat; again, the product became contaminated with *L. monocytogenes* after processing (USDA-FSIS 2001). More recently, the largest product recall ever issued, 27.4 million pounds of fresh and frozen RTE turkey and chicken products, was linked to another major outbreak of listeriosis emanating from a manufacturer in Pennsylvania. Each year, approximately 2,300 cases of foodborne listeriosis are reported in the United States at an estimated cost of $2.33 billion (~$1 million per case), making *L. monocytogenes* the second costliest foodborne pathogen after *Salmonella* (2.38 billion) (USDA-FSIS 2001).

Product slicing and packaging operations are points at which both pathogenic and spoilage organisms can be introduced to cooked RTE foods. In commercial manufacturing facilities, slicing of RTE meat products can easily increase microbial population 100-fold or greater (Cagri et al. 2004). Heightened consumer demand for enhanced keeping quality and freshness of RTE foods has given rise to the concept of active packaging – a type of packaging that alters conditions surrounding the food to maintain product quality and freshness, improve sensory properties, or enhance product safety and shelf life. A specific function of active packaging includes restriction of antimicrobial activity through controlled diffusion of one or more antimicrobial agents from packaging material into the product. Antimicrobial edible films and coatings can potentially serve as active food packaging materials by altering permeability of a product to water vapor and oxygen, as well as by minimizing growth of surface contaminants during refrigerated storage, providing an alternative to postprocess pasteurization for inactivation of surface contaminants.

Some of the more commonly used preservatives and antimicrobials in edible films and coatings include benzoates, proprionates, sorbates, parabens, acidifying agents (e.g., acetic and lactic acids), curing agents (e.g., sodium chloride and sodium nitrite) bacteriocins, and natural preservatives (e.g., natural oils, lysozyme, liquid smoke) (Cagri et al. 2004). Antifungal compounds, organic acids, potassium sorbate, or the bacteriocin, nisin, were reported to be more effective in reducing levels of foodborne microorganisms when immobilized or incorporated into edible gels (i.e. starch, carrageenan, waxes, cellulose ethers, or alginate) and applied to meat surfaces than when these agents were used alone (Cutter and Sumner 2002).

Incorporation of antibiotic and antifungal compounds into carrageenan films was reported by Meyers et al. (1959), who showed a two log reduction in bacterial counts at the surfaces of coated poultry products. Hotchkiss (1995) reported incorporating antimycotics into wax and cellulose-based coatings. Siragusa and Dickson (1993) reported that organic acids were more effective in reducing *L. monocytogenes, S. typhimurium* and *E. coli* O157:H7 levels on beef carcasses when they were immobilized in calcium alginate gels (and applied), compared to when they were applied alone. Baron (1993) verified that potassium sorbate and lactic acid incorporated into an edible cornstarch film inhibited growth of *S. typhimurium* and *E. coli* O157:H7 on surfaces of poultry. Incorporation of the bacteriocin, nisin, into calcium alginate gels was investigated by Cutter and Siragusa (1996, 1997). They showed greater reductions in bacterial populations and greater retention of nisin activity on both lean and adipose beef surfaces when nisin was immobilized in an alginate gel matrix, compared to controls that were treated with only nisin. Natrajan and Sheldon (2000) reported that incorporation of nisin and chelators into protein- and polysaccharide-based films inhibited growth of *Salmonella* on poultry skin. Control of *L. monocytogenes* at the surface of refrigerated RTE chicken was achieved with edible zein film coatings containing nisin and/or calcium propionate (Janes et al. 2002). McCormick et al. (2005) reported that wheat gluten films containing nisin were effective in reducing populations of *L. monocytogenes*, but were not effective against *S. typhimurium*. Inhibition of *L. monocytogenes* was demonstrated on turkey frankfurters coated with soy protein films containing nisin combined with grape seed and green tea extracts (Theivendran et al. 2006). Miller and Cutter (2000) incorporated nisin into collagen-based films to provide protection against pathogenic and spoilage organisms. A sausage casing containing sorbate and glycol was commercially marketed by Union Carbide (Labuza and Breene 1989).

Incorporation of essential oils has also been investigated in production of antimicrobial edible films and coatings. Although antimicrobial properties of essential oils have been recognized for centuries, there has been renewed interest in their use because of consumer demand for natural ingredients and additives. Essential oils are responsible for odor, aroma and flavor of spices and herbs. These compounds are added to edible films to modify flavor, aroma and odor, as well as to impart antioxidant and antimicrobial properties. Films containing these ethanol-soluble compounds show activity against both Gram-positive and Gram-negative bacteria. Essential oils of angelica, anise, carrot, cardamom, cinnamon, cloves, coriander, dill weed, fennel, fenugreek, garlic, nutmeg, oregano, parsley, rosemary, sage and thymol are inhibitory to various spoilage and/or pathogenic bacteria, as well as molds. Oussalah et al. (2004) produced milk protein-based edible films containing oregano, pimiento and oregano-pimiento mixtures. The oregano-containing films provided the greatest antimicrobial activity against *E. coli* O157:H7 and *Pseudomonas* spp. on beef muscle slices, whereas pimiento-containing films were least effective against these two bacteria. More recently, Zivanovic et al. (2005) reported that chitosan films enriched with essential oils (anise, basil, coriander and oregano) had strong antimicrobial effects on *L. monocytogenes* when applied to inoculated bologna samples.

Other antimicrobial films based on whey protein isolate (WPI) containing sorbic acid or *p*-aminobenzoic acid were developed by Cagri et al. (2001). Both films reportedly inhibited *L. monocytogenes*, *E. coli* O157:H7 and *typhimurium* DT104 on a nonselective plating medium. Subsequently, these films were tested with beef bologna and summer sausage slices that were surface-inoculated with these same pathogens at levels of 10^6 CFU/g (Cagri et al. 2002). WPI films containing sorbic or *p*-aminobenzoic acid decreased *L. monocytogenes*, *E. coli* O157:H7 and *S. typhimurium* DT 104 populations by 3.4–4.1, 3.1–3.6 and 3.1–4.1 log, respectively, on bologna and sausage slices after 21 days of aerobic storage at refrigerated temperature. Growth of mesophilic aerobic bacteria, lactic acid bacteria, mold, and yeast on meat slices was also inhibited by the same antimicrobial WPI films, compared to those coated with antimicrobial-free control films.

The same research group subsequently developed heat-sealable WPI casings containing sorbic acid, *p*-aminobenzoic acid, or sorbic acid–*p*-aminobenzoic acid (1:1 ratio) for hot dog manufacture (Cagri et al. 2003). WPI casings containing *p*-aminobenzoic acid inhibited *L. monocytogenes* on hot dogs for 42 days of refrigerated storage; however, films containing sorbic acid or sorbic acid–*p*-aminobenzoic acid were less effective. Sensory (texture, flavor, juiciness, overall acceptability), chemical (thiobarbituric acid; pH; moisture, fat, and protein contents), physical (purge, color) and mechanical (shear force) characteristics of hot dogs processed within WPI casings containing *p*-aminobenzoic acid were comparable to those prepared with collagen and natural casings. Thus, WPI casings containing *p*-aminobenzoic acid may eventually provide a viable alternative to postprocess pasteurization of hot dogs for minimization of *Listeria* risk.

Incorporation of other antimicrobials into edible films and coatings such as liquid smoke, sodium chloride and nitrites, has not been extensively studied. Liquid smoke, which contributes antimicrobial, antioxidant, color and flavor attributes, is also a potentially attractive film additive. Liquid smoke has been studied in conjunction with edible collagen casings, where liquid smoke is introduced into the acid-swollen collagen mass before extrusion as a casing or film (Miller 1975). Because liquid smoke is generally very acidic (pH 2.5 or less), it is compatible with the gel system, and can replace a portion of acid normally added to induce swelling of collagen. Sodium chloride (common salt) has been recognized as a food preservative since ancient times, and can be used alone or in combination with other preservation techniques, such as pasteurization or fermentation. Most bacterial foodborne pathogens are susceptible to elevated concentrations of salt, particularly in the presence other preservatives. Incorporation of salt into protein-based films (e.g., whey protein) as an antimicrobial agent is of limited use, since physical properties of protein films are altered with increasing ionic strength. At high ionic strength, proteins aggregate to form turbid, opaque gels with high water-holding capacity (Smith and Culbertson 2000). Nitrite has not yet been studied as an edible film additive, although it appears to be suitable for production of antimicrobial edible films. Application of films containing nitrite to RTE meat products may not only help prevent growth of *L. monocytogenes* and spoilage bacteria that can contaminate such products after processing, but might also improve meat surface color.

Release of antimicrobial agents from an edible film and controlling the conditions for release are very important considerations for effectiveness of antimicrobial films and coatings. Release of antimicrobial substances from edible films is dependent on many factors, including electrostatic interactions between the antimicrobial agent and polymer chains, ionic osmosis and structural changes induced by the presence of the antimicrobial and environmental conditions. Diffusion of antimicrobials through the edible film is also influenced by film (type, manufacturing procedures), food (pH, water activity), hydrophilic characteristics and storage conditions (temperature, relative humidity, duration).

8.3.6 Incorporation of Other Additives

Many potential benefits of edible films as carriers of other additives (e.g., flavors, antioxidants, coloring agents, vitamins, probiotics and nutraceuticals) justify continued research in the field of active packaging. However, only antioxidants have been investigated extensively. Although oxygen-barrier properties of edible films and coatings may reduce need for oxidative stabilization, antioxidants have been used in edible coatings to provide further protection to meat and poultry products.

Incorporation of antioxidants such as gallic or ascorbic acids into carrageenan coatings to extend shelf life of poultry products has been reported as early as 1948 (Stoloff et al. 1948; Allingham 1949). Antioxidants were incorporated into a mixture of lard and tallow coating containing lactic acid–fatty acid triacylglycerol, which was used to coat freeze-dried and fresh meats, including beef steaks, pork chops and beef cubes (Sleeth and Furgal 1965). Thiobarbituric acid levels were significantly reduced in meats to which antioxidant-containing coatings were applied, compared to those of non-coated controls (Sleeth and Furgal 1965). Pork chops treated with alginate–starch coatings containing tocopherol were reported to be juicier and less susceptible to lipid oxidation, compared to the untreated controls. However, they still developed off-flavors (Hargens-Madsen et al. 1995). Herald et al. (1996) reported that turkey breasts wrapped in corn zein films containing butylated hydroxyanisole (BHA) had a lower hexanal content than samples packaged in polyvinylidene chloride (PVDC).

Essential oils from oregano, sage rosemary, thyme and pimiento are reported to possess antioxidant properties comparable to or greater than BHA or butylated hydroxytoluene (BHT) (Wu et al. 1982; Shelef 1983; Howard et al. 2000; Bendini et al. 2002). Oussalah et al. (2004) have reported on antioxidant properties of milk protein-based edible films containing oregano, pimiento and oregano-pimiento mix. Pimiento-containing films provided the highest antioxidant activity on beef muscle slices; oregano-based films were also able to inhibit lipid oxidation in beef muscle samples. Armitage et al. (2002) evaluated egg albumen coatings with natural antioxidants, fenugreek, rosemary and vitamin E, for antioxidant activities in diced raw and diced cooked poultry breast meat. In both raw and cooked experiments, coatings with no added antioxidant showed greatest inhibition of lipid oxidation as monitored by thiobarbituric acid reactive substances (TBARS).

Although some edible films have received consumer acceptance, further research is needed to develop cost-effective production and application methods for continuous extrusion and application of films and coatings as flat sheets or casings.

8.4 Film-forming Techniques and Application

Several different techniques, including solvent removal, thermal gelation and solidification of melt, have been developed for forming edible films and coatings. Solvent removal has been used to produce hydrocolloid edible films. In this process a continuous structure is formed and stabilized by chemical and physical interactions between polymer molecules. Macromolecules in film-forming solution are dissolved in a solvent, such as water, ethanol or acetic acid, which contains other functional additives (plasticizers, cross-linking agents, solutes). Film-forming solution is then cast in a thin layer, dried and peeled from the surface. Freestanding films are formed by a casting technique in which film-forming solutions are poured onto a smooth, flat, level surface, and allowed to dry, usually within a defined surface (Donhowe and Fennema 1994).

In preparing some types of protein films (e.g., whey protein, casein, soy protein, wheat gluten), solution is heated to promote protein gelation and coagulation, which involves denaturation, gelation, or precipitation, followed by rapid cooling. Intramolecular and intermolecular disulfide bonds in the native protein complex are cleaved and reduced to sulfhydryl groups during protein denaturation (Okamoto 1978). When film-forming solution is cast, disulfide bonds re-form to link polypeptide chains together to produce the film structure, which is further stabilized by hydrogen and hydrophobic bonding.

Melting followed by solidification is a means for producing lipid-based films. Casting molten wax on dried methylcellulose films followed by solubilization of the methylcellulose can also be used to form wax films (Donhowe and Fennema 1993).

Edible films and coatings may be applied to meat and poultry products by foaming, dipping, spraying, casting, brushing, individual wrapping, or rolling (Donhowe and Fennema 1994; Grant and Burns 1994). Foam application is used for emulsion coatings, where a foaming agent is added to the coating or foam is created by compressed air; meat may be coated with protective foam as it moves over rollers. Flaps or brushes may also be used to distribute coating over the product (Grant and Burns 1994). Irregular surfaces of food products present particular challenges to uniform application of coatings. Thus, several applications may be needed to obtain a uniform coating on a food surface. Emulsions may work in case of multiple applications where the coating will need to set (or solidify). This is done typically after dipping product that is being coated, letting the coating drain and set prior to application of the next coating layer (Donhowe and Fennema 1994). Spraying is most suitable where thinner films are needed, or when a coating is applied to only one side of a product. Spraying is also used when dual applications are used to promote cross-linking (e.g., with alginate–Ca^{2+} coatings) (Donhowe and Fennema 1994). Early procedures for coating meat and poultry products involved spraying materials onto rollers or brushes, and allowing meat to be coated by the tumbling action of the product (Grant and Burns 1994). Brushing and/or rolling the coating directly onto meat surfaces may also be used to apply edible coatings. Edible films and coatings applied to food products will need to set or solidify at the food surface. This requirement can be accomplished at ambient temperatures or with the use of heat. Shorter drying times tend to provide for more uniformly distributed coatings (Grant and Burns 1994).

Other than on collagen films and casings, only limited information is available on continuous extrusion of freestanding films for meat applications.

8.5 Regulatory Aspects

The process of determining acceptability of most ingredients used in edible films and coatings is similar to that used for food formulations (Heckman and Heller 1986). If the edible film or coating is made from materials that have been generally recognized as safe (GRAS) and its use in the film or coating is in accordance with good manufacturing practices, it will be considered GRAS. If materials are not GRAS, but safety of the film or coating can be demonstrated, the manufacturer may file a GRAS Affirmation Petition with FDA or proceed without FDA concurrence on the basis of self-determination. The manufacturer may not need to establish that materials used are GRAS if the film or coating materials have received a pre-1958 FDA clearance or "prior sanction". If materials used in films and coatings are not GRAS or have not received prior sanction, then a food additive petition to the FDA must be filed (Krochta and De Mulder-Johnston 1997).

With the exception of chitosan, polysaccharides including cellulose and its approved derivatives (CMC, MC, HPMC), starches and their approved derivatives, and seaweed extracts (agar, alginates, carrageenan), are either approved food additives or GRAS substances. Regulations regarding their specific use are described in the Code of Federal Regulations (CFR) Title 21 (Nisperos-Carriedo 1994). Many of the lipid compounds used are classified either as GRAS or food additives (FDA 1991). Beeswax, carnauba wax, candelilla wax, carnauba wax, mono-, di-, and triacylglycerols, glyceryl monooleate, glyceryl monostearate and steric acid are GRAS substances. Acetylated monoacylglycerol is classified as a multipurpose additive; fatty acids are approved as lubricants and defoamers (CFR Title 21; Baldwin et al. 1997). Proteins that are most commonly used for edible films and coatings, such as corn zein, wheat gluten, soy protein and milk proteins, have GRAS status. However, some concerns have been raised due to allergenicity or intolerance some consumers have to wheat proteins or lactose. Since edible films and coatings become part of the food to be consumed, all materials used in these products must be appropriately declared on the label. Regulations regarding incorporation of antimicrobials, antioxidants, essential oils, color and other additives are the same as those that will be applicable to food formulations.

8.6   Conclusions

Edible films and coatings provide numerous opportunities to meat processors as well as to consumers. Alternatives to collagen casings are feasible. Incorporation of antimicrobials, antioxidants, flavors and colors provide for additional opportunities. Once thermoplastic processing of other proteins and materials are refined, large-scale production of these films will be viable. Development and application of these materials should improve quality and safety of meat products.

References

Allen L, Nelson AI, Steinberg MP, McGill JN (1963) Edible corn carbohydrate food coatings. II. Evaluation of fresh meat products. Food Technol. 17, 1442–1446

Allingham WJ May 17, 1949. U.S. patent 2,470,281

Amin S, Ustunol Z (2007) Solubility and mechanical properties of heat cured whey protein-based edible films compared to collagen collagen and natural casings. Int. J. Dairy Tech. 60(2), 149–153

Anderson TR August 9, 1960. U.S. patent 2,948,623

Anderson TR June 20, 1961a. U.S. patent 2,989,401

Anderson TR June 20, 1961b. U.S. patent 2,989,402

Armitage DB, Hetiarachchy NS, Monsoor MA (2002) Natural antioxidants as a component of an egg albumin film in the reduction of lipid oxidation in Cooked and uncooked poultry. J. Food Sci. 67, 631–638

Anonymous (1967) Edible packaging offers pluses for frozen meat, poultry. Quick Frozen Foods 29, 165–167, 213–214

Ayers JC (1959) Use of coating materials or film impregnated with chlortetracy-cline to enhance color and shelf life of fresh beef. Food Technol. 13, 512–515

Baker RA, Baldwin EA, Nisperos-Carriedo MO (1994) Edible coatings and films for processed foods. In: JM Krochta, EA Baldwin, MO Nisperos-Carriedo (eds.) Edible Coatings and Films to Improve Food Quality. Technomic Publishing, Lancaster, PA, pp 89–104

Balasubramaniam VM, Chinnan MS, Mallikarjunan P, Phillips RD (1997) The effect of edible film oil uptake and moisture retention of a deep-fat Fried poultry product. J. Food Process Eng. 20, 17–20

Baldwin EA (1994) Edible coatings for fresh fruits and vegetables: past, present and future. In: JM Krochta, EA Baldwin, MO Nisperos-Carriedo (eds) Edible Coatings and Films to Improve Food Quality. Technomic Publishing, Lancaster, PA, pp 25–64

Baldwin EA, Nisperos NO, Baker RA (1995) Use of edible coatings to pre-Serve quality of lightly (and slightly) processed products. Crit. Rev. Food Sci. Nutr. 35, 509–524

Baldwin EA, Nisperos NO, Hagenmaier RD, Baker RA (1997) Use of lipids in coatings for food products. Food Technol. 51(6), 56–62, 64

Baron JK (1993) Inhibition of Salmonella typhimurium and Escherichia coli O157:H7 by an antimicrobial containing edible film. MS Thesis, University of Nebraska, Lincoln

Bauer CD, Neuser GL December 9, 1969. U.S. patent 3,483,004

Bauer CD, Neuser, GL, Pinkalla HA October 15, 1968. U.S. patent 3,406,081

Becker OW October 6, 1936. U.S. patent 2,056,595

Becker OW April 26, 1938. U.S. patent 2,115,607

Becker OW June 13, 1939. U.S. patent 2,161,908

Ben A, Kurth LB (1995) Edible film and coatings for meat cuts and primal. Meat'95. The Australian Meat Industry Research Conference. CSIRO, September 10–12 1995

Bendini A, Gallina TG, Lercker G (2002) Antioxidant activity of oregano (Origanum vulgare L.) leaves. Ital. J. Food Sci. 14, 17–24

Boni KA December 27, 1988. U.S. patent 4,794,006

Brault D, D'Aprano G, Lacroix M (1997) Formation of free-standing steri-Lized edible films from irradiated caseinates. J. Agric. Food Chem. 45, 2964–2969

Burke NI January 13. 1976. U.S. patent 3,932,677

Cagri A, Ustunol Z, Ryser E (2001) Antimicrobial, mechanical and moisture barrier properties of low pH whye protein-based edible films containing p-aminobenzoic acid and sorbic acids. J. Food Sci. 66, 865–871

Cagri A, Ustunol Z, Ryser E (2002) Inhibition of three pathogens on bologna and summer sausage slices using antimicrobial edible films. J. Food Sci. 67, 2317–2324

Cagri A, Ustunol Z, Osburn WN, Ryser E (2003) Inhibition of Listeria monocytogenes on hot dogs using antimicrobial whey protein based edible casings. J. Food Sci. 68, 291–299

Cagri A, Ustunol Z, Ryser E (2004) Antimicrobial edible films and coatings: A review. J. Food Prot. 67, 833–848

Centers for Disease Control and Prevention (2000) Multi-state outbreak of listeriosis. Morb. Mortal. Wkly. Rep. 49, 1129–1130

Cherian G, Gennadios A, Weller C, Chinachoti P (1995) Thermo-mechanical behavior of wheat gluten films: effect of sucrose, glycerin, and sorbitol. Cereal Chem. 72, 1–6

Childs WH October 29, 1957. U.S. patent 2,811,453

Cuq B, Gontard N, Cuq JL, Guilbert S (1997) Selected functional properties of fish myofibrillar protein-based films as effected by hydrophilic plasticizers. J. Agric. Food Chem. 45, 622–626

Cutter CN, Siragusa GR (1996) Reduction of *Bronchotrix thermosphacta* on beef surfaces following immobilization of nisin in calcium alginate gels. Lett. Appl. Microbiol. 23, 9–12

Cutter CN, Siragusa GR (1997) Growth of *Bronchotrix thermosphacta* in ground beef following treatments with nisin and calcium alginate gels. Food Microbiol. 14, 425–430

Cutter CN, Sumner SS (2002) Application of edible coatings in muscle foods. In: Protein-Based Films and Coatings. A Gennedios (ed). CRC, Boca Raton, FL

Daniels R (1973) Edible Coatings and Soluble Packaging. Park Ridge, NJ, Noyes DataCorp

Deacon MJ, Kindleysides L (1973) U.S. patent 3,767,821. In Sausage casing technology, Food Technology, Review No. 14. Karmas E (ed) Park Ridge NJ: Noyes Data Corp. p. 88–89

Donhowe IG, Fennema O (1993) Water vapor and oxygen permeability of Wax films. J. Am. Oil Chem. Soc. 70, 867–873

Donhowe IG, Fennema O (1994) Edible films and coatings: Characteristics, formation, definitions and testing methods. In: Edible Coatings and Films to Improve Food Quality. JM Krochta, EA Baldwin, MO Nisperos-Carriedo (eds). Technomic Publishing, Lancaster, PA, pp 1–24

Earle RD July 30, 1968. US patent 3,395,024

Farouk MM, Price JF, Salih AM (1990) Effect of an edible collagen film overwrap on edudation and lipid oxidation in beef round steaks. J. Food Sci. 55:1510–1515

Feuge RO (1955) Acetoglycerides – new fat products of potential value for the Food industry. Food Technol. 9, 314–318

Galietta G, Di-Gioia L, Guilbert S, Cuq B (1998) Mechanical and ther-momechanical properties of films based on whey proteins as affected by plasti-izer and crosslinking agents. J. Dairy Sci. 81:3123–3130

Gennadios A, Hanna MA, Kurth LB (1997) Application of edible coatings on meats, poultry and seafoods: A review. Lebensm. Wiss. Technol. 30, 337–350

Gennadios A, McHugh T, Weller CL, Krochta JM (1994) Edible coatings and films based on proteins. In: Edible Coatings and Films to Improve Food *Quality* p. 201–277. JM Krochta, EA Baldwin. MO Nisperos-Carriedo (eds). Technomic Publishing, Lancaster, PA

Gennadios A, Weller CL (1990) Edible films and coatings from wheat and corn proteins. Food Technol. 44, 63–69

Glicksman M (1983) Food Hydrocolloids. Vol. III. CRC, Boca Raton, FL

Gontard NG, Guilbert S, Cuq JL (1993) Water and glycerol as plasticizers affect mechanical and water vapor barrier properties of an edible wheat gluten film. J. Food Sci. Technol. 58, 201–211

Grant LA, Burns J (1994) Application of coatings. In: Edible Coatings and Films to Improve Food Quality. JM Krochta, EA Baldwin, MO Nisperos-Carriedo (eds) Technomic Publishing, Lancaster, PA, pp 189–200

Griffin DB, Keeton JT, Savell JW, Leu R, Vanderzant C, Cross HR (1987) Physical and sensory characteristics of vacuum packaged beef steaks and roasts treated with an edible acetylated monoglyceride. J. Food Prot. 50, 550–553

Guilbert S (1986) Technology and application of edible protective films. In: Packaging and Preservation – Theory and Practice. M Mathlouthi (ed). Applied Science, Elsevier, New York, NY, p 371

Handley D, Ma-Edmonds M, Hamouz F, Cuppett S, Madigo R, Schnepf M (1996) Controlling oxidation and warmed over flavor in precooked pork chops with rosemary oleoresin and edible film. In: Natural antioxidants chemistry, health effects and applications. Shahidi F. (ed). AOCS Champaign, IL, pp 311–318

Hargens-Madsen M, Schnepf M, Hamouz F, Weller C, Roy S (1995) Use of edible films and tocopherol in the control of warmed over flavor. J. Am. Diet. Assoc. 95, A-41

Hartal D (1966) The development and evaluation of carbohydrate-alginate food coatings. MS Thesis, University of Illinois, Urbana

Heine C, Wust R, Kamp B January 39, 1979. US patent 4,137,334

Heckman JH, Heller IR (1986) Food, drug and cosmetic packaging regulations. In: The Wiley Encyclopedia of Packaging Technology ed. M Bakker. Wiley, New York, NY, pp 351–359

Herald TJ, Hachmeister KA, Huang S, Bowers JR (1996) Corn zein pack-ageing materials for cooked turkey. J. Food Sci. 61:415–417,421

Harvard C, Harmony MX June 1869. U.S. patent 90,944

Hernandez E (1994) Edible coatings from lipids and resins. In: Edible Coatings and Films to Improve Food Quality. JM Krochta, EA Baldwin, MO Nisperos-Carriedo (eds). Technomic Publishing, Lancaster, PA, pp 279–303

Hirasa K (1991) Moisture loss and lipid oxidation in frozen fish-effect of a casein-acetylated monoglyceride edible coating. M.S. Thesis, University of California, Davis

Hood LL (1987) Collagen in sausage casings. In: Advances in Meat Research Vol. 4. AM Pearson, TR Dutson and AJ Bailey (eds) New York, NY: Van Nostrand Reinhold p. 109–129

Holownia KI, Chinnan MS, Erickson MC, Mallikarjunan P (2000) Quali-Ty evaluation of edible film-coated chicken strips and frying oils. J. Food Sci. 65, 1087–1093

Hotchkiss JH (1995) Safety considerations in active packaging. In: Active Food Packaging. Rooney M.L. (ed). Blackie Academic and Professional, New York, NY, pp 238–255

Howard LR, Talcott TS, Brenes CH, Villalon B (2000) Changes in phytochemical and antioxidant activity of selected pepper cultivars (Capsicum species) as influenced by maturity. J. Agric. Food Chem. 48, 1713–1720

Janes ME, Kooshesh S, Johnson MG (2002) Control of Listeria monocyto-genes on surface of refrigerated, ready to eat chicken coated with edible zein film coatings containing nisin and/or calcium proprionate. J. Food Sci. 67, 2754–2757

Keil HL February 14, 1961. U.S. patent 2,971,849

Keil HL, Hagen RF, Flaws RW September 20, 1960. U.S. patent 2,953,462

Kester JJ, Fennema OR (1986) Edible films and coatings: a review. Food Technol. 40, 47–59

Kim SJ, Ustunol Z (2001a) Thermal properties, heat sealability and seal attributes of whey protein isolate/lipid emulsion edible films. J. Food Sci. 66, 985–990

Kim SJ, Ustunol Z (2001b) Sensory attributes of whey protein isolate and candelilla wax edible films. J. Food Sci. 66, 909–911

Kim SJ, Ustunol Z (2001c) Solubility and moisture sorption isotherms of Whey protein based edible films as influenced by lipid and plasticizer type. J. Agric. Food Chem. 49, 4388–4391

Klose AA, Mecchi EP, Hanson HL (1952) Use of antioxidants in the fro-Zen storage of turkeys. Food Technol. 6, 308–311

Kobussen J, Kobussen M, Basile II VL March 30, 1999. U.S. patent 5,888,131

Kroger M, Igoe RS (1971) Edible containers. Food Prod. Dev. 5:74, 76, 78–79,82

Krochta JM (2002) Proteins as raw materials for films and coatings: definitions, current status, and opportunities, p.1. In: Gennadios A (ed), Protein-based Films and Coatings. CRC, Boca Raton, FL

Krochta JM, De Mulder-Johnston C (1997) Edible and biodegradable polymer films: challenges and opportunities. Food Technol. 51, 2:61–74

Kunte LA (1996) Effectiveness of 7S soy protein and edible film in controlling Lipid oxidation in chicken. M.S. Thesis. University of Nebraska, Lincoln

Krumel KL, Lindsay TA (1976) Nonionic cellulose ethers. Food Technol. 30, 36–38, 40, 43

Labell F (1991) Edible packaging. Food Process. 52, 24

Labuza TP, Breene WM (1989) Application of active packaging for improvement of shelf life and nutritional quality of fresh extended shelf life foods. J. Food Proc. Pres. 13, 1–69

Lazarus CR, West RL, Oblinger JL, Palmer AZ (1976) Evaluation of a calcium alginate coating and a protective plastic wrapping for the control of lamb carcass shrinkage. J. Food Sci. 41, 639–641

Letney LR January 24, 1958. U.S. patent 2,819,975

Leu R, Keeton JT, Griffin DB, Savell JW, Vanderzant C (1987) Microflora of vacuum packaged beef steaks and roasts treated with an edible acetylatedmonoglyceride. J. Food Prot. 50:554–556

Lieberman ER, Guilbert SG (1973) Gas permeation of collagen films as affected by cross-linkage, moisture, and plasticizer content. J. Polymer Sci. 41, 33–43

Love JD (1988) Sensory analysis of warmed-over flavor in meat. Food Technol. 42, 6:140–143

Ma-Edmonds M, Hamouz F, Cuppett S, Madigo R, Schnepf M (1995) Use of rosemary oleoresin and edible film to control warmed-over flavor in pre-cooked beef patties. Abstract No. 50-6 IFT Annual Meeting. Anaheim, CA

Marquie C, Aymard C, Cuq JL, Guilbert S (1995) Biodegradable packa-ging made from cottonseed flour: formation and improvement by chemical treatments with gossypol, formaldehyde, and glutaraldehyde. J. Agric. Food Chem. 43, 2762–2767

McCormick KE, Han IY, Acton JC, Sheldon BW, Dowson PL (2005) In-package pasteurization combined with biocide-impregnated films to inhibit *Listeria monocytogenes* and *Salmonella* Typhimurium in turkey bologna. J. Food Sci. 70, M52–7

McGarth EP (1955) Packaging costs cut, quality protected by wax-coating frozen meats. Food Engr. 27, 50–51, 77

McKnight JT March 3, 1964a. U.S. patent 3,123,483

McKnight JT October 6, 1964b. U.S. patent 3,151,990

Meyers RC, Winter AR, Weister HH (1959) Edible protective coating for extending the shelf life of poultry. Food Technol. 13, 146–148

Meyers MA (1990) Functionality of hydrocolloids in batter coating systems. In: Batters and Breadings in Food Processings. K Kulp, R Loewe (eds). American Association of Cereal Chemists, St. Paul MN, pp 117–141

Miller AT January 16 1975. US patent 3,894,158

Miller AT June 14, 1983. U.S. patent 4,388,311

Miller AT, Marder B October 13, 1998. U.S. patent 5,820,812

Miller BJ, Cutter CN (2000) Incorporation of nisin into a collagen films retains antimicrobial activity against *Listeria monocytogenes* and *Brochothrix thermosphacta* associated with ready-to-eat meat products

Morgan BH (1971) Edible packaging update. Food Prod. Dev. 5, 6:75–77, 108

Morris A, Parker JA March 17, 1896. U.S. patent 556,471

Mountney GJ, Winter AR (1961) The use of calcium alginate film for coat-Ing cut-up poultry. Poultry Sci. 40, 28–34

Natrajan N,Sheldon BW (1995) Evaluation of bacteriocin-based packaging and edible film delivery systems to reduce Salmonella in fresh poultry. Poultry Sci. 74(Suppl. 1):31

Natrajan N, Sheldon BW (2000) Inhibition of Salmonella on poultry skin Using protein and polysaccharide-based films containing nisin formulation. J. Food Prot. 63, 1268–1272

Nisperos-Carriedo MO (1994) Edible coatings and films based on polysaccharides. In:Edible Coatings and Films to Improve Food Quality. JM Krochta, EA Baldwin MO and Nisperos-Carriedo (eds) Technomic Publishing, Lancaster, PA, pp 305–335

Okamoto S (1978) Factors affecting protein film formation. Cereal Foods World. 23, 256–262

Osburn WN (2002) Collagen casings. In: Protein-Based Films and Coatings. Gennadios A (ed). CRC, Boca Raton, FL

Oussalah M, Caillet S, Salmieri S, Saucier L, Lacroix M (2004) Antimi-Crobial and antioxidant effects on milk protein-based films containing essential oils for preservation of whole beef muscle. J. Agric. Food Chem. 52, 5598–5605

Pearce JA, Lavers CG (1949) Frozen storage of poultry. V. Effect of some processing factors on quality. Can. J. Res. 27, 253–265

Rhim JW, Gennadios A, Handa A, Weller CL, Handa MA (2000) Solubility, tensile and color properties modified soy protein isolate films. J. Agric. Food Chem. 48, 4937–4941

Riaz MN (1999) Processing biodegradable packaging material from starches using extrusion technology. Cereal Food World. 705–709

Rose HJ August 12, 1968. U.S. patent 3,413,130

Sacharow S (1972) Edible films. Packaging 43, 6, 9

Schmidt GR (1986) Processing and Fabrication in Muscle as Food. Bech-tel PJ (ed). Academic, London, pp 201–238

Sleeth RB, Furgal HP January 12, 1965. U.S. patent 3,165,416

Smith DM, Culbertson JD (2000) Proteins: Functional properties. In: Food Chemistry: Principles and Applications. Christen GL and Smith JS (eds). Science Technology System, West Sacramento, CA, p 131

Smits JW (1985) The sausage coextrusion process. In Trends in Modern Meat Technology. B Krol, PS van Roon, JA Hoeben (eds). Center for Agricultural Publishing and Documentation, Wageningen, The Netherlands, pp 60–62

Schneide J June 6, 1972. U.S. patent 3,667,970

Shelef LA (1983) Antimicrobial effects of spices. J. Food Saf. 6, 29–44

Simelane S, Ustunol Z (2005) Effect of meat processing conditions on mechanical properties of heat cured whey protein-based edible films: A comparison to collagen casing. J. Food Sci. 70, 2:E131–134

Siragusa GR, Dickson JS (1993) Inhibition of *Listeria monocytogenes, Salmonella* Typhimurium and *Escherichia coli* O157:H7 on beef muscle tissue by lactic or acetic acid contained in calcium alginate gels. J. Food Safety. 13, 147–158

St. Angelo AJ, Bailey MF (1987) Warmed Over Flavor in Meats. Academic, Orlando, FL, 294p

St. Angelo AJ, Vercellotti JR, Legendre MG, Vinnett CH, Kuan JW, James C, Dupuy HP (1987) Chemical and instrumental analysis of warmed-over flavor in beef. J. Food Sci. 52:1163–1168

Stemmler M, Stemmler H (1976) US Patent 3, 936, 312. Feb. 3

Stoloff LS, Puncochar JF, Crowther HE (1948) Curb mackerel fillet ran-cidity. Food Ind. 20, 1130–1132, 1258

Stubbs CA, Cornforth DP (1980) The effect of an edible pectinate film on Beef carcass shrinkage and microbial growth. In: Proceeding of the 26th Euro-Pean Meeting of Meat Research Workers. American Meat Science Association, Kansas City, MO, pp 276–278

Suderman DR, Wiker J, Cunningham FE (1981) Factors affecting adhesion of coating to poultry skin: Effects of various protein and gum sources in the coating composition. J. Food Sci. 46, 1010–1011

Theivendran S, Hettiarachchy NS, Johnson MG (2006) Inhibition of *List-eria monocytogenes* by nisin combined with grape seed extract or green tea extract in soy protein films coated on turkey frankfurters. J. Food Sci. 71, M39–44

Tims MJ, Watts BM (1958) Protection of cooked meats with phosphates. Food Technol. 12, 240–243

Tsuzuki T, Lieberman ER August 11, 1972. U.S. patent 3,681,093

USDA-FSIS (1998–2002) Recall information center. http://www.fsis.usda.gov/OA/recalls.

Ustunol ZB, Mert (2004) Water solubility, mechanical, barrier and thermal properties of cross-linked whey protein isolate based films. J. Food Sci. 69(3), 129–133

Villegas R, O'Connor TP, Kerry JP, Buckley DJ (1999) Effect of gelatin dip on the oxidative and colour stability of cooked ham and bacon pieces during frozen storage. Int. J. Food Sci. Tech. 34, 385–389

Waldman RC (1985) Co-extrusion – high tech innovation. The National Provisioner. Jan 12 pp 13–16

Wang PY (1986) Meat processing I. In: Encyclopedia of Food Engineering 2nd edition. CW Hall, AW Frall, AL Rippen (eds) AVI Publishing, Westport, CT, pp 545–550

Wanstedt KG, Seideman SC, Donnelly LS, Quenzer NM (1981) Sensory attributes of precooked, calcium alginate-coated pork patties. J. Food Prot.44, 732–735

Whistler RL, Daniel JL (1990) Functions of polysaccharides in foods.In: Food Additives. AL Branen, PM Davidson and SY Salminen (eds) Marcel Dekker, New York, NY, pp 395–424

Whitman GR, Rosenthal H January 19, 1971. U.S. patent 3,556,814

Williams SK, Oblinger JL, West RL (1978) Evaluation of a calcium algi-nate film for use on beef cuts. J. Food Sci. 43, 292–296

Wilson JR, Burke NI December 13, 1977. U.S. patent 4,062,980

Wu Y, Rhim JW, Weller CL, Lamouz F, Cuppet S, Schnepf M (2000) Moisture loss and lipid oxidation for precooked beef patties stored in edible coatings and films. J. Food Sci. 65, 300–304

Wu JW, Lee MH, Ho CT, Chang SS (1982) Elucidation of the chemical structures of natural antioxidants isolated from rosemary. J. Am. Oil Chem. Soc. 59, 339–345

Wu Y, Weller CL, Hamouz F, Cuppet S, Schnepf M (2001) Moisture loss and lipid oxidation for precooked ground-beef patties packaged in edible starch alginate-based composite films. J. Food Sci. 66, 486–492

Zabik ME, Dawson LE (1963) The acceptability of cooked poultry protected by an edible acetylated monoglyceride coating during fress and frozen storage. Food Technol. 17, 87–91

Ziolko FJ November 29, 1977. U.S. patent 4,060,361

Ziolko FJ February 6, 1979. U.S. patent 4,138,503

Zivanovic S, Chi S, Draughon AF (2005) Antimicrobial activity of chitosan films enriched with essential oils. J. Food Sci. 70, M45–52

Chapter 9
Edible Films and Coatings for Flavor Encapsulation

Gary A. Reineccius

9.1 Introduction

An important use of edible films is for encapsulation of flavorings. The edible film serves many purposes including permitting production of a dry, free-flowing flavor (most flavors are liquids), protection of flavoring from interaction with food or deleterious reactions such as oxidation, confinement during storage, and, finally, controlled release. The degree to which the edible film meets these requirements depends upon the process used to form the film around the flavoring and the film composition itself. This chapter will discuss major processes used in manufacturing encapsulated flavorings, materials commonly used as the encapsulation matrix, and factors determining the efficacy of the edible film (hereafter termed the *encapsulation matrix*).

9.2 Historical

If we consider encapsulation processes used in the food industry as a whole, pan coating is likely the earliest example of such a process. A patent dating from 1875 describes a process for pan coating of almonds (with sugar), confectionery products, pills, etc. Focusing on food flavorings, the earliest reference would be to spray drying. Accidental discovery that acetone, added to tomato puree to help maintain color and flavor of tomato powder, was lost to only a small extent in the spray-drying process started the development of encapsulated flavorings (A. Boake Roberts Co, 1932). Since that time, spray drying has maintained a leadership position in the flavor industry. With time, other processes have entered the flavor industry. Coacervation as a process was first studied by Bungenberg de Jong (1949)

G.A. Reineccius
Department of Food Science and Nutrition, 1334 Eckles Avenue,
St. Paul, MN 55108-1038, USA
e-mail: greinecc@umn.edu

M.E. Embuscado and K.C. Huber (eds.), *Edible Films and Coatings for Food Applications*, 269
DOI 10.1007/978-0-387-92824-1_9, © Springer Science+Business Media, LLC 2009

and Kruyt (1949). In 1957, the coacervation process was patented by Green and Schleicher (1957) as applied to carbonless paper. Soon after, claims were made for delivery of flavorings and fragrances via this process (e.g., Brynko and Scarpelli 1965). About the same time, an "extrusion" process was developed to deliver flavoring materials. This process originated from the basic ideas of Schultz et al. (1956), who simply added citrus oils to a molten carbohydrate mass (like hard candy), agitated it to form a crude emulsion, and then allowed it to cool and solidify as a mass. The mass was ground to the desired particle size and sold as an encapsulated flavoring material. Swisher (1957) used a similar formulation, but added extrusion to the process. The final process of significance for flavor delivery is use of cyclodextrins (CyDs). CyDs have been known since 1891 when they were first described by Villars (1891). Over the years work, progressed on characterizing this family of compounds, and Szejtli et al. (1979) were the first to find applications for them in the flavor industry.

If we consider technologies popular today, they have very old origins and largely represent improvements of old processes.

9.3 Technical Developments

While focus of this chapter is on edible films, one cannot separate the film from the process. Therefore, we will first present an overview of the processes used in the encapsulation of flavorings. There are numerous processes described for flavor encapsulation in both the scientific and patent literature; however, only four processes find significant commercial usage: spray drying, extrusion, coacervation and CyDs, and our discussion will be limited to them.

9.3.1 Encapsulation Processes

9.3.1.1 Spray Drying

Spray drying is used to produce close to 90% of encapsulated flavorings on the market. Spray drying is accomplished by initially making an emulsion of the flavoring (20% on a dry basis) in an encapsulation matrix (80% on a dry weight basis) using a minimal amount of water (preferably <50% water wet basis). An oil-in-water emulsion is formed via homogenization to obtain a particle size averaging about 1 μm, and the emulsion is pumped into a spray dryer. The flavor emulsion is atomized into a hot air stream where evaporative cooling brings inlet air temperature down from 200–325°C to 80–90°C (Fig. 9.1). During drying, the flavoring particle never exceeds the exit air temperature.

This process requires the encapsulating matrix to meet several requirements. The matrix must be water-soluble, since that is the solvent used in spray drying. The infeed

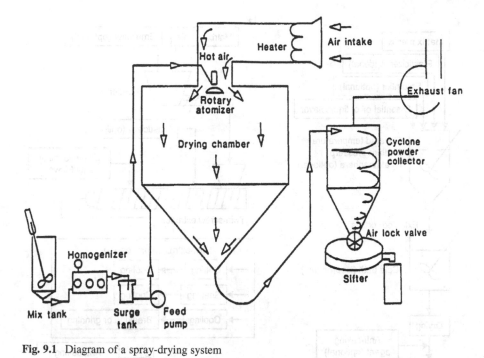

Fig. 9.1 Diagram of a spray-drying system

emulsion must be low enough in viscosity with high solids concentrations (50–70%) to permit pumping and, more restrictive, atomization. While the final product need not necessarily be a stable emulsion on reconstitution (e.g., cake mix or frosting flavorings), the encapsulation matrix must yield at least a temporary emulsion, which should not break from the time of homogenization until it is atomized in the spray dryer. Breakage of the emulsion may result in a fire or explosion in the spray dryer. The encapsulation matrix must dry well in a normal spray dryer (e.g., not become sticky at high temperatures) so that a good yield is obtained, and then not be hygroscopic following drying. The matrix must ultimately produce a good-quality flavoring (i.e., it must retain volatile flavor constituents), protect it from degradation or evaporation during storage, and finally provide release in the finished product.

Requirements discussed above largely limit matrix materials to maltodextrins, corn syrup solids, modified starches and gum acacia, the traditional encapsulation matrix. Some new materials have become available that will be discussed later.

9.3.1.2 Extrusion Process

The extrusion process traditionally consisted of initially making a low-moisture carbohydrate melt (ca. 15% moisture), adding flavor (10–20% on a dry weight basis), forming an emulsion, and then extruding the melt through a die (1/64 in. holes)

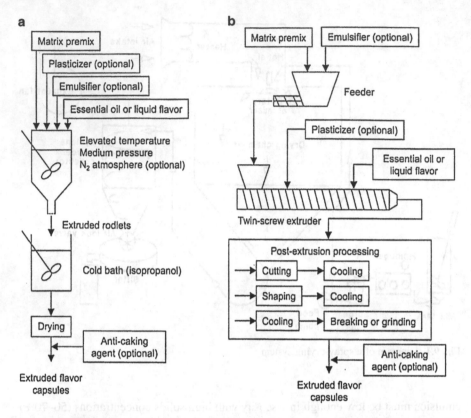

Fig. 9.2 (a–b) Processes for making "extrusion" flavorings: (a) traditional batch process; (b) modern extruder (Ubbink and Schoonman 2005)

under low pressure. The extruded flavoring dropped into a cold isopropanol bath where it immediately formed an amorphous glass structure, was broken into small pieces by a mixer, and was then further dried in hot air to yield the finished flavoring material (Fig. 9.2).

Significant developments have occurred in this process, notably, (1) use of scraped surface heat exchangers in combination (or a twin screw extruder) to permit continuous processing, (2) use of low moisture contents, and (3) alternatives to cooling in isopropanol solutions. While there have been numerous patents outlining the progress, the following provide the key references: Benczedi and Bouquerand (2001); McIver et al. (2004); Subramaniam and McIver (2007); Porzio and Popplewell (2001); Porzio and Zasypkin (2004). The evolution in this process has made it very competitive costwise with spray drying. If we consider these changes, going to a scraped surface heat exchanger (twin screw extruder) was a major innovation (Fig. 9.2). This permitted use of a continuous process that offerred inherent advantages in cost, process control and quality. Quality improvements came partly in being able to add flavoring late in the process, thereby exposing it to less heat. Quality improvements also came in being able to use much lower amounts of water in the system: the scraped surface heat exchangers could work at higher viscosities, as could the extrusion process. Thus, in more recent forms, there was no water to be removed from the system, eliminating a processing step (drying) and improving flavor quality through better retention of flavorings (improved retention from 75–85% to >90%) (Subramaniam and McIver 2007). The last innovation in the process was eliminating the washing of the formed particles in cold isopropanol. This step traditionally accomplished washing the surface to remove any surface flavoring, rapidly solidified the molten extrudate (rapidly moving it in the glassy state), and reduced the moisture content (to some extent). It was problematic in that it was slow and costly, and involved working with an organic solvent (legal issues from residue and hazards in handling). Solidifying extrudate in liquid nitrogen or a nitrogen-cooled environment offered significant improvement.

The encapsulating material used in this process has similar requirements as that used in spray drying. The encapsulating matrix does not have to be atomized but it must be somewhat workable (soluble) at 85% or higher solids at 110°C or higher. Emulsification is also less of an issue since synthetic emulsifiers are added to the carbohydrate melt (ca. 2% on a total weight basis), and there is little opportunity for phase separation during manufacturing. This is due to the short time between emulsion formation and solidification, and the extremely high viscosity of the carbohydrate melt. The matrix must form a good amorphous structure and be nonhygroscopic. It must also result in good retention of flavor compounds during manufacturing, offer protection during storage, and release flavor when placed in water.

9.3.1.3 Coarcervation

While coacervation has been known for many years, it has only recently been used to any significant extent in the food flavor industry. Complex coacervation is used in the food industry: "complex" coacervation refers to two hydrocolloids (one polycationic and the other polyanionic) interacting typically through ionic bonding to form a complex. This complex forms on the surface of hydrophobic droplets (flavoring), thereby serving as the capsule wall. This process has changed very little from the time it was patented by Green and Schleicher (1957).

As is shown in Fig. 9.3, the first step is to dissolve one of the hydrocolloids in water and add flavoring (hydrophobic liquid). With mixing, the system forms an emulsion. Since the particle is the template for formation of the capsule wall, the size of this emulsion is the primary determinant of the final capsule size. While the literature suggests that these particles can be extremely small (in the nanometer range), in practice it is difficult to make food-grade coacervates of less than 100 μm without severe agglomeration. The second hydrocolloid is then added as a solution. The system is adjusted to a pH where the two hydrocolloids carry a net opposite charge and diluted with water. The water serves to dilute the system, so that when it is cooled to gel, the entire mass does not form a gel. At this time, the two

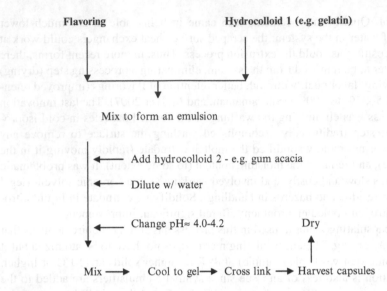

Fig. 9.3 Process for the formation of coacervates

hydrocolloids interact through ionic bonding to form the final wall. Capsules may be chemically cross-linked, which hardens the wall structure, making it more durable in further handling and insoluble in application. Cross-linking may be done with glutaraldehyde, transglutaminase (enzyme), or other cross-linking chemicals. The final step is harvesting the capsules and drying them. They are typically harvested by filtering, and dried by freeze-drying or mixing with a drying agent such as silicates. An excellent detailed discussion of this process is provided by Thies (2007).

Coacervation has found recent application primarily because of its insolubility: All other major encapsulation systems are soluble (or slowly soluble) in water and, therefore, their flavoring is released quickly on hydration. The slow-release property of coacervates (typically diffusion through the insoluble capsule wall on hydration) makes it attractive where a slow release is required, e.g., in thermally processed foods. The greatest innovations in this process have been in the choices of food polymers as will be discussed later.

9.3.1.4 Inclusion Complex Formation

Inclusion complexes can be formed between flavor compounds and some food components. The most notable are the interactions with starch and CyDs.

Starches, primarily those high in amylose, will form a helical structure that can complex small molecules. Some early work by Wyler and Solms (1981) on complexation of model volatiles (limonene, menthone and decanal) with gelatinized potato starch demonstrated that these compounds could be readily complexed. Time for maximum complexation was variable, ranging from 1 min (decanal) to several days (limonene), depending upon the compound. The amount complexed was also both compound and temperature dependent: less was complexed at higher temperatures. A subsequent study by Wyler and Solms (1982) demonstrated that the same volatiles were stable to evaporative losses and oxidation when included in the starch complex. Saldarini and Doerig (1981) were awarded a patent on flavor encapsulation in starch hydrolysates (10-13 DE). They found that they could produce a powder containing 4.6% acetaldehyde in this manner which retained 4.1% after 1 week storage at 40°C.

There has been continued interest in starch–flavor complexation (e.g., Delarue and Giampaoli 2000; Rutschmann and Solms 1990; Arvisenet et al. 2002; Conde-Petit et al. 2006; Tapanapunnitikul et al. 2008). Numerous authors have investigated starch interactions between several classes of compounds, as well as temperatures, starch treatment, starch source, etc. For example, Wulff et al. (2005) optimized these interactions and found that 5.7–10.4% loading was possible. On the basis of this loading, they proposed that segments of 8–16 anhydroglucose units were involved in binding. The association constant of a given volatile compound was found to be strongly dependent on origin and chain length of amylose.

CyDs are well known for their ability to form inclusion complexes. In aqueous solution, CyDs orient such that they have a hydrophobic cavity that is very suitable for the inclusion of flavor compounds (Fig. 9.4). CyDs include molecules generally on an equimolar molar basis: one mole of CyD (host) will include 1 mol of a "guest" molecule. This means that the weight of flavor molecules included in a mole of CyD flavor:complex varies with the molecular weight of the guest. Normally, the final inclusion complex to contains 9–15% guest molecules (by weight).

Pagington (1986) listed several methods of preparing flavor/CyD inclusion complexes, including stirring or shaking a solution of CyD with the guest and filtering off the precipitated complex; blending solid CyD with the guest in a mixer and drying; and passing the vapor of a guest flavor through a CyD solution. Qi and Hedges (1995) provide experimental details of a coprecipitation method deemed most suitable for laboratory evaluation. Large-scale production is generally accomplished by the paste method (may be called a *slurry method*) since less water must subsequently be removed during drying.

9.3.2   Edible Films Formed in Encapsulation

9.3.2.1   Materials

Just as the encapsulation process is chosen on the basis of several factors, so are materials used in this process. Materials are chosen on the basis of compatibility with a process, cost, flavor type, stability during storage, legal constraints, religious

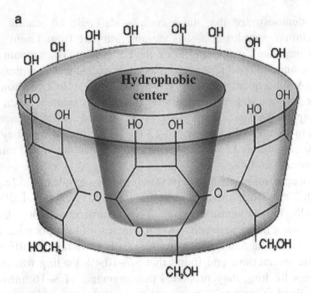

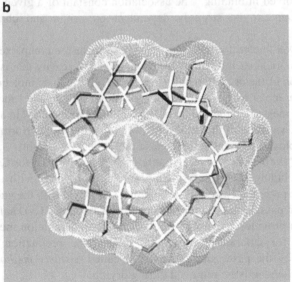

Fig. 9.4 (a–b) Cyclodextrin structures ((a) compliments of Dr. S. Patil, S.K. Patil and Associates; (b) compliments of Wacker Chemie AG, the picture owner)

constraints and functionality in the final application. For example, one may wish to have the flavoring used in an all-natural food and therefore have to use a natural wall material such as gum acacia or pectin. If flavoring is water soluble, one can use a nonemulsifying wall material such as a maltodextrin. However, if flavoring is insoluble, an emulsion must be made, thereby dictating use of an emulsifying food polymer such as modified starch, gum acacia, or protein. If encapsulated

Table 9.1 Food polymers used for encapsulation

Encapsulation material	Compound class	Primary function	Encapsulation process
Gum acacia	Polysaccharide	Emulsification (film former) polyanion polymer	Spray drying and coacervation
Maltodextrin	Starch hydrolysate	Bulking agent	Spray drying and extrusion
Chemically modified starch	Polysaccharide	Emulsification (film former)	Spray drying
Corn syrup solids	Starch hydrolysate	Bulking agent (oxygen barrier)	Spray drying
Glucose, sucrose, sugar alcohols	Mono and disaccharides, and alcohols thereof	Filler (oxygen barrier)	Spray drying and extrusion
Gelatin	Hydrolyzed protein	Polycation polymer	Coacervation
Whey proteins, caseinates	Proteins	Polycation polymer and emulsification	Coacervation and spray drying
Carrageenan, alginates, pectins, etc	Polysaccharides	Polycation polymer	Coacervation
Chitosan	Polysaccharide	Polyanionic polymer	Coacervation and spray drying
Cyclodextrins	Oligosaccharides	Host molecule	Inclusion complex formation

flavoring is to be used in a kosher or Halal product, then gelatin is not acceptable. Generally, numerous considerations must be taken into account in this selection.

Of the materials used in flavor encapsulation, water-soluble carbohydrates are most widely used. These carbohydrates serve different functions in different encapsulation processes. The most commonly used carbohydrates are listed in Table 9.1. Proteins have found very minimal use except in the coacervation process. These edible encapsulating materials will be discussed individually.

Gum Acacia

Gum acacia is the material traditionally used for the encapsulation of food flavors. It is derived from the gum acacia tree, which is grown in the semidesert region of North Central Africa (Thevenet 1986). While there are over 1350 species of the acacia tree (Al-Assaf et al. 2005), only a few species are used for gum manufacture (primarily *Acacia senegal* and *Acacia seyal*). *A. senegal* is produced by the tree in response to artificial injury (Fig. 9.5). *A. seyal* produces tears (droplets) naturally which are gathered from under the trees. Each tree will produce about 300 g of gum, which is allowed to partially dry on the tree before hand-harvesting, sorting, grading and distributing to processors. Processors most often grind the gum, solubilize it in water, filter or centrifuge to remove foreign materials, pasteurize and then spray-dry it. While gum may be sold in the crude form,

Fig. 9.5 Acacia senegal "tear" (compliments of TIC Gums, Belcamp, MD)

conditions of harvesting and storage result in contamination by microbes and foreign materials; so it must be cleaned prior to use as a food ingredient.

Gum acacia is a polymer made up primarily of D-glucuronic acid, L-rhamnose, D-galactose and L-arabinose (Fig. 9.6). It also contains a small amount of protein (from traces to >2% depending upon species), which is believed to be primarily responsible for its emulsification properties (Thevenet 1986). Gum acacia is unique among the plant gums since it exhibits low viscosity at relatively high solids (ca. 300 cps at 30–35% solids), and is a very good emulsifier.

Since gum acacia is a relatively expensive ingredient, there has been substantial work put in to making the gum more functional. This effort typically was aimed at improving performance of lesser quality gums so they may be used in place of the expensive *A. senegal*. A series of articles from the Phillips Hydrocolloids Research Centre (UK) have described a method to age gums to increase their emulsification performance (Al-Assaf et al. 2005). They state that aged gums or gums from older tress offer better performance, and they propose in these articles and related patent that this can be done artificially to all gums (Al-Assaf et al. 2003). The authors did not evaluate their products for an encapsulation application.

TIC Gums has also approached this goal through the chemical modification of *A. seyal* (Ward 2002). Essentially, they have chosen to add an octenylsuccinate anhydride to the gum in a way similar to how the starches are modified. This has been proposed to offer better emulsification and good encapsulation performance compared to the original gum.

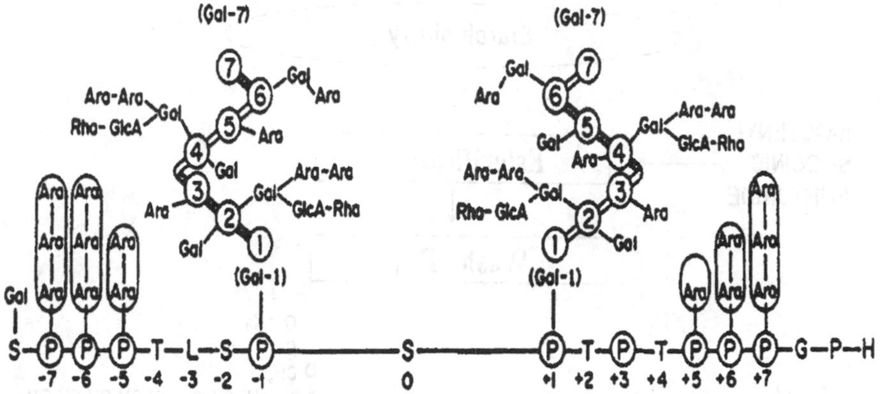

Fig. 9.6 Structure of gum acacia

Modified Food Starch

Starches have virtually no intrinsic emulsification properties. If they make a contribution to emulsion stability, it is only through their effects on viscosity. The need
for emulsification properties in encapsulation matrices, as well as other food applications (e.g., beverage flavor emulsions or cloud emulsions), prompted development of starches which have been chemically modified for this purpose (Trubiano
and Lacourse 1986). Emulsification properties are attained by chemical addition of
octenylsuccinate to partially hydrolyzed starch (Fig. 9.7). Octenyl succinate may be
used at a 0.02 degree of substitution on the starch polymer and still be approved for
food use. However, octenyl succinate is not considered to be a natural product and
this may affect claims for food labels (e.g., one cannot put a "100% natural" label
on the product).

Octenyl succinate-derivatized starches have been chemically engineered to be
excellent emulsifiers, and exhibit low viscosity at high solids (40–50% solids).
They spray-dry exceptionally well and, therefore have found wide use in the encapsulation of flavorings.

Maltodextrins and Corn Syrup Solids

Both maltodextrins and corn syrup solids are made from treatment of starches using
acid, enzyme or acid/enzyme. If product has a dextrose equivalent (DE) of less than
20, it is a maltodextrin. Products with a DE equal to or greater than 20 are classified
as corn syrup solids.

Neither maltodextrins nor corn syrup solids have any inherent emulsification
properties. This is problematic for their use in the encapsulation of lipophilic
flavorings via spray drying, but not an issue in encapsulation of hydrophilic

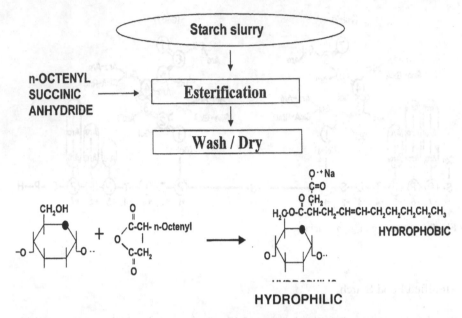

Fig. 9.7 Manufacture of chemically modified starches (Kettlitz et al. 2005)

flavorings. In fact, they are often the wall material of choice when spray drying water soluble flavorings. For water insoluble flavorings, they may be used at sufficiently high solids levels (and therefore high viscosities) during preparation of "emulsions" for spray drying so that they will temporarily maintain an emulsion. However, they offer no emulsion stability to the finished product. Their use in encapsulation of water insoluble flavorings can be dangerous since, if an emulsion breaks in manufacturing and the processor ends up feeding pure flavoring into a spray dryer, an explosion may occur in the dryer.

Considering their use in flavor encapsulation via extrusion, emulsification, when required, is generally provided by a secondary emulsifier (not the basic wall material). Thus, maltodextrins can make up a substantial proportion of wall material and perform the needed function.

Maltodextrins and corn syrup solids are inexpensive compared to either gum acacia or the modified food starches and, therefore, are commonly used as nonfunctional "fillers": basically materials that offer a cost reduction. One finds many references to using blends of either gum acacia or modified starch with maltodextrins or corn syrup solids to take advantage of emulsification properties of gum acacia or modified starch, and then accomplishing a cost reduction through use of the maltodextrins/corn syrup solids. One has to keep in mind that higher DE products do not dry well (spray drying) and are prone to caking on storage. Thus, selection of product DE, as well as substitution level, must be chosen carefully.

Mono- and Disaccharides

Simple sugars (e.g., glucose, sucrose, lactose and maltose, and hydrogenated versions thereof) also find use in encapsulation of flavorings (extrusion and spray-drying processes). They are used to some extent as fillers (like starch hydrolysates), but more importantly are used to confer increased oxidative stability to encapsulated flavorings (to be discussed later). They might be used as 10–40% of wall solids; higher levels result in poor drying properties and caking during storage.

Other Polysaccharides

Other polysaccharides (e.g. pectin, carrageenan, alginate and chitosan) find limited use in flavor encapsulation. These materials have been reported to be used in flavor encapsulation via coacervation (Thies 2007). Pectin, carrageenan and alginate find use as anionic hydrocolloids (in place of gum acacia), while chitosan serves as the cationic hydrocolloid (in place of protein).

These hydrocolloids find little use in other flavor encapsulation technologies because of their high viscosity. High viscosity materials are difficult to atomize in a spray-drying operation at solids level required for economics in operation. Basic information on each of the polysaccharides is provided in the book by Williams and Phillips (2000).

The most unique, and likely least known, of these materials is chitosan. It is a polysaccharide derived from alkaline deacetylation of chitin (1-4)-linked 2-acetamido-2-deoxy-β-D-glucan): a widely distributed polysaccharide found in invertebrates (Terbojevich and Muzzarelli 2000, Fig. 9.8). Chitosans are a heterogeneous group having >7% nitrogen and a degree of acetylation <0.40. The amino nitrogen sets chitosans apart from other polysaccharides. Chitosans have

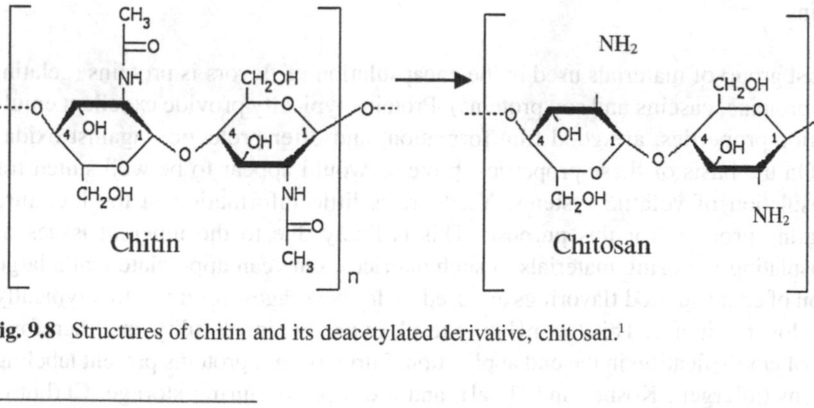

Fig. 9.8 Structures of chitin and its deacetylated derivative, chitosan.[1]

---

[1] (http://upload.wikimedia.org/wikipedia/commons/6/66/Chitin_fixed.png)

a pKa of 6.3, and are solubilized by addition of salts such as formate, acetate, lactate, malate, citrate, glyoxylate, pyruvate, glycolate and ascorbate (Terbojevich and Muzzarelli 2000).

While there have been a substantial number of publications evaluating chitosans for drug delivery, only one article was found on its use for flavor encapsulation (Borognoni et al. 2006). This paper reported on encapsulation of limonene in N-succinyl chitosan. While authors noted good stability of encapsulated limonene on storage, there was inadequate information in the abstract (the article was in Portuguese) to draw any conclusions.

CyDs

CyDs (α, β, γ) are cyclic structures with 6, 7, or 8 glucose members in the ring, respectively (Fig. 9.9). As noted earlier, in an aqueous system, glucose molecules orient such that the hydroxyl groups orient towards the solvent (water) and the C–H bonds towards the center. This creates a hydrophobic center in the CyD. Dimensions of the inner cavity of α-, β-, and γ-CyDs are 5.7, 7.8, and 9.5 Å, respectively, thus potentially accommodating a range of sizes of guest molecules (Hedges et al. 1995). Other measurements of inner cavity have also been published, as shown in Fig. 9.9b, which are close to those by Hedges et al. (1995). For a substance included within a CyD, an equilibrium exists between free and complexed guest molecules; the equilibrium constant depends on nature of the CyD and guest molecule, as well as on factors such as temperature, moisture level and other aspects of food composition. Generally speaking, however, presence of water or very high temperature is required to liberate guest molecules once complexed (Reineccius et al. 2002–2005). Goubet et al. (2001) have addressed a very interesting topic of competitive binding by CyDs. Since flavors are composed of numerous compounds, the issue of competitive binding is relevant.

Proteins

The last group of materials used in the encapsulation of flavors is proteins (gelatin, whey proteins, caseins and soy proteins). Proteins typically provide excellent emulsification properties, are good film formation, and offer protection against oxidation. On the basis of these properties, proteins would appear to be well suited for encapsulation of volatile systems. Yet there is little information in the literature evaluating proteins for this purpose. This is likely due to the inherent issues of encapsulating flavoring materials in such matrices. One can appreciate that a large portion of encapsulated flavorings are used in dry beverages: products that typically have a low pH in use. This low pH may result in protein insolubility and, therefore, a loss of emulsification in the end application. Furthermore, proteins present labeling concerns (allergen, Kosher and Halal), and are reactive during storage. Carbonyl compounds are important components in most flavorings, and these compounds

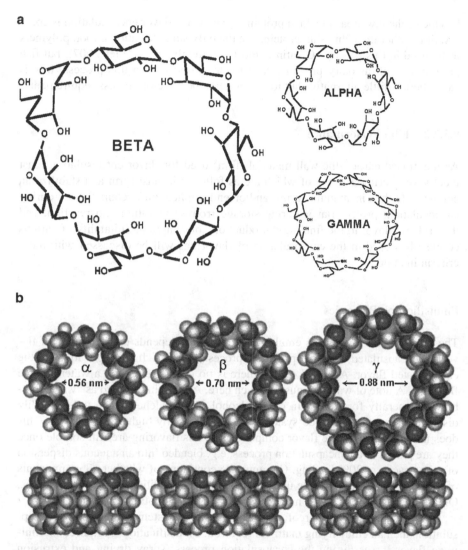

Fig. 9.9 (**a–b**) Cyclodextrin structures ((**a**) compliments of Dr. S. Patil, S.K. Patil and Assoc.; (**b**) compliments of Wacker Chemie AG, the picture owner)

readily react the amino groups of proteins, thereby resulting in Schiffs base formation: a loss of flavor and development of colored compounds in the product. Proteins also present economic issues. First, they are expensive relative to the competing encapsulation ingredients. Additionally, they generally have sufficiently high viscosity in solution that they result in significant flavor losses during spray drying and low productivity during manufacturing.

One of the few areas in which proteins are used for flavor encapsulation is coacervation. Other than chitosan, proteins are the only suitable cationic food polymers to be used in this process. Gelatin is most commonly used (Thies 2007), but fish protein, as well as dairy proteins, have found application in this process. In this case, there is little alternative but to use proteins because of process requirements.

9.3.2.2 Film Performance

As mentioned earlier, the wall material (film) used for flavor encapsulation must meet several criteria, some of which are as follows: it must form and stabilize an emulsion (either in manufacturing and/or end application), retain flavors during encapsulation, protect flavor during storage from evaporation and reaction, and then release flavor to the final food product on consumption. Suitability of various edible films used in the encapsulation of flavorings will be discussed with these criteria in mind.

Emulsification Properties

The importance of imparting emulsifying properties depends upon the type of flavoring encapsulated, the encapsulation process, and the final application of the encapsulated flavor. As noted earlier, there is no need to emulsify a water soluble flavoring. A note of warning must be given here. A flavoring labeled as "water soluble" is generally formulated in either alcohol or propylene glycol and is readily dispersed in an aqueous food system at 0.1% or slightly higher usage levels. This does not mean that all the flavor components in this flavoring are still soluble once they are placed in an encapsulation process, e.g. blended into an aqueous dispersion of gum acacia, at 20% loading. One must be conscious of whether all components of a flavoring are truly soluble under manufacturing conditions and concentrations. Only then can one use a wall matrix that has no emulsification properties.

If flavors, or any part thereof, are insoluble in the system being used for encapsulation, then an emulsifying matrix is required. Emulsification is required to minimize flavor losses during the encapsulation process (spray drying and extrusion processes). There is ample data in the literature showing that retention of water insoluble flavorings are substantially improved if a good-quality emulsion is prepared and used during the encapsulation process (Risch and Reineccius 1988). The work of Soottitantawat et al. (2003) illustrates the benefit of a good-quality emulsion to retention of water insoluble volatiles (i.e., limonene) during spray drying (Fig 9.10). Good-quality emulsions are those with a mean particle size of ca. 1 μm. These emulsions are prepared by using an emulsifying wall material such as gum acacia, a modified starch or protein. One notes that retention is not necessarily improved by small mean particle sizes for more water-soluble volatiles.

In terms of the final application, products such as dry savory mixes (e.g., gravies or sauces), baked goods (e.g., cake mixes or frostings), or confectionery products

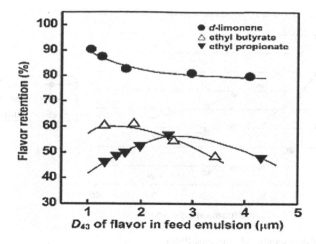

Fig. 9.10 The effect of infeed emulsion droplet size and properties of model flavors on retention during spray drying. Wall materials: blend gum acacia and maltodextrin blend (Soottitantawat et al. 2003)

(e.g., candies) do not require the edible film to provide significant emulsification properties in end use. However, a substantial portion of dry flavorings is used in dry beverage mixes. Dry beverage mixes must give a 1 week shelf life (no visible ring of separated flavor) following reconstitution. Therefore, the edible film used as an encapsulant for this application must provide good emulsification properties to these beverage products.

As noted earlier, maltodextrins and corn syrup solids impart no emulsion stability other than that due to their high viscosity when initially manufactured. When an encapsulated flavor is based on maltodextrin or corn syrup solids, it cannot be used for a dry beverage flavor application unless a secondary emulsifier is incorporated. Secondary emulsifiers are not commonly used with spray-dried flavors (although an emulsifying agent such as a modified starch or gum acacia may be used in combination with the maltodextrin or corn syrup solid flavor carrier), but are essential for manufacture and stability of extruded flavorings. Both gum acacia and the modified food starches are excellent emulsifiers. The relative ability of corn syrup solids, traditional gum acacia and modified food starch to stabilize an orange oil emulsion is shown in (Fig. 9.11). This figure gives an initial absorption for the emulsion (time = 0), which indicates the particle size of flavor droplets and resultant cloudiness.

Modified food starch formed an emulsion that was substantially cloudier than either gum acacia or corn syrup emulsions (0.8 vs. 0.6 or 0.3 absorbance units, respectively). Upon centrifugation (500 × g), all emulsions cleared to some extent as a result of separation of the emulsion. However, the modified food starch remained more stable than gum acacia or corn syrup emulsions. An advantage of using modified food starches for flavor encapsulation is their ability to form a stable flavor emulsion.

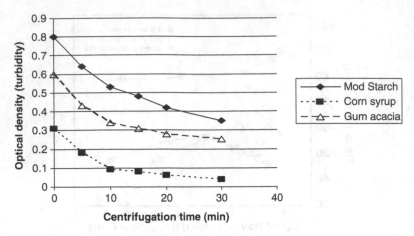

Fig. 9.11 Emulsion stability (Reineccius 1991)

Flavor Retention

The ability of an edible film to "capture" or hold onto flavor compounds during the drying process is critical. If flavors are lost during the drying process, the resultant flavor will lose strength and potentially become imbalanced in character. Lighter, more volatile constituents will be preferentially lost during the process, with the resulting dry flavor lacking in the very volatile light fresh notes. A secondary concern centering on retention is that a flavor compound not retained in the powder is lost to the processing environment. In the spray-drying process, lost flavorings enter the dryer exit air, which then must be removed by a costly scrubbing process to protect the environment. In extrusion, lost volatiles enter the workplace environment or the cold solvent used to solidify particles. Therefore, encapsulation materials that offer poor retention result in increased processing costs and decreased product quality.

Of the common edible films used in flavor encapsulation, modified food starches excel in terms of flavor retention during drying (Table 9.1). As noted earlier, modified food starches are excellent emulsifiers, and emulsion quality has a strong influence on flavor retention during spray drying (e.g., Risch and Reineccius 1988; Baranauskiene et al. 2007). Additionally, modified food starches can typically be used at 50–55% solids levels vs. 30–35% solids for gum acacia. Thus, although both gum acacia and modified food starches will yield good emulsions (i.e., they should yield good flavor retention), modified food starches yield the best flavor retention, partially because they can be used at higher infeed solids levels, which also improves flavor retention (Reineccius 1989).

Maltodextrins, corn syrup solids, or simple sugars (and their alcohols) typically can be used at high infeed solids levels, but their poor emulsification properties result in poor flavor retention during drying.

A few authors have reported the use of proteins for flavor encapsulation. For example, Kim and Morr (1996) evaluated sodium caseinate, whey protein isolate, soy protein isolate and gum acacia (control) for encapsulation (spray drying) of orange oil. They found soy protein isolate to be the best at retaining orange oil during spray drying (85.7% retention), followed by sodium caseinate (81.5%) and gum acacia (75.9%), with whey protein isolate being the poorest (72.7%). One must note that authors chose to use all materials at 10% solids (they started by making an aqueous solution of each carrier at this concentration and then added orange oil at 10, 20, and 30% of carrier solids). While this choice allows a comparison of the inherent carrier's ability to retain volatile substances, it is an impractical approach from a commercial standpoint. Manufacturers use a carrier at maximum solids content that permits adequate atomization. Thus, they would use each of these carriers at different solids levels based on the viscosity of the infeed slurry. Since solids content of the infeed slurry is *the* major determinant of flavor retention during drying, one would get a totally different ranking of carriers and retention if the experimental design involved use of carriers at constant viscosity.

Dronen (2004) evaluated the retention of a volatile model flavor system (water soluble components) during spray drying prepared on a constant viscosity basis. She used 10 DE maltodextrin (MD) at 40% infeed solids (IS); 25 DE corn syrup solids (CSS) at 40% IS; gum acacia (GA) at 35% IS; modified starch (MS) at 45% IS; soy protein concentrate (SPC) at 15% IS; and whey protein isolate (WPI) at 30% IS. These solids levels corresponded to nearly equivalent viscosities, which were low enough to permit atomization during spray drying (ca. 250 cp). She found the 10 DE MD to give 55% retention (R) ~ equal to GA (55% R) ~ equal to MS (51% R), but >25 DE CSS (40% R), >WPI (25% R), and >SPC (13% R). It is relevant that using formulations consistent with industry practice, order of retention was quite different from that reported by Kim and Morr (1996).

Storage Stability

The most common problem occurring during storage of flavors is deterioration due to oxidation. Any flavoring containing citrus oils or oils based on aldehydes is susceptible to oxidative reactions and development of off-flavors. Thus, an important function of an edible film is protection of flavoring from oxygen.

Films used for flavor encapsulation differ greatly in their ability to protect a flavoring from oxygen. Maltodextrins provide varying protection depending upon their dextrose equivalent (DE) (Anandaraman and Reineccius 1986). In general, the higher the DE, the better the protection. This presents a problem since higher DE materials tend to be difficult to dry, yield poor flavor retention and are very hygroscopic. Thus, one can get excellent shelf life using a high DE maltodextrin (more correctly, a corn syrup solid), but have an otherwise unsuitable product. This observation also holds true for use of mono and disaccharides.

The observation that high DE starch hydrolysates (corn syrup solids) offered greater oxidative stability during storage has led to a number of commercial products

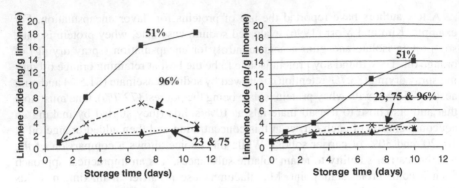

Fig. 9.12 Rate of limonene oxidation as influenced by RH of storage (50°C); *left* – Gum acacia:MD blend (1:3); *right* – Hi-Cap (adapted from Soottitantawat et al. 2004)

that are blends of emulsifying wall materials and higher DE maltodextrins, corn syrup solids, or simple sugars. Sufficient emulsifying wall materials are used to impart necessary emulsifying capacity and corn syrup solids or simple sugars are added to impart oxidative stability. One might use as little as 20% of an emulsifying wall material and 80% of a wall material that imparts oxidation stability. Fortunately, starch hydrolysates are very inexpensive, thereby lowering product costs.

While modified food starches produce an encapsulated flavoring that has excellent flavor retention and emulsion stability, they have traditionally provided very poor protection against oxidation (e.g., Reineccius 1991; Krishnan et al. 2005). As new generations of modified starch have been developed, their performance in this respect has improved (e.g., Hi-Cap-100, National Starch Corp.). If one compares the two parts of (Fig. 9.12), one can see that there is little difference between the oxidation profile of gum acacia–maltodextrin blend (left) and Hi-Cap (right) samples.

Data in (Fig. 9.12) also give the reader an idea of the importance of RH on oxidative stability imparted by these two materials. Interestingly, the least stable RH for both materials is 51% (evaluated 23, 51, 75, and 96% RH). For the Hi-Cap samples, 96% RH is the next least stable, and there is no significant difference between the 23 and 75% RH. For gum acacia–maltodextrin samples, there was little difference in stability when the samples were stored at 12, 75, or 96% RH. There is substantial information in the literature suggesting that the ability of a given wall material to protect a sample against oxidation is not easily predicted on the basis of structure or water activity.

Acacia products vary greatly in their suitability as encapsulating materials. As was mentioned earlier, there are several hundred different species of acacia trees, many of which yield a gum that may inadvertently enter commercial products. Thus, dilution of quality gums, or other natural variability in preferred species, may result in variable stability when using this product. Thus, one will find some gums that

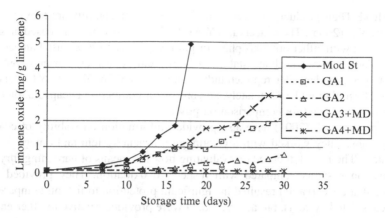

Fig. 9.13 Limonene oxidation when spray-dried in different commercial gum acacias and stored at 37°C (numbers after GA in legend specify different gum acacias; Risch and Reineccius 1990)

offer excellent protection against oxidation, while another samples offer little protection. Plots shown in Fig. 9.13 illustrate how stability varies with gum type and the addition of maltodextrin.

There are a few publications on the effectiveness of proteins to protect flavorings from oxidation. Of these publications, Kim and Morr (1996) spray-dried orange oil in different proteins (WPI, SPI, and SC), as well as in gum acacia to serve as a control, stored the resultant powders, and monitored oxidation in each. They found WPI and SPI offered the best protection to orange oil, and gum acacia and SC the poorest. There is a great deal more information on using proteins as carriers to protect lipids from oxidation. Of these papers, most have shown proteins to be very effective for this purpose (e.g., Rosenberg and Moreau 1996; Jimenez et al. 2004; Robert et al. 2003; Bylaite et al. 2001). Rosenberg has postulated that this protection comes from protein moving to the lipid–carrier interface, and forming an oxygen impermeable membrane.

One of the major advantages of CyDs is their ability to protect their guests from oxidation (Szente and Szejtli 2004). This capability is so well known that it has become common knowledge. While there is much less information on stabilizing properties of inclusion complexes formed between volatiles and amylose, it appears that these complexes are also stable to oxidation (e.g., Wulff et al. 2005).

There is little information in the public domain on the ability of coacervate systems to protect flavorings against oxidation. The few papers in this topic area are focused on preservation of highly oxidizable oils (fish, algal). For example, Barrow et al. (2007) commented on the stability of highly unsaturated oils (fish, algal, etc.) when coacervated using their proprietary method of encapsulation. While no data were provided in this article, they claimed stability for 18 months while achieving a 60% load. Their product is described and some data on stability are provided in a patent by Yan (2005). The coacervates Yan (2005) worked with were based on gelatin (Type A) with either polyphosphate or gum acacia. Yan (2005) used an accelerated oxidation test to evaluate stability of different coacervate formulations (all containing antioxidant). He reported induction periods of 26–38 h at 65°C, depending upon formulation. Unfortunately, no controls or competing encapsulation techniques were provided for comparison purposes.

Lamprecht et al. (2001) have also provided information on stability of coacervate systems. They worked with a gelatin–gum acacia system and omega-3 fatty acid esters. Their work focused on evaluating how the means of hardening (drying) coacervation systems compared with respect to oxidation. They provided data showing that coacervation resulted in significant protection to fish oil compared to nonencapsulated systems, but again no data were provided relative to other encapsulation systems.

Flavor Release

Flavor release can have several meanings. As used in this chapter, it will refer to how an encapsulated flavoring gives up its contents. A shortcoming of the major process used for flavor encapsulation (spray drying) is its inability to use edible films that are water-insoluble. The spray-drying process requires that an aqueous soluble, edible film must be used in manufacturing and, thus, they release flavor upon contact with water.

While in a sense all encapsulated flavorings offer controlled release (e.g., release only after contact with water), one can appreciate that additional controlled release properties may be desirable in certain product applications. For example, one may desire that an encapsulated flavoring does not release its contents during early stages of thermal processing. A delayed release may result in less flavor loss, since flavor would be protected from heat until late in processing. Here a slow or delayed release is desirable. In the case of an encapsulated flavoring for a dry beverage mix, one desires a rapid release on reconstitution. Thus, the desired flavor release will be dependent upon the application. Currently, these controlled release properties can be attained directly through the use of coacervation, extrusion and inclusion complex formation. Since spray-dried particles are water soluble, controlled release properties may be imparted to them through application of secondary coatings, e.g., coating with a fat or shellac. Secondary coatings (e.g., fats, oils and shellacs) are costly and problematic to apply, and therefore it is desirable to accomplish controlled release by choosing the appropriate encapsulation technique. The topic of controlled release deserves a chapter of its own, so the following discussion will be an overview in nature.

As mentioned above, coacervation, extrusion and CyDs all may be used to impart controlled release properties to a flavoring. Coacervation does this through use of capsule materials that may be insoluble in the final application. Cross-linked capsules will be insoluble irrespective of the food system, and non-cross-linked capsules may be "soluble" or insoluble depending upon sample environment. If capsules are insoluble and retain their integrity in a food application, release is governed by diffusion through the capsule wall. For example, Yeo et al. (2005) have provided information on the controlled release of a flavor oil prepared by complex coacervation. They reported on how formulation, freeze–thaw cycles and ionic strength will affect release of an encapsulated oil. In essence, dissolution of the capsule (in an ionic solution) resulted in rapid release, while systems in which the wall matrix remained intact gave slower, diffusion-controlled release. Weinbreck et al. (2004) reported on the encapsulation of sunflower oil, lemon oil, and orange oil in a gum acacia–whey protein coacervate. They found that larger capsules gave strongest flavor release (during eating). Glutaraldehyde cross-linked batches (small and large capsules) gave the lowest release intensity.

Controlled release properties may be imparted to flavorings encapsulated by extrusion processes by using some portion of the wall material that is somewhat insoluble. This approach is illustrated in a patent issued to McIver et al. (2002), who included agar (7%) in a typical extrusion formulation. They produced an extruded capsule initially containing 13.6% cinnamic aldehyde. This capsule still contained 13.1% cinnamic aldehyde after being in water for 4 days.

CyDs tend to produce controlled release since they must be hydrated to release any flavor, and the extent of release is based on partitioning of the guest between the food system and the CyD cavity. As one would expect, this is based on affinity of the host for the guest. As an example, Reineccius et al. (2003) demonstrated that as little as 10% menthol is released from β-CyD when placed in an aqueous solution, but more than 70% of ethyl butyrate is released under similar conditions. Mourtzinos et al. (2008) demonstrated a similar disparity in release of geraniol and thymol from β-CyD: geraniol was rapidly released (nearly 100%), while thymol was only slowly and incompletely released (30% into aqueous solution).

9.4 Summary

Properties of the edible films used for the encapsulation of flavoring materials are critical to the efficacy of the flavoring material. The edible film must often serve as an emulsifier, film former (trap flavoring during the dehydration process), oxygen barrier (to protect against flavor deterioration) and controlled release agent. The materials available serve each of these functions to varying degrees.

Maltodextrins are relatively inexpensive, but lack most of the functional properties desired in an encapsulation material. Corn syrup solids are also inexpensive and give excellent protection against oxidation. However, they also give poor retention during drying and no emulsification. Modified food starches are moderately priced excellent emulsifiers, and give excellent flavor retention during drying. These materials may give only limited protection to oxidation unless they are blended with other materials (e.g., mono- and disaccharides, maltodextrins, or corn syrup solids). Gum acacias vary greatly in function properties. They generally are slightly more expensive than the modified food starches and somewhat inferior in emulsification properties and flavor retention. They offer excellent protection against oxidation, and therefore can be an excellent overall choice as an edible film for encapsulation of flavors. The gum acacias have the additional advantage of being "natural," which is important to some product labels. Proteins offer excellent emulsion stability and oxidative stability, but they are reactive with flavorings containing any carbonyls.

Selection of these materials is dependent upon the process being used, the performance desired (emulsification or oxidation issues) and the controlled release properties desired in a finished application. These is no question that cost always enters the decision-making process.

9.5 Future trends

There is little probability that new encapsulation processes will be introduced in the near future. Processes in use today reflect minor improvements in old methodologies as opposed to new or novel approaches. One might look to the pharmaceutical or agrochemical fields for innovation, but constraints imposed by cost and legal use in the food industry generally make adoption of their methodologies not feasible. New materials are needed that meet our requirements from performance, cost, and legal use aspects.

References

Arvisenet G, Le Bail P, Voilley A, Cayot N (2002) Influence of physicochemical interactions between amylose and aroma compounds on the retention of aroma in food-like matrices. J Agric Food Chem 50:7088–93

Baranauskiene R, Venskutonis PR, Dewettinck K, Verhe R (2006) Properties of oregano (Origanum vulgare L.), citronella (Cymbopogon nardus G.) and marjoram (Majorana hortensis L.) flavors encapsulated into milk protein-based matrices. Food Res Int 39:413–425

Baranauskiene R, Bylaite E, Zukauskaite J, Venskutonis RP (2007) Flavor retention of peppermint (Mentha piperita L.) essential oil spray-dried in modified starches during encapsulation and storage. J Agric Food Chem 55:3027–3036

Barrow CJ, Nolan C, Jin Y (2007) Stabilization of highly unsaturated fatty acids and delivery into foods. Lipid Technol 19:108–111

Benczedi D; Bouquerand PE (2001) Process for the preparation of granules for the controlled release of volatile compounds. Firmenich Inc, WO 01/17372 A1

Brynko C, Scarpelli JA (1965) Production of microscopic capsules. National Cash Register Co, US 3190837

Bungenberg de Jong HG (1949) In:Kruyt HR (ed) Colloid Science. Elsevier Publishing Co, Amsterdam. pp 335–384

Bylaite E, Rimantas Venskutonis P, Mapdpieriene R (2001) Properties of caraway (Carum carvi L.) essential oil encapsulated into milk protein-based matrices. Eur Food Res Technol 212:661–670

Conde-Petit B, Escher F, Nuessli J (2006) Structural features of starch-flavor complexation in food model systems. Trends Food Sci Technol 17:227–235.

Delarue J, Giampaoli P (2000) Study of interaction phenomena between aroma compounds and carbohydrate matrixes by inverse gas chromatography. J Agric Food Chem 48:2372–2375

Dronen D (2004) Characterization of volatile loss from dry food polymer materials. Ph.D. Thesis, University of Minnesota, Minneapolis, MN.

Fulger CV, Popplewell LM (1999) Flavor encapsulation. McCormick & Co Inc US 5958502

Goubet I, Dahout C, Semon E, Guichard E, Le Que're' JL, Voilley A (2001) Competitive binding of aroma compounds by α-cyclodextrin. J Agric Food Chem 49:5916–5922

Green BK, Schleicher L (1957) Oil containing microscopic capsules and method of making them. National Cash Register, Dayton, OH, US Patent 2800457

Hedges AR, Shieh WJ, Sikorski CT (1995) Use of cyclodextrins for encapsulation in the use and treatment of food products. In: Reineccius GA, Risch SJ (eds) Encapsulation and Controlled Release of Food Ingredients. Amer. Chem Soc, Washington DC. p 60

Jimenez M, Garcia HS, Beristain CI (2004) Spray-drying microencapsulation and oxidative stability of conjugated linoleic acid. Eur. Food Res Technol 219:588–592

Kim YD, Morr CV (1996) Microencapsulation properties of gum Arabic and several food proteins: spray-dried orange oil emulsion particles. J Agric Food Chem 44:1314–1320

Kettlitz B, Pereman J, Fonteyn D, Sips N. (2005) Characterization and Application of a New Emulsifying Food Starch. Lectures of 56 Starch Convention

Kruyt HR (1949) In: Kruyt HR (ed) Colloid Science, vol. II, Elsevier, Amsterdam

Lamprecht A, Schafer U, Lehr CM (2001) Influences of process parameters on preparation of microparticle used as a carrier system for Ω- 3 unsaturated fatty acid ethyl esters used in supplementary nutrition. J Microencap 18:347–357

McIver R, Vlad F, Golding RA, Leichssenring TD (2002) Encapsulated flavor and/or fragrance composition. Firmenich SA, Geneva, US Patent 2002/0187223 A1

McIver RC, Leresche JP, Neffah B, Kelly R (2004) Continuous process for the incorporation of a flavor or fragrance ingredient or composition into a carbohydrate matrix. Firmenich Inc, WO 2004/082393

Mourtzinos I, Kalogeropoulos N, Papadakis SE, Konstantinou K, Karathanos VT (2008) Encapsulation of nutraceutical monoterpenes in β-cyclodextrin and modified starch. J Food Sci 73:S89–S94

Pagington J (1986) Beta-cyclodextrin. Perfum. Flavorist, 11:49

Porzio MA, Popplewell LM (2001) Encapsulation compositions. McCormick & Co. Inc, EP1123660

Porzio MA, Zasypkin D (2004) Encapsulation compositions and process for preparing the same. McCormick & Co Inc, US 2004241444

Qi Z, Hedges A (1995) Use of cyclodextrins for flavors. In: Ho C-T, Tan C-T, Tong C-H (eds) Flavor Technology: Physical Chemistry, Modification and Process. American Chemical Society, Washington, DC. pp 231

Reineccius GA (1989) Flavor encapsulation. Food Rev Int 5(2):147

Reineccius GA (1991) Carbohydrates for flavor encapsulation. Food Technol 45:144

Reineccius TA, Reineccius GA, Peppard TL (2002) Encapsulation of flavors using cyclodextrins: Comparison of flavor retention in α, β, and γ types. J Food Sci 67:3271–79

Reineccius TA, Reineccius GA, Peppard TL (2003) Comparison of flavor release from alpha-, Beta- and Gamma-cyclodextrins. J Food Sci 68:1234–1239

Reineccius TA, Reineccius GA, Peppard TL (2004) Beta-cyclodextrin as a partial fat replacer. J Food Sci 69:334–341

Reineccius TA, Reineccius GA, Peppard TL (2005) The effect of solvent interactions on alpha-, beta-, and gamma-cyclodextrin/flavor molecular inclusion complexes. J Agric Food Chem 53:388–92.

Risch SJ, Reineccius GA (1988) Effect of emulsion size on flavor retention and shelf-stability of spray dried orange oil. In: Reineccius GA, Risch SJ (eds) Encapsulation and Controlled Release of Food Ingredients. Am. Chem Soc, Washington, DC. p 67.

Risch SJ, Reineccius GA (1990) Difference between gum acacias for the spray drying of citrus oils. Perfum Flavorist 15:55–58.

Robert P, Carlsson RM, Romero N, Masson L (2003) Stability of spray-dried encapsulated carotenoid pigments from rosa mosqueta (Rosa rubiginosa) oleoresin. J Am Oil Chem Soc 80:1115–1120

Rosenberg M, Moreau DL (1996) Oxidative stability of anhydrous milkfat microencapsulated in whey proteins. J Food Sci 61:39–43

Rutschmann MA, Solms J (1990) Formation of inclusion complexes of starch with different organic compounds IV. Ligand binding and variability in helical conformations of amylose complexes. Lebensm Wiss Technol 23:84–87

Saldarini A, Doerig R (1981) Fixing a volatile flavoring agent in starch hydrolysate. Norda Inc, USA US 4285983

Schultz IH, Dimick RP, Mackower B (1956) Incorporation of natural fruit flavors into fruit juice powders. Food Technol 10:57

Soottitantawat A, Yoshii H, Furuta T, Ohkawara M, Linko P (2003) Microencapsulation by spray drying: influence of emulsion size on the retention of volatile compounds. J Food Sci 68:2256–2262

Soottitantawat A, Yoshii H, Furuta T, Ohgawara M, Forssell P, Partanen R, Poutanen K, Linko P (2004) Effect of water activity on the release characteristics and oxidative stability of D-limonene encapsulated by spray drying. J Agric Food Chem 52:1269–1276

Subramaniam A, McIver R (2007) Process for the incorporation of a flavor or fragrance ingredient or composition into a carbohydrate matrix. Firmenich Inc, CN101005771

Swisher HE (1957) Solid essential oil containing compositions, US Patent 2,809,895

Szejtli J, Szente L, Banky-Elod E (1979) Molecular encapsulation of volatile, easily oxidizable labile flavor substances by cyclodextrins. Acta Chimica Academiae Scientiarum Hungaricae 101:27–46.

Szente L, Szejtli J (2004) Cyclodextrins as food ingredients. Trends Food Sci Tech 15:137–142.

Tapanapunnitikul O, Chaiseri S, Peterson DG, Thompson DB (2008) Water solubility of flavor compounds influences formation of flavor inclusion complexes from dispersed high-amylose maize starch. J Agric Food Chem 56:220–226

Ubbink J, Schoonman A (2005) Flavor delivery systems. In: Seidel A (ed) Kirk-Othmer Encyclopedia of Chemical Technology (5th Edition). Wiley, Hoboken. pp 527–563

Weinbreck F, Minor M, de Kruif G (2004) Microencapsulation of oils using whey protein/gum arabic coacervates. J Microencap 21:667–679

Wulff G, Avgenaki G, Guzmann MSP (2005) Molecular encapsulation of flavours as helical inclusion complexes of amylose. J Cereal Sci 41:239–249

Wyler R, Solms J (1982) Starch -flavor complexes III. Stability of dried starch -flavor complexes and other dried flavor preparations. Lebensmittel-Wissenschaft und Technologie 15:93–96

Wyler R, Solms J (1981) Inclusion complexes of potato starch with flavor compounds. In: Schreier P (ed) Flavour. de Gruyter, Berlin. pp 693–9

Yan N (2005) Encapsulated agglomeration of microcapsules and method for the preparation thereof. Ocean Nutrition Canada Limited (Dartmouth, CA) US Patent 6974592

Yeo Y, Bellas E, Firestone W, Langer R, Kohane DS (2005) Complex coacervates for thermally sensitive controlled release of flavor compounds. J Agric Food Chem 53:7518–7525

Chapter 10
Delivery of Flavor and Active Ingredients Using Edible Films and Coatings

Olga Martín-Belloso, M. Alejandra Rojas-Graü, and Robert Soliva-Fortuny

10.1 Introduction

Edible films and coatings are promising systems for improvement of food quality, shelf life, safety, and functionality. They can be used as individual packaging materials, food coating materials, and active ingredient carriers. They can also be used to separate the compartments of heterogeneous ingredients within foods. In fact, edible films and coatings can incorporate food additives, such as anti-browning agents, antimicrobials, flavors, colorants, and other functional substances. Enhanced sensory properties of a food can be achieved by adding flavoring agents to an edible film or coating, leading to development of new flavor delivery systems that improve food quality and utility. Currently, there are numerous applications of edible films and coatings whose main purpose is to impart desirable mouthfeel to the coated product; this is especially true for snacks (e.g., popcorn, corn chips, and potato chips), nuts, meat, fish, and poultry. Further, incorporation of active ingredients can enhance functionality of edible films and coatings, thereby providing health benefits to consumers. For example, addition of probiotic organisms to films and coatings could open up opportunities to develop new, health-enhancing products. Edible films and coatings are promising delivery systems for flavor and active ingredients that improve food quality and functionality.

This chapter will focus on use of edible films and coatings as carriers of flavor and active ingredients in foods. It will also discuss incorporation of some flavoring and active substances into edible films and coatings, which can improve quality and functionality of foods. It will also provide insights about recent advances in this area.

O. Martín-Belloso (✉), M.A. Rojas-Graü, and R. Soliva-Fortuny
Department of Food Technology, University of Lleida,
Alcalde Rovira Roure, 191, Lleida, 25198, Spain
e-mail: omartin@tecal.udl.es

M.E. Embuscado and K.C. Huber (eds.), *Edible Films and Coatings for Food Applications*, 295
DOI 10.1007/978-0-387-92824-1_10, © Springer Science+Business Media, LLC 2009

10.2 Historical Background

Application of edible films and coatings to foods to extend their shelf life is not a new practice. Edible films and coatings have been used for centuries to prevent moisture migration, improve food appearance, and increase product shelf life. Wax coatings on whole fruits and vegetables have been used since the 1800s. In fact, coating of fresh citrus fruits (oranges and lemons) with wax to retard desiccation was practiced in China in the twelfth and thirteenth centuries (Hardenburg 1967). Currently, edible coatings are widely used on whole fruits, such as apples, pears, oranges, lemons, and grapefruits, to reduce water loss, improve appearance by imparting sheen to the food surface, provide a carrier for fungicides or growth regulators, and create a barrier to gas exchange between the commodity and external atmosphere. In fact, application of a coating to tropical fruits is essential because they are susceptible to weight loss and, in some cases, to physiological breakdown. Coatings for tropical fruits such as mangos, avocados, papayas, melons, and pineapples usually consist of emulsions composed of carnauba wax (Grant and Burns 1994).

Edible films and coatings have also been used to preserve meats, nuts, snacks, and candies. A good example of this is M&M chocolate candies. The practice of applying edible coatings to candy was originally introduced in 1941 to overcome lack of chocolate sales during warm summer months. The coating application kept chocolate from melting during storage and handling.

Nevertheless, edible films and coatings have since been recognized for even more innovative uses and applications beyond their historical usage. They have a great potential to deliver functional compounds as carriers of special ingredients. For example, incorporation of probiotic and nutraceutical compounds represents a new advancement in the concept of edible films and coatings. The term "nutraceutical" has become synonymous with dietary supplements. It was originally defined in 1995 as any substance that may be considered a food or part of a food that provides medical or health benefits, including prevention and treatment of disease (Gulati and Ottaway 2006). The term, nutraceutical, is used to describe substances that are not traditionally recognized as foods (e.g., vitamins and minerals), but have positive physiological effects on the human body (Gulati and Ottaway 2006). In addition, probiotics have been defined as "live microbial feed supplements that have beneficial effects on the host by improving their intestinal microbial balance" (Adhikari et al. 2000). Potential health benefits and biological functions of bifidobacteria in humans include intestinal production of lactic and acetic acids, inhibition of pathogens, reduction of colon cancer risks, reduction of serum cholesterol, improved calcium absorption, and activation of the immune system, among others (Mitsuoka 1990; Gibson and Roberfroid 1995; Kim et al. 2002). A viable bifidobacteria population of 5 cfu/g in the final product has been identified as the therapeutic minimum to attain aforementioned health benefits (Naidu et al. 1999). In order to confer health benefits to humans, the viable count of bifidobacteria at the time of consumption should be 10^6 cfu/g (Samona and Robinson 1991). Ingestion of 10^6 cells per gram has been recommended for classic probiotic foods, such as yogurt (Kurman and Rasic 1991). It is also important for manufacturers and retailers to be able to confirm viable counts of these organisms in bifidus-containing products.

10.3   Edible Films and Coatings as Carriers of Flavors, Colorants and Spices

As previously outlined, edible films and coatings can deliver and maintain desirable concentrations of color, flavor, spiciness, sweetness, saltiness, etc. Several commercial films, especially Japanese pullulan-based films, are available in a variety of colors, with spices and seasonings included (Guilbert and Gontard 2005). Owing to its excellent oxygen barrier properties, pullulan films can be used to entrap flavors and colors and to stabilize other active ingredients within the film.

Currently, the US Department of Agriculture's (USDA's) Agricultural Research Service (Albany, California), in cooperation with Origami Foods (Pleasanton, California), has developed vegetable and fruit edible films as alternatives to the seaweed sheets (nori) traditionally used for sushi and other Asian cuisine. These wraps, produced as soft and pliable sheets using infrared drying, can be made from broccoli, tomato, carrot, mango, apple, peach, pear, as well as a variety of other fruit and vegetable products. They can also be used to contain spices, seasonings, colorants, flavors, vitamins, and other beneficial plant-derived compounds. These food films are made commercially available by California-based Origami Foods and the USDA for use in a growing number of food applications, such as a bright orange, carrot-based wrap to encircle a cucumber, garlic, and rice filling; a deep red, tomato and basil-based wrap to hold a spicy tuna and rice filling; a blueberry or strawberry-based wrap to cover creamy cheesecake in mini-desserts; a pineapple–apricot–ginger–based wrap to enclose rice and diced roast pork in elaboration of a sushi; and a broccoli-based wrap to encircle sushi of carrots, onions, and asparagus. Other uses might include snack crackers wrapped with fruit and vegetable films, apple wedges wrapped in peach film and tempura strawberry wrapped bananas.

In a recent study, Laohakunjit and Kerdchoechuen (2007) coated nonaromatic milled rice with 30% sorbitol–plasticized rice starch, containing 25% natural pandan leaf extract (*Pandanus amaryllifolius* Roxb.). This extract is primarily responsible for the jasmine aroma of aromatic rice. The rice starch coating containing natural pandan extract produced nonaromatic rice with aroma compounds similar to that of aromatic rice. Additionally, coating treatment also reduced *n*-hexanal content of storage grains. This coating technique represents a promising approach for improving rice aroma and, at the same time, for reducing potential for oxidative rancidity during grain storage.

10.4   Edible Films and Coatings Carrying Nutraceuticals

Edible films and coatings also have capacity to hold many active ingredients that could be used to enhance nutritional value of food products. However, few studies have actually reported integration of nutritional or nutraceutical ingredients into edible films or coatings of foods, though there is a recent growing interest in this area. For such applications, the concentration of nutraceuticals added to the films or coatings must be carefully studied in relation to basic properties (e.g., barrier and mechanical) of carrier films.

Mei and Zhao (2003) evaluated feasibility of milk protein-based edible films to carry high concentrations of calcium and vitamin E. They used calcium caseinate (CC) and whey protein isolate (WPI) films containing 5 or 10% Gluconal Cal (GC), a mixture of calcium lactate and gluconate, or 0.1 or 0.2% of α-tocopheryl acetate (VE), respectively. Both CC and WPI films were capable of carrying high concentrations of calcium or vitamin E, though film functionality can be compromised in the process. For example, vitamin E incorporation at tested levels increased elongation at break and reduced tensile strength of films. In contrast, incorporation of calcium into CC films reduced tensile strength for both levels of GC addition, and decreased both elongation at break and water vapor permeability values at 10% GC addition level.

In subsequent studies, Park and Zhao (2004) evaluated functionality of chitosan-based films containing a high concentration of calcium (Gluconal Cal- GC), 5–20% zinc lactate (ZL), vitamin E (5–20% α-tocopheryl acetate) and acetylated monoglyceride (AM). Addition of GC significantly increased the pH and decreased the viscosity of film-forming solutions; however, addition of ZL or VE resulted in no such effects. Water barrier properties of the films were improved as either the concentration of mineral or vitamin E component within the film matrix increased. Nevertheless, tensile strength of films was affected by the incorporation of high concentrations of GC or VE, although other mechanical properties such as film elongation, puncture strength, and puncture deformation were not affected. They concluded that both milk protein and chitosan films may be used as carriers of nutraceuticals.

Some studies have reported that edible coatings might serve as excellent carriers of low levels of nutrients for fruits and vegetables, thereby improving nutritional value. Mei et al. (2002) developed xanthan gum coatings, containing high concentrations of calcium and vitamin E (5% Gluconal Cal and 0.2% α-tocopheryl acetate, respectively), for the purpose of enhancing nutritional and sensory qualities of fresh baby carrots. Calcium and vitamin E contents of the coated carrot samples increased from 2.6 to 6.6% and from 0 to 67% of the Dietary Reference Intakes (DRI) values per serving (85 g), respectively. Furthermore, these nutrient levels were achieved without affecting fresh aroma, flavor, sweetness, crispness, or β-carotene levels of the carrots. Because peeled baby carrots are poor sources of calcium and vitamin E, incorporating both nutrients into edible coating is an excellent approach for enhancing nutritional qualities of this type of product.

Application of edible coatings carrying nutraceutical compounds has also been studied with fruits. In fact, Han et al. (2004) added calcium and vitamin E to chitosan-based coatings to improve shelf life and nutritional properties of fresh and frozen strawberries and red raspberries. The addition of high concentrations of Gluconal Cal or α-tocopheryl acetate into chitosan-based coatings does not alter native antifungal and moisture barrier functions associated with chitosan. Coatings significantly reduced incidence of decay and weight loss and delayed changes in color, pH and titratable acidity of strawberries and red raspberries during cold storage. In addition, chitosan-based coatings demonstrated great capacity to hold high concentrations of calcium or vitamin E, thereby increasing significantly the content of these nutrients in both the fresh and frozen fruits. For one serving (100 g), coated fruits contained 34–59 mg of calcium and 1.7–7.7 mg of vitamin E, depending on the type of fruit and time of storage, whereas uncoated fruits contained only 19–21 mg and 0.25–1.15 mg of calcium and vitamin E, respectively. Similarly, Hernández-Muñoz et al. (2006) observed that the amount of calcium (3,079 g/kg dry matter) retained by coated fruit was greater than that obtained with calcium dips alone (2,340 g/kg). On the other hand, addition of 1% calcium gluconate to the chitosan coating formulation (1.5% in 0.5% acetic acid) did not extend shelf life of coated strawberries.

10.5   Edible Films and Coatings to Carry Probiotic Organisms

Incorporation of probiotics into functional edible films and coatings has been scarcely studied. Recently, Tapia et al. (2007) developed probiotic edible films for coating fresh-cut fruits. In this work, feasibility of alginate (2% w/v) and gellan (0.5%) based edible coatings as carriers of organisms, such as bifidobacteria, was investigated in an attempt to obtain functional probiotic-coated fruits. Fresh-cut apples and papayas were successfully coated with alginate or gellan film-forming solutions containing viable bifidobacteria. In fact, values higher than 10^6 cfu/g *Bifidobacterium lactis* Bb-12 were maintained on both papaya and apple pieces for up to 10 days of refrigerated storage (Fig. 10.1). This observation demonstrated that alginate- and gellan-based edible coatings could feasibly carry and support viable probiotics on fresh-cut fruits. Furthermore, water vapor permeability for alginate (6.31 and 5.52 $\times 10^{-9}$ g m/Pa s m^2) and gellan (3.65 and 4.89 $\times 10^{-9}$ g m/Pa s m^2) probiotic coatings of papayas and apples, respectively, was higher than in the corresponding cast

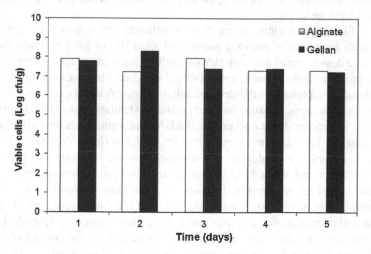

Fig. 10.1 Viable cells (log ufc/g) of *Bifidobacterium lactis* Bb-12 in alginate and gellan coatings, containing sunflower oil, glycerol, and *N*-acetylcysteine applied on fresh-cut apples, followed by refrigerated (2°C) storage (adapted from Tapia et al. 2007)

films. This work represents a recent advancement in the use of edible films and coatings as carriers of diverse food additives, and opens new possibilities for development of probiotic film and coating products.

10.6 Edible Films and Coatings to Carry Antimicrobial Agents

Active compounds, such as antimicrobials, can be incorporated into edible films and coatings. These antimicrobial edible films and coatings inhibit spoilage and pathogenic bacteria by maintaining effective concentrations of active compounds on food surfaces (Gennadios and Kurth 1997). There are several categories of antimicrobials that can be potentially incorporated into edible films and coatings: organic acids (acetic, benzoic, lactic, propionic, sorbic); fatty acid esters (glyceryl monolaurate); polypeptides (lysozyme, peroxidase, lactoferrin, nisin); plant essential oils (cinnamon, oregano, lemongrass); and nitrites and sulfites (Franssen and Krochta 2003). Within these categories, plant essential oils are outstanding alternatives to chemical preservatives, and their use in foods meets consumer demands for minimally processed natural products (Burt 2004). Essential oils are designated as "Generally Recognized as Safe" (GRAS) (Burt 2004), and are used as flavoring agents in various foods (Fenaroli 1995). These compounds can also be added to edible films and coatings to modify food flavor, aroma, and odor (Cagri et al. 2004). In two recent studies, Rojas-Graü et al. (2006a, 2007c) compared effects of oregano, cinnamon, and lemongrass oils, incorporated into apple puree and alginate–apple puree edible films, for effectiveness against *Escherichia coli* O157:H7. Both works showed that edible films, as carriers of antimicrobial compounds, constitute a feasible approach for incorporating plant essential oils onto fresh food surfaces.

Edible films and coatings, as carriers of antimicrobial compounds, provide a novel means by which to improve safety and shelf life of food systems. Several studies have demonstrated that antimicrobial edible films can reduce bacterial levels on meat products. For instance, antimicrobial chitosan films, containing acetic or propionic acid, reportedly inhibited growth of *Enterobacteriaceae* and *Serratia liquefaciens* on bologna, cooked ham, and pastrami (Ouattara et al. 2000). Cagri et al. (2002) also incorporated *p*-aminobenzoic (PABA) and sorbic acids into whey protein isolate films to inhibit *Listeria monocytogenes*, *Escherichia coli* O157:H7 and *Salmonella enterica* on bologna and summer sausage. Oussalah et al. (2004) applied milk protein-based edible films containing oregano, pimento, or oregano–pimento essential oil mix onto beef muscle slices to control growth of *E. coli* and *Pseudomonas* spp., thereby increasing product shelf life during refrigerated storage. The same authors also applied alginate-based edible films, containing Spanish oregano or Chinese cinnamon essential oils, onto beef muscle slices for control of *E. coli* O157:H7 and *Salmonella typhimurium*. Incorporation of essential oils into films helped to reduce growth of both bacterial species on beef muscle during 5 days of storage (Oussalah et al. 2006).

10.7 Edible Films and Coatings to Carry Antioxidant Agents

Antioxidants are used to protect against oxidative rancidity, degradation, and enzymatic browning in fruits and vegetables. Ascorbic acid (Soliva-Fortuny et al. 2001; Son et al. 2001), 4-hexylresorcinol (Dong et al. 2000; Luo and Barbosa-Canovas 1997; Monsalve-Gonzalez et al. 1993), and some sulfur-containing amino acids, such as cysteine and glutathione (Gorny et al. 2002; Molnar-Perl and Friedman 1990; Nicoli et al. 1994; Son et al. 2001) have been widely studied, individually and in combination, for their ability to prevent enzymatic browning, to serve as sulfite substitutes, and to improve shelf life of minimally processed fruits (Pizzocaro et al. 1993; Rojas-Graü et al. 2006a, b). Several authors have studied incorporation of many of these antibrowning agents into edible films used to coat minimally-processed fruits (Wong et al. 1994; Baldwin et al. 1996; Lee et al. 2003; Perez-Gago et al. 2006). Rojas-Graü et al. (2007a) and Tapia et al. (2005) applied alginate- and gellan-based coatings to fresh-cut apples and papayas, demontrating that coatings were good carriers for antioxidant agents, including cysteine, glutathione, and ascorbic and citric acids. Most of these antibrowning agents, however, are hydrophilic compounds, and may increase water vapor transmission rate and induce water loss when incorporated into films and coatings (Ayranci and Tunc 2004).

10.8 Technical Development

Edible films and coatings can be applied by different methods such as brushing, wrapping, spraying, dipping, casting, panning, or rolling. Dipping is advantageous when a product requires several applications of a coating to obtain uniformity on an irregular surface. After dipping the product and draining away the excess coating, the film is allowed to set or solidify on the product (Donhowe and Fennema 1994).

Casting is another technique used to apply edible coatings to food. For casting, film-forming solutions are poured onto a level surface and allowed to dry, usually within a confined space. Casting produces freestanding films that exhibit a specified thickness, smoothness, and flatness (Donhowe and Fennema 1994). Depending on firmness and flexibility, cast films can then be used to wrap surfaces. This wrapping technique allows films to be cut to any size, and serves as an innovative and easy method for carrying and delivering a wide variety of ingredients such as flavorings, spices and seasonings that can later be used to cover foods. This method is especially useful when applied to highly spicy materials that need to be separated from the food product.

Another technique, spraying, provides a more uniform coating. Spraying is desirable when a coating application is needed only on one side of the food product or when a dual application must be used to achieve the desired function (e.g., cross-linking of alginate or pectin coatings by subsequent coating with $CaCl_2$) (Donhowe and Fennema 1994). Sometimes it is advisable to heat product after spraying to hasten the drying process and to improve uniformity of film solution on the food surface (Grant and Burns 1994).

The type of food product dictates the most appropriate method for coating application. For example, the most effective method for coating nuts is panning; cheese, on the other hand, is often dipped in or brushed with wax. Spraying is the conventional method for applying most coatings to fruits and vegetables. Meat products, however, are coated using any of several techniques such as foaming, dipping, spraying, casting, brushing, individual wrapping, or rolling (Donhowe and Fennema 1994).

Coatings are more prevalently used than films. For example, collagen films are used for sausage casings; some hydroxymethyl cellulose films, such as soluble pouches, are used for dried food ingredients. Shellac and wax coatings are most commonly used on fruits and vegetables, zein coatings on candies, and sugar coatings on nuts (Krochta and De Mulder-Johnston 1997). Another typical example of using coatings is a type of Japanese polysaccharide film that is used for meat products, such as ham and poultry packaging, before they are smoked and/or steamed. The film dissolves during the process, and coated meat exhibits improved yield, structure, and texture, as well as reduced moisture loss (Labell 1991; Stollman et al. 1994). Use of carboxymethylcellulose and methylcellulose as matrix in elaboration of coatings is predominant with fruits, vegetables, meats, fish, nuts, confectionery products, baked goods, grains and other agricultural products (Nussinovitch 2003).

The first edible films made from fruit purees were developed by McHugh et al. (1996), who also characterized their water vapor and oxygen permeability properties. Soon afterwards, a patent application was filed on products and processes for wrapping foods with fruit- and/or vegetable-based edible films (McHugh and Senesi 1999). After that, they developed a novel method (apple wraps) to extend shelf life and improve quality of fresh-cut apples. These wraps were made from an apple puree that contained various concentrations of fatty acids, fatty alcohols, beeswax and vegetable oil. Fruit and vegetable wraps can be used to extend the shelf life of fresh-cut fruits and vegetables, as well as to enhance nutritional value and increase consumer appeal. Additionally, fruit and vegetable wraps can be used as barriers for other food systems such as nuts, baked goods and confectionery products (McHugh and Senesi 2000).

10.9 Properties and Functions

Potential properties and applications of edible films and coatings have been extensively reviewed (Bravin et al. 2006; Jagannath et al. 2006; Min et al. 2005; Serrano et al. 2006). Edible films and coatings are known to improve product shelf life and food quality, as they are selective barriers to moisture transfer, oxygen uptake, and loss of volatile aromas and flavors (Kester and Fennema 1986). When used to coat fresh-cut fruits and vegetables, edible films may reduce deleterious effects concomitant with minimal processing. Moisture barrier properties of edible films and coatings have been extensively studied by measuring their water vapor properties, because of water's key role in deteriorative reactions.

Edible films coatings made from naturally occurring polymers, such as polysaccharides and proteins are regarded as good oxygen barriers because of their hydrogen-bonded network structures, which are very tightly packed and ordered (McHugh and Krochta 1994). Major drawbacks of such films and coatings are their relatively low water resistance and poor vapor barrier properties (Yang and Paulson 2000). Incorporation of lipids, either in an emulsion or as a layer coating of film formulations, greatly improves their water vapor barrier properties (García et al. 2000; Yang and Paulson 2000). In addition to the increased barrier properties, edible films and coatings control adhesion, cohesion and durability, and they improve appearance of coated foods (Krochta and De Mulder-Johnston 1997).

Besides physical and chemical quality enhancements, edible films and coatings contribute to better visual quality, surface smoothness, color and, in addition, serve as carriers of various active agents such as emulsifiers, antioxidants, antimicrobials, nutraceuticals, flavors and colorants. These active compounds, which serve to enhance food quality and safety, may be added up to the levels in which they could interfere with physical and mechanical properties of films (Kester and Fennema 1986; Gennadios and Weller 1990; Guilbert and Gontard 1995; Krochta and De Mulder-Johnston 1997; Miller et al. 1998).

For instance, edible coatings can improve effectiveness of the popping process for popcorn (Wu and Schwartzberg 1992), acting as adhesion agents between heterogeneous food ingredients (Anonymous 1997). Furthermore, edible coatings are applied to surfaces of snack foods and crackers, as the base or adhesive for seasonings. For example, fat and oil have traditionally been used for adhering seasonings and flavorings to surfaces of cereal-based snack foods. However, as a result of recent market demands for reduced-fat and fat-free snack foods, many companies have introduced other edible coatings that are especially useful in low-fat applications as a replacement for oil-based adhesives. For example, oil-roasted and dry-roasted peanuts, which require an adhesive to act as a coating or bonding agent for salt and/or seasonings, now utilize a coating solution made from modified food starch, corn syrup, water and glycerol. This solution is applied to peanuts during tumbling, after which seasoning or salt may be added.

In addition, edible films and coatings allow the incorporation of preservation agents as well as nutrients and nutraceuticals. This practice improves product quality, as is the case with colored/flavored confectionery goods, glazed bakery goods, flavored nuts and vitamin-enriched rice. Active compounds can be added to food coatings to extend shelf life, preserve color, and improve nutritional value of foods. Multifunctional edible films and coatings can be utilized for value-added confections, medicinal and therapeutic foods, pharmaceutical products, and other nutraceuticals, as well as conventional perishable foods (Han 2002).

In addition to improving health and nutritional value, incorporated flavors and colorants can improve taste and visual appeal of coated products. For example, edible coatings can be sprayed or dipped onto surfaces of snack foods and crackers to serve as adhesives for flavorings (Druchta and De Mulder Johnston 1997). They can also be used as carriers of flavors in precooked meats. The practice of coating is gaining popularity in development of a variety of ready-to-heat products for quick meals. Candies are often coated with edible films to improve their texture by reducing stickiness (McHugh and Senesi 2000). Because of the various chemical characteristics of these active additives, the film composition should be modified to maintain a homogeneous film structure as these additives are incorporated into the film structure (Debeaufort et al. 1998).

10.10 Sensory Evaluation of Edible Films and Coatings with Active Ingredients

For the majority of cases, edible coatings with little or no taste are desired to prevent their detection during consumption (Contreras-Medellin and Labuza 1981). To accomplish this purpose, films must have neutral organoleptic properties (e.g., clarity, transparency, lack of discernible odor and taste, etc.). Since edible films and coatings are intended to be consumed with their respective products, the incorporation of compounds such as nutraceuticals, antioxidants and antimicrobial agents should not adversely affect consumer acceptance.

Several studies have investigated sensory characteristics of coated food products when nutraceutical ingredients are incorporated. The taste contributed by these ingredients has been considered a particularly important aspect, since many nutraceutical compounds have natural bitter, astringent, or other undesirable off-flavors (Drewnowski and Gomez-Carneros 2000) that can lead to unacceptable attributes in these products (LeClair 2000). Han et al. (2005) evaluated sensory attributes of fresh strawberries treated with chitosan-based edible coating material, with and without addition of vitamin E incorporated into the coating. Coated strawberries were evaluated by consumers for acceptance attributes and by trained panelists for descriptive analysis of their appearance, texture and flavor. Results from consumer testing 1 week after coating application indicated that all chitosan coatings evaluated increased the appearance scores of strawberries. On the other hand, the incorporation of vitamin E reduced the glossiness of coated strawberries, which could affect consumer acceptance.

Other studies have illustrated effects of incorporating other compounds, such as antibrowning agents, on sensory qualities of coated products. The type and concentration of different antioxidant compounds incorporated into coatings of whey protein concentrate (WPC)/beeswax (BW) and their effect on the color and sensory quality of fresh-cut apples have been studied (Perez-Gago et al. 2006). Results showed that incorporation of ascorbic acid (AA-0.5% and 1%), cysteine (Cyst-0.1%, 0.3% and 0.5%) or 4-hexylresorcinol (0.005% and 0.02%) as antioxidant agents into coating formulations reduced apple browning. The most effective WPC/BW coatings were those containing 1% AA or 0.5% Cyst. A sensory panel was able to discriminate between samples coated with WPC/Cyst solution and those just dipped in Cyst aqueous solution. Notably, panelists were also able to detect a smell of sulfur in Cyst-coated samples, which supports the conclusion that use high concentrations of sulfur-containing compounds, such as *N*-acetylcysteine and glutathione, as dipping agents may produce an unpleasant odor in fruits and vegetables (Iyidogan and Bayindirli 2004; Richard et al. 1992; Rojas-Graü et al. 2007a, 2006b). Interestingly, no differences were noted between coated and uncoated samples containing AA (Perez-Gago et al. 2006), which make the WPC/AA coating a good alternative to application of AA alone.

Lee et al. (2003) studied the effect of two edible coatings (carrageenan and whey protein concentrate) in combination with antibrowning agents on fresh-cut apple slices, and observed that the incorporation of ascorbic acid, citric acid and oxalic acid did not negatively affect their sensory characteristics. Hence, addition of antibrowning agents to these coating solutions was advantageous in maintaining color over a 2-week time span. Also, in quality testing, apples coated with any of the assayed formulations had higher sensory scores than uncoated apples. Whey protein concentrate coating solutions (5 g/100 mL), containing ascorbic acid (1 g/100 mL) and $CaCl_2$ (1 g/100 mL), were most effective in preserving sensory quality of apples.

However, incorporation of some antimicrobial agents into edible coatings might be anticipated to alter original flavors of foods as a result of the strong flavors associated with them, especially when plant essential oils are used. Indeed, essential oils have been extensively evaluated for their ability to protect foods against pathogenic bacteria, in addition to being used as flavoring agents in baked goods, sweets, ice cream, beverages and chewing gum (Fenaroli 1995). Currently, little is known about sensory characteristics of edible films and coatings that include essential oils in their formulations. Recently, Rojas-Graü et al. (2007b) evaluated sensory quality of coated fresh-cut apples containing plant essential oils, such as lemongrass, oregano oil, and vanillin. Essential oils were incorporated into apple puree–alginate edible coatings used to surface-coat fresh-cut "Fuji" apples. Taste evaluations indicated that coated fresh-cut apples containing incorporated vanillin (0.3% w/w) were the most acceptable in terms of sensory quality after 2 weeks of storage. In contrast, the lowest overall preference was observed for coated apple samples containing oregano oil (Fig. 10.2). Despite the low concentration of

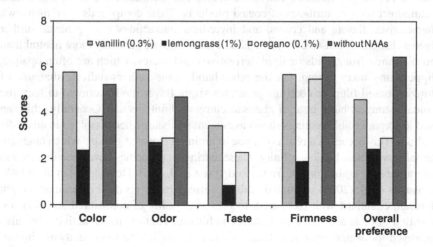

Fig. 10.2 Sensory attribute scores for coated fresh-cut apple, with or without antimicrobials agents, after 2 weeks of storage (adapted from Rojas et al. 2007)

oregano oil used (0.1% w/w), some consumers detected a residual aromatic herbal taste that diminished overall preference for these cut apples.

In contrast with sensory changes observed when plant essential oils are incorporated into coatings, several studies report good results when other antimicrobial compounds are used. The effects of a casein coating on properties of Kashar cheese, and its effectiveness in carrying natamycin to prevent mold growth, were studied by Yildirim et al. (2006). Five cheese groups were evaluated: uncoated, vacuum-wrapped, coated with casein, coated with casein containing natamycin, and dipped in natamycin solution. Sensory evaluation showed no significant differences among cheese samples, which support the thought that casein coating with natamycin can suppress mold growth for about 1 month without any adverse effects to cheese sensory quality.

Eswaranandam et al. (2006) incorporated malic and lactic acids into soy protein coatings, with the purpose of evaluating their effect on the sensory characteristics of whole apples and fresh-cut cantaloupe. Organic acids incorporated into films did not adversely impact sensory properties of either whole apples or cantaloupe cubes, as evaluated by trained sensory panelists.

Chen et al. (1999) developed edible coatings composed of methylcellulose, glycerol and fatty acid (stearic acid or palmitic acid) as carriers of preservatives to inhibit microbial growth on Taiwanese-style fruit preserves. In this study, two types of fruit (Ten-shing mei and Ching-shuan mei) prepared from plums were covered with edible coatings containing benzoic acid. Two osmophilic yeasts (*Zygosaccharomyces rouxii* and *Z. mellis*) were used as test microorganisms. Results indicated that yeast growth was inhibited in both coated fruit preserves containing 50–100 μg of benzoic acid per gram of coating solution. Furthermore, coated fruits were not significantly different from uncoated fruits in terms of sensory quality, suggesting that the edible coating could serve as a food additive carrier without modifying sensory properties.

It is important to highlight that many compounds used in development of edible films and coatings – including edible matrices, plasticizers and other active ingredients – can affect sensory attributes of coated products. These compounds have their own characteristic flavors and colors, and ingredient interactions may generate further diverse changes. For instance, hydrocolloid-based films are generally more neutral than those formed from lipids or lipid derivatives and waxes, which are often opaque, slippery and waxy tasting. On the other hand, some authors indicate that use of chitosan-based films or coatings generates slight flavor modifications; in fact, the typical astringent/bitter taste of chitosan currently limits its use. Generally, chitosan used to prepare edible coatings shows astringent attributes when dissolved in an acidic medium. This occurs as a result of a rise in amine protonated groups, which augment the salivary protein binding affinity. These astringent properties have been detected in several sensory evaluations of fruits (Rodriguez et al. 2003; Devlieghere et al. 2004).

Vargas et al. (2006) evaluated edible coatings based on high molecular weight chitosan, combined with oleic acid, to preserve quality of cold-stored strawberries. Addition of oleic acid not only enhanced chitosan's antimicrobial activity, but also improved the water vapor resistance of coated samples. Sensory analysis showed that coating application led to a significant decrease in strawberry aroma and flavor, especially when the ratio of oleic acid to chitosan was high in the film. Aroma and flavor of coated strawberries were considered less intense than those of uncoated samples. Panelists also detected an atypical oily aroma in samples coated with the formulation containing the highest level of oleic acid. Additionally, strawberries coated with 1% chitosan edible coatings were more astringent than the uncoated fruit, and were significantly more astringent than those coated with a mixture of chitosan and 4% oleic acid. Consequently, astringency detected in chitosan films appeared to be reduced by addition of oleic acid.

In contrast, Han et al. (2005) evaluated sensory quality of fresh strawberries coated with a 1% chitosan solution containing vitamin E, and observed that application of the coating resulted in no perception of astringency. Similarly, Chien et al. (2007) observed that low molecular weight chitosan (LMWC) did not adversely influence the natural taste of sliced red pitayas (dragon-fruit), a medium-large berry. The chitosan coating helped maintain the soluble solids content, titratable acidity, and ascorbic acid content and, in addition, inhibited microbial growth. Sensorial analysis revealed superior results for the LMWC-coated samples after 7 days of storage.

Flavor and coloring agents may also be added to an edible film or coating matrix to improve the sensory attributes of wrapped or coated products. Unfortunately, few studies have been reported regarding this topic.

10.11 Current State and Recent Advances

There are multiple applications of edible films and coatings in the food industry. These include (1) oxygen-sensitive foods, such as nuts, to extend shelf life and reduce packaging; (2) nuts, to prevent oil migration into surrounding food components (e.g., nuts in chocolate); (3) fragile foods, such as breakfast cereals and freeze-dried foods, to improve integrity and reduce loss due to breakage; (4) fresh fruits and vegetables, whole and pre-cut, to extend product shelf life by reducing moisture loss, respiration and color change; (5) moisture-sensitive foods or inclusions, (e.g., nuts, cookies and/ or candies in ice cream) to provide a moisture barrier to keep products and inclusions crisp; (6) low-fat and non-fat snack foods, (e.g., chips) to keep seasonings adhered to products; (7) frozen foods, to prevent oxidation, as well as to prevent moisture, aroma or color migration; film separation layers for heterogeneous foods; and film pouches for dry food ingredients. Within these applications, use of edible films and coatings as carriers of active substances stands out as a promising application of active food packaging (Cuq et al. 1995; Han 2000).

10.12 Regulatory Status

Edible films and coatings can be classified as food products, food ingredients, food additives, food contact substances, or food packaging materials (Debeaufort et al. 1998). Therefore, edible films and coatings should follow all required regulations pertinent to food ingredients, since they are an integral part of the edible portion of food products (Guilbert and Gontard 1995). To maintain product safety and eating quality, all film-forming components, as well as any functional additives in the film-forming materials, should be food-grade, nontoxic materials; further, all process facilities should be acceptable for food processing and should strictly observe current Good Manufacturing Practice (cGMP) (Guilbert et al. 1996; Guilbert and Gontard 1995; Han 2002; Nussinovitch 2003). Ingredients acceptable for use in edible films and coatings should be GRAS, and used within any limitations specified by the Food and Drug Administration (FDA).

Since edible films and coatings may have ingredients included for functional effect, inclusion of such ingredients on product labels would be required. In Europe, the intended use of various food additives must always be included on packaging according to the specific functional category (antioxidant, preservative, color, etc.), with either their name or E-number. In European Community Legislation, a food additive is defined as "any substance not normally consumed as a food in itself and not normally used as a characteristic ingredient of food whether or not it has nutritive value, the intentional addition of which to food for a technological purpose results in it or its by-products becoming directly or indirectly a component of such foods". Food additives are intentionally added to foods to perform certain technological functions, for example, color, sweeten, or to preserve. Additives that are most likely to be found on food labels are antioxidants, colorants, emulsifiers, stabilizers, gelling agents, thickeners, flavor enhancers, preservatives and sweeteners. Flavorings, or substances used to give taste and/or smell to food, also must be present in the ingredient list on packaging of food products.

In the case of nutraceutical incorporation, specific regulations regarding its use must be applied. Within European Union Law, the legal categorization of a nutraceutical is, in general, made on the basis of its accepted effects on the body. Thus, if the substance contributes only to maintenance of healthy tissues and organs, it may be considered a food ingredient.

In food regulations of most countries, chemical substances added as antimicrobials are regarded as food additives, if the primary purpose of the substances is preservation of food to prolong its shelf life. For example, and according to U.S. regulations, organic acids including acetic, lactic, citric, malic, propionic and tartaric, as well as their salts, are GRAS for miscellaneous and general purpose usage (Doores 1993). On the other hand, many plant essential oils are used widely in food, health and personal care industries, and are also classified as GRAS substances or permitted as food additives (Kabara 1991).

Besides ingredient regulations, there is another important topic within the regulatory statutes. Many edible films and coatings are made with ingredients that can cause allergic reactions in some consumers. Allergens inherent to milk, soybeans, fish, peanuts, nuts, and wheat are the most important (Anonymous 2001). Several edible films and coatings are formed from milk (whey, casein), wheat (gluten), soy and peanut proteins. Therefore, a coating containing a known allergen must also be clearly labeled (Franssen and Krochta 2003).

10.13   Future Trends

Development of new technologies to improve carrier properties of edible films and coatings is a major issue for future research. Currently, use of such edible films and coatings is limited. One of the main obstacles is cost, restricting their application to products of high value. Besides cost, other limiting factors for commercial use of edible films and coatings are lack of materials with desired functionalities, cost of investment for the installation of new film production or coating equipment, difficulty of the production process and strictness of regulations.

In spite of these limitations, the food industry is looking for edible films and coatings that can be used on a broad spectrum of foods, add value to their products, increase product shelf life, and/or reduce packaging. However, more studies are yet necessary to develop new edible films and coatings containing active ingredients to understand interactions among components used in their production. When flavorings and active compounds (e.g., antimicrobials, antioxidants, and nutraceuticals) are added to edible films and coatings, mechanical properties can be dramatically affected. Studies addressing this subject are still very limited, and more information is needed to understand this behavior.

Acknowledgments This work was supported by the Ministry of Science and Technology, Spain (AGL2003-09208-C03-01).

References

Adhikari K, Mustapha A, Grun IU, Fernando L (2000) Viability of microencapsulated bifidobacteria in set yogurt during refrigerated storage. J Dairy Sci 83:1946–1951

Anonymous (1997) Edible films solve problems. Food Technol 51:60

Anonymous (2001) Food Safety and Food Labeling; Presence and Labeling of Allergens in Food, Rockville, Food and Drug Administration, HHS

Ayranci E, Tunc S (2004) The effect of edible coatings on water and vitamin C loss of apricots (*Armeniaca vulgaris* Lam) and green peppers (*Capsicum annuum* L). Food Chem 87:339–342

Baldwin EA, Nisperos MO, Chen X, Hagenmaier RD (1996) Improving storage life of cut apple and potato with edible coating. Postharvest Biol Technol 9:151–163

Bravin B, Peressini D, Sensidoni A (2006) Development and application of polysaccharide-lipid edible coating to extend shelf-life of dry bakery products. J Food Eng 76:280–290

Burt S (2004) Essential oils: their antibacterial properties and potential applications in foods: a review. Int J Food Microbiol 94:223–53

Cagri A, Ustunol Z, Ryser ET (2002) Inhibition of three pathogens on bologna and summer sausage slices using antimicrobial edible films. J Food Sci 67:2317–2324

Cagri A, Ustunol Z, Ryser E (2004) Antimicrobial edible films and coating. J Food Prot 67:833–848

Chen MJ, Weng YM, Chen WL (1999) Edible coating as preservative carriers to inhibit yeast on Taiwanese-style fruit preserves. J Food Safety 19:89–96

Chien PJ, Sheu F, Lin HR (2007) Quality assessment of low molecular weight chitosan coating on sliced red pitayas. J Food Sci 79:736–740

Contreras-Medellin R, Labuza TP (1981) Prediction of moisture protection requirements for foods. Cereal Food World 26:335–342

Cuq B, Gontard N, Guilbert S (1995) Edible films and coatings as active layers. In: Rooney M (ed) Active Food Packaging. Blackie, Glasgow, pp 111–142

Debeaufort F, Quezada-Gallo JA, Voilley A (1998) Edible films and coatings tomorrow's packaging: a review. Crit Rev Food Sci Nutr 38:299–313

Devlieghere F, Vermeulen A, Debevere J (2004) Chitosan: antimicrobial activity, interactions with food components and applicability as a coating on fruit and vegetables. Food Microbiol 21:703–714

Donhowe IG, Fennema O (1994) Edible films and coatings: Characteristics, formation, definitions, and testing methods. In: Krochta JM, Baldwin EA, Nispero-Carriedo MO (eds) Edible Coatings and Films to Improve Food Quality. Technomic Publishing Company, Lancaster, pp 1–24

Dong X, Wrolstad RE, Sugar D (2000) Extending shelf life of fresh-cut pears. J Food Sci 65:181–6

Doores S (1993) . Organic acid. In Davidson PM, Branen AL (eds) Antimicrobials in Foods. Marcel Dekker, New York, NY, pp 95–136

Drewnowski A, Gomez-Carneros C (2000) Bitter taste, phytonutrients, and the consumer: a review. Am J Clin Nutr 72:1424–35

Druchta J, De Mulder Johnston C (1997) Edible films solve problems. Food Technol 51:61–74

Eswaranandam S, Hettiarachchy NS, Meullenet JF (2006) Effect of malic and lactic acid incorporated soy protein coatings on the sensory attributes of whole apple and fresh-cut cantaloupe. J Food Sci 71:S307–S313

Fenaroli G (1995) Fenaroli's Handbook of Flavor Ingredients. CRC, Boca Raton, FL

Franssen LR, Krochta JM (2003) Edible coatings containing natural antimicrobials for processed foods. In: Roller S (ed) Natural Antimicrobials for Minimal Processing of Foods. CRC Cambridge, Boca Raton, FL, pp 250–262

García MA, Martinó MN, Zaritzky NE (2000) Lipid addition to improve barrier properties of edible starch-based films and coatings. J Food Sci 65:941–947

Gennadios A, Kurth LB (1997) Application of edible coatings on meats, poultry and seafoods:a rewiev. Lebensm Wiss U Technol 30:337–350

Gennadios A, Weller CL (1990) Edible films and coatings from wheat and corn proteins. Food Technol 44:63–69

Gibson GR, Roberfroid MB (1995) Dietary modulation of the human colonic microbiota: introducing the concept of probiotics. J Nutr 125:14011412

Gorny JR, Hess-Pierce B, Cifuentes RA, Kader AA (2002) Quality changes in fresh-cut pear slices as affected by controlled atmospheres and chemical preservatives. Postharvest Biol Technol 24:271–8

Grant LA, Burns J (1994) Application of coatings. In: Krochta JM, Baldwin EA, Nispero-Carriedo MO (eds) Edible Coatings and Films to Improve Food Quality. Technomic Publishing Company, Lancaster, pp 189–200

Guilbert S, Gontard N (1995) Edible and biodegradable food packaging. In: Ackermann P, Jägerstad M, Ohlsson T (eds) Foods and Packaging Materials – Chemical Interactions. The Royal Society of Chemistry, England, pp 159–168

Guilbert S, Gontard N (2005) Agro-polymers for edible and biodegradable films: review of agricultural polymeric materials, physical and mechanical characteristics. In: Han JH (ed) Innovation in Food Packaging. Elsevier, CA, pp 263–276

Guilbert S, Gontard N, Gorris LGM (1996) Prolongation of the shelf life of perishable food products using biodegradable films and coatings. Lebensm Wiss U Technol 29:10–17

Gulati OP, Ottaway PB (2006) Legislation relating to nutraceuticals in the European Union with a particular focus on botanical-sourced products. Toxicology 221:75–87

Han JH (2000) Antimicrobial food packaging. Food Technol 54:56–65

Han J (2002) Protein-based edible films and coatings carrying antimicrobial agents. In: Gennadios, A (ed) Protein-Based Films and Coatings. CRC, Boca Raton, FL, pp 485–498

Han C, Zhao Y, Leonard SW, Traber MG (2004) Edible coatings to improve storability and enhance nutritional value of fresh and frozen strawberries (*Fragaria x ananassa*) and raspberries (*Rubus ideaus*). Postharvest Biol Technol 33:67–78

Han C, Lederer C, McDaniel M, Zhao Y (2005) Sensory evaluation of fresh strawberries (*Fragaria ananassa*) coated with chitosan-based edible coatings. J Food Sci 70:S172–178

Hardenburg RE (1967) Wax and related coatings for horticultural products. A bibliography. Agr Res Bul – USDA 51–15

Hernández-Muñoz P, Almenar E, Ocio MJ, Gavara R (2006) Effect of calcium dips and chitosan coatings on postharvest life of strawberries (Fragariaxananassa). Postharvest Biol Technol 39:247–53

Iyidogan NF, Bayindirli A (2004) Effect of l-cysteine, kojic acid and 4-hexylresorcinol combination on inhibition of enzymatic browning in Amasya apple juice. J Food Eng 62:299–304

Jagannath JH, Nanjappa C, Das Gupta D, Bawa AS (2006) Studies on the stability of an edible film and its use for the preservation of carrot (*Daucus carota*). Int J Food Sci Technol 41:498–506

Kabara JJ (1991) Phenols and chelators. In: Russell NJ, Gould GW (eds) Food Preservatives. Blackie, London, pp 200–214

Kester JJ, Fennema O (1986) Edible Films and Coatings: a review. Food Technology 40:47–59

Kim W, Tanaka T, Kumura H, Shimazaki K (2002) Lactoferrin-binding proteins in *Bifidobacterium bifidum*. Biochem Cell Biol 80:91–94

Krochta JM, De Mulder-Johnston C (1997) Edible and biodegradable polymer films: challenges and opportunities. Food Technol 51:61–74

Kurman JA, Rasic JL (1991) The health potential of products containing bifidobacteria. In: Robinson RK (ed) Therapeutic Properties of Fermented Milks. Elsevier Applied Food Sciences, London, pp 117–158

Labell F (1991) Edible packaging. Food Process Eng 52:24

Laohakunjit N, Kerdchoecuen O (2007) Aroma enrichment and the change during storage of non-aromatic milled rice coated with extracted natural flavour. Food Chem 101:339–344

LeClair K (2000) Breaking the sensory barrier for functional foods. Food Prod Design 7:59–63

Lee JY, Park HJ, Lee CY, Choi WY (2003) Extending shelf-life of minimally processed apples with edible coatings and antibrowning agents. Lebensm-Wiss Technol 36:323–329

Luo Y, Barbosa-Canovas GV (1997) Enzymatic browning and its inhibition in new apple cultivars slices using 4-hexylresorcinol in combination with ascorbic acid. Food Sci Tech Int 3:195–201

McHugh TH and Krochta JM (1994) Water vapour permeability properties of edible whey protein-lipid emulsion films. J Am Oil Chem Soc 71:307–312

McHugh TH, Senesi E (2000) Apple wraps: a novel method to improve the quality and extend the shelf life of fresh-cut apples. J Food Sci 65:480–5

McHugh TH, Huxsoll CC, Krochta JM (1996) Permeability properties of fruit puree edible films. J Food Sci 61:88–91

McHugh TH, Senesi E, inventors; USDA-ARS-WRRC assignee. Filed 1999 June 11. Fruit and vegetable edible film wraps and methods to improve and extend the shelf life of foods. U.S. Patent Application Serial No. 09/330,358

Mei Y, Zhao Y (2003) Barrier and mechanical properties of milk protein-based edible films incorporated with nutraceuticals. J Agric Food Chem 51:1914–1918

Mei Y, Zhao Y, Yang J, Furr HC (2002) Using edible coating to enhance nutritional and sensory qualities of baby carrots. J Food Sci 67:1964–1968

Miller KS, Upadhyaya SK, Krochta JM (1998) Permeability of d-limonene in whey protein films. J Food Sci 63:244–247

Min S, Harris LJ, Han JH, Krochta JM (2005) *Listeria monocytogenes* inhibition by whey protein films and coatings incorporating lysozyme. J Food Prot 68:2317–2325

Mitsuoka T (1990) Bifidobacteria and their role in human health. J Ind Microbiol 6:263–267

Monsalve-Gonzalez A, Barbosa-Canovas GV, Cavalieri RP, McEvily AJ, Iyengar R (1993) Control of browning during storage of apple slices preserved by combined methods. 4-hexyl-resorcinol as anti-browning agent. J Food Sci 58:797–800

Molnar-Perl I, Friedman M (1990) Inhibition of browning by sulfur amino acids. 3. Apples and potatoes. J Agric Food Chem 38:1652–6

Naidu AS, Bidlack WR, Clemens RA (1999) Probiotic spectra of lactic acid bacteria (LAB). Crit Rev Food Sci Nutrition 38:13–126

Nicoli MC, Anise M, Severinc C (1994) Combined effects in preventing enzymatic browning reactions in minimally processed fruit. J Food Qual 17:221–9

Nussinovitch A (2003) Water Soluble Polymer Applications in Foods. Blackwell Science, Oxford

Ouattara B, Simard RE, Oiette G, Begin A, Holley RA (2000) Inhibition of surface spoilage bacteria in processed meats by application of antimicrobial films prepared with chitosan. Int J Food Microbiol 62:139–148

Oussalah M, Caillet S, Salmieri S, Saucier L, Lacroix M (2004) Antimicrobial and antioxidant effects of milk protein-based film containing essential oils for the preservation of whole beef muscle. J Agric Food Chem 52:5598–5605

Oussalah M, Caillet S, Salmieri S, Saucier L, Lacroix M (2006) Antimicrobial effects of alginate-based film containing essential oils for the preservation of whole beef muscle. J Food Prot 69:2364–2369

Park S, Zhao Y (2004) Incorporation of a high concentration of mineral or vitamin into chitosan-based films. J Agric Food Chem 52:1933–1939

Perez-Gago MB, Serra M, del Rio MA (2006) Color change of fresh-cut apples coated with whey protein concentrate-based edible coatings. Postharvest Biol Technol 39:84–92

Pizzocaro F, Toregiani D, Gilardi G (1993) Inhibition of apple polyphenol oxidase (PPO) by ascorbic acid, citric acid and sodium chloride. J Food Proces Preser 17:21–30

Richard FC, Goupy PM, Nicolas JJ (1992) Cysteine as an inhibitor of enzymatic browning. 2. Kinetic studies. J Agric Food Chem 40:2108–2114

Rodriguez MS, Albertengo LA, Vitale I, Agullo E (2003) Relationship between astringency and chitosan-saliva solutions turbidity at different pH. J Food Sci 68:665–667

Rojas-Graü MA, Avena-Bustillos R, Friedman M, Henika P, Martín-Belloso O, McHugh T (2006a) Mechanical, barrier and antimicrobial properties of apple puree edible films containing plant essential oils. J Agric Food Chem 54:9262–9267

Rojas-Graü MA, Sobrino-López A, Tapia MS, Martín-Belloso O (2006b) Browning inhibition in fresh-cut 'Fuji' apple slices by natural antibrowning agents. J Food Sci 71:S59–S65

Rojas-Graü MA, Tapia MS, Rodríguez FJ, Carmona AJ, Martín-Belloso O (2007a) Alginate and gellan based edible coatings as support of antibrowning agents applied on fresh-cut Fuji apple. Food Hydrocolloids 21:118–127

Rojas-Graü MA, Raybaudi-Massilia RM, Soliva-Fortuny RC, Avena-Bustillos RJ, McHugh TH, Martín-Belloso O (2007b) Apple puree-alginate edible coating as carrier of antimicrobial agents to prolong shelf-life of fresh-cut apples. Postharvest Biol Technol 45:254–264

Rojas-Graü MA, Olsen C, Avena-Bustillos RJ, Friedman M, Henika PR, Martín-Belloso O, Pan Z, McHugh TH (2007c) Effects of plant essential oils and oil compounds on mechanical, barrier and antimicrobial properties of alginate-apple puree edible films. J Food Eng 81:634–641

Samona A, Robinson RK (1991) Enumeration of Bifidobacteria in dairy products. J Soc Dairy Technol 44:64–66

Serrano M, Valverde JM, Guillen F, Castillo S, Martínez-Romero D, Valero D (2006) Use of Aloe vera gel coating preserves the functional properties of table grapes. J Agric Food Chem 54:3882–3886

Soliva-Fortuny RC, Grigelmo-Miguel N, Odriozola-Serrano I, Gorinstein S, Martin-Belloso O (2001) Browning evaluation of ready- to- eat apples as affected by modified atmosphere packaging. J Agric Food Chem 49:3685–90

Son S, Moon K, Lee C (2001) Inhibitory effects of various antibrowning agents on apple slices. Food Chem 73:23–30

Stollman U, Hohansson F, Leufven A (1994) Packaging and food quality. In: Man CMD, Jones AA (eds) Shelf life Evaluation of Foods. Blackie Academic and Professional, New York, NY, pp 52–71

Tapia MS, Rodríguez FJ, Rojas-Graü MA, Martín-Belloso O (2005) Formulation of alginate and gellan based edible coatings with antioxidants for fresh-cut apple and papaya. IFT Annual Meeting. New Orleans, LA. Paper 36E–43.

Tapia MS, Rojas-Graü MA, Rodríguez FJ, Ramírez J, Carmona A, Martin-Belloso O (2007) Alginate- and Gellan-based edible films for probiotic coatings on fresh-cut fruits. J Food Sci 72:E190–E196

Vargas M, Albors A, Chiralt A, Gonzalez-Martinez C (2006) Quality of cold-stored strawberries as affected by chitosan-oleic acid edible coatings. Postharvest Biol Technol 41:164–171

Wong WS, Tillin SJ, Hudson JS, Pavlath AE (1994) Gas exchange in cut apples with bilayer coatings. J Agric Food Chem 42:2278–2285

Wu PJ, Schwartzberg HG (1992) Popping behaviour and zein coating of popcorn. Cereal Chem 69:567–573

Yang L, Paulson AT (2000) Effects of lipids on mechanical and moisture barrier properties of edible gellan film. Food Res Int 33:571–578

Yildirim M, Gulec F, Bayram M, Yildirim Z (2006) Cheese properties of kashar cheese coated with casein as a carrier of natamycin. Italian J Food Sci 18:127–138

Chapter 11
Delivery of Food Additives and Antimicrobials Using Edible Films and Coatings

Jesus-Alberto Quezada-Gallo

11.1 Introduction

Functional efficiency of edible films and coatings strongly depends on the nature of film components and physical structure. Choice of a film-forming substance and/or active additive depends on the desired objective, nature of the food product, and specific application. Thus, lipids or hydrophobic substances such as resins, waxes or some insoluble proteins are most efficient for retarding moisture transfer. On the contrary, water-soluble hydrocolloids, like polysaccharides and proteins, are not very efficient barriers against water transfer. However their permeability to gases is often lower than those of plastic films. Moreover, hydrocolloids usually provide better mechanical properties to edible films and coatings than lipids and hydrophobic substances. Therefore, combinations of desirable properties can be obtained through collective use of hydrocolloids and lipids to create composite films. Film-forming substances, particularly proteins, require film additives such as plasticizers to improve film resistance and elasticity, or emulsifiers, to increase hydrophobic particle distribution in composite emulsion-based edible films (Debeaufort et al. 1998).

One of the important emerging functions of edible films and coatings is their use as carriers of antimicrobials and antifungal agents to increase shelf life of foods. They may also be used as carriers of nutrients to increase nutritional value of processed food products. Use of edible films and coatings as active delivery systems in foods is a relatively recent concept, although this function has been widely studied and used for pharmaceuticals. This chapter will discuss edible films and coatings as delivery matrices for food additives and antimicrobials.

J.-A. Quezada-Gallo
Departamento de Ingeniera y Ciencias Qumicas, Universidad Iberoamericana,
Ciudad de Mexico, Edificio F segundo piso, Prol. Paseo de la Reforma 880,
Lomas de Santa Fe, 01210 Mexico, D.F., Mexico
e-mail: jesus.quezada@uia.mx

M.E. Embuscado and K.C. Huber (eds.), *Edible Films and Coatings for Food Applications*, 315
DOI 10.1007/978-0-387-92824-1_11, © Springer Science+Business Media, LLC 2009

11.2   History of Incorporating Additives into Edible Films and Coatings

Though use of edible films in food systems seems somewhat new, food products were covered with films and coatings over hundreds of years ago. Wax has been used to delay dehydration of citrus fruits in China since the twelfth and thirteenth centuries. Application of lipid-based coatings to meats to prevent shrinkage has been a traditional practice since at least the sixteenth century, while later in the last century, meat and other foodstuffs were preserved by coating them with gelatin films. Yuba, a proteic edible film obtained from the skin of boiled soymilk, was traditionally used in Asia to improve appearance and aid preservation of foods since the fifteenth century. In the nineteenth century, sucrose solution was applied as an edible protective coating on nuts, almonds and hazelnuts to prevent their oxidation and rancidity during storage.

One of the most important advancements in edible film and coating technology since the 1930s involves use of an emulsion made of waxes, oil and water. Such emulsions are applied to fruits to improve appearance (gloss and color) and to prevent softening and onset of mealiness. They are also used for delivery of fungicides, to control ripening, and retard moisture loss in fruit. A number of edible polysaccharide coatings including alginates, carrageenans, cellulose ethers, pectin, and starch derivatives have been used to improve stored meat quality. Over the last 40 years, a great number of studies have investigated formulation, application and characterization of edible films and coatings; evidence of such efforts can be found in both the scientific and patent literature (Debeaufort et al. 1998).

11.3   Food Additives

11.3.1   Antimicrobials

The main cause of spoilage for many food products is surface microbial growth. Reduction of water activity (a_w) and protection with moisture-proof packaging are common methods used to prevent spoilage in food products. Nevertheless, a reorganization of water inside the package, due to temperature changes, can induce condensation of moisture on the food surface, increasing the possibility for microbial growth (Torres et al. 1985a, b). Growth of yeasts, molds, and bacteria during product storage and distribution can drastically reduce food quality and food safety.

Use of antimicrobials such as benzoic acid, sodium benzoate, sorbic acid, potassium sorbate, and/or propionic acid represents an additional means of food preservation. Edible coatings have been studied as antimicrobial carriers because of their effectiveness in retaining additives on food surfaces. Within the last decade, several groups have evaluated diffusivity of sorbic acid and potassium sorbate in model systems (Guilbert 1988; Vodjani and Torres 1990; Torres and Karel 1985).

Early research studies applied antimicrobials and fungicides by dipping food in additive solution. This method was initially effective in reducing the total number of viable microorganisms, but on storage, preservative diffused into the food, allowing surface spoilage to occur. Application of fungicides in wax-based emulsions or water suspension has been studied on citrus fruit. The primary factor responsible for their effectiveness is dependent on method of application. The emulsified method was less effective of the two treatments, possibly due to encapsulation of the additive in the lipid phase (Ecker and Kolbezen 1977).

Chitosan, as a matrix material for edible films and coatings, has been studied extensively. Chitosan films can prevent growth of *Listeria monocytogenes* (Jeon et al. 2002) and *Aspergillus niger* (Sebti et al. 2005). Chitosan used in combination with pectin improves mechanical properties of edible films (Hoagland and Parris 1996). Chitosan has been investigated as a dough ingredient in precooked pizza to protect against *Alternaria sp, Penicillium sp, Aspergillus sp and Cladosporium sp* (Rodriguez et al. 2003). When combined with hydroxypropyl methylcellulose on fresh strawberries, it can control *Cladosporium sp and Rhizopus sp* growth (Park et al. 2005).

Interesting results have been obtained from enriched chitosan films with essential oils. These additives can protect skinless pink salmon fillets against *Listeria monocytogenes* (Zivanovic et al. 2005), while also preventing moisture loss and lipid oxidation (Sathivel 2005). Upon irradiation, chitosan coating on intermediate-moisture meat products had antioxidant and antimicrobial roles (Rao et al. 2005). When applied on cooked ground beef and turkey, it had a protective effect against *Clostridium perfringens* spores during chilling (Juneja et al. 2006).

11.3.2   Antioxidants

Antioxidants increase stability of food components, especially lipids, and maintain nutritional value and color by preventing oxidative rancidity, degradation and discoloration. Acid or phenolic compounds act as antioxidants. Acid compounds, such as citric and ascorbic acid, are metal chelating agents. Phenolic compounds, such as butylated hydroxyanisole (BHA), butylated hydroxytoluene (BHT), tertiary butylated hydroxyquinone (TBHQ), propyl gallate and tocopherols inhibit lipid oxidation. These antioxidants can be incorporated into edible coatings, thus being retained on the food surface where they are most effective, since oxidation is a surface-air phenomenon. Edible coatings can also reduce enzymatic processes such as enzymatic oxidation (Nisperos-Carriedo et al. 1990).

11.4 Components of the Film Matrix

11.4.1 Plasticizers

Plasticizers such as glycerol, sorbitol, acetylated monoglyceride, polyethylene glycol, sucrose and edible oil are used to modify mechanical properties of films and coatings. Incorporation of small molecules into the polymer network also alters film barrier properties.

Physicochemical and barrier properties of polymeric networks are affected by plasticizer concentration and molecular weight, water content of the film and strength and type of polymer-polymer and polymer-plasticizer interactions. For example, solubility of fish myofibrillar protein-based films in water is not significantly affected by plasticizer concentration, whether the plasticizer is glycerol, sorbitol or sucrose. In general, no significant difference was observed in of functional properties as a function of plasticizer type (Cuq et al. 1997). On the other hand, incorporation of sorbitol or glycerol into whey protein isolate films affected their solubility and equilibrium moisture contents, which significantly increased at $a_w \geq 0.65$ (Kim and Ustunol 2001). A combination of plasticizers is used when one plasticizer is not capable of individually performing this function satisfactorily. When sodium dodecyl sulfate was added, at a concentration of <16% (w/w) as a co-plasticizer, whey protein films exhibited better mechanical properties, without change in water vapor permeability. However, at concentrations higher than 23% (w/w), an antiplasticization effect was observed, resulting in a more brittle and breakable film (Fairley et al. 1996).

11.4.2 Surfactants

Water-soluble films and coatings generally do not have good water barrier properties, due to the solubility of their matrix components in water. Composite coatings are heterogeneous films of hydrophobic particles within a hydrophilic matrix. In this case, surfactants may be used to stabilize the dispersed phase in a polymeric solution before drying the film or prior to applying it to food surfaces. Emulsifiers can also be used to improve adhesion at the interface between the food and coating, or between two layers of films of differing polarity within a multilayer film. In multilayer films or coatings, the good water barrier properties of lipids and the good cohesion and gas permeability properties of hydrophilic polysaccharide or protein can be combined as separate layers to create materials with enhanced properties.

Emulsifiers are selected for use based on their hydrophilic-lipophilic balance (HLB) and phase inversion temperature (PIT). HLBs range from values of 1 to 40 for emulsifiers, and are calculated based on the specific polar and hydrophobic groups present on the surfactant molecule. Very hydrophilic surfactants have high HLB values, while those with a low HLB values are more generally lipophilic.

Surfactants with HLB values <7 are recommended for water-in-oil (w/o) emulsions; those with HLB values >7 are generally appropriate for oil-in-water (o/w) emulsions. PIT indicates the temperature at which a w/o emulsion turns into an o/w emulsion. Many food emulsifiers are derivatives of glycerol and fatty acids (e.g., polyglycerols-polystearates, polyoxyethylene sorbitan derivatives (TWEEN) and lecithin).

11.5   Improvement of Sensory Attributes

Edible coatings can be used to carry color and aroma compounds with the aim of improving sensory qualities such as appearance and flavor, of processed foods. The ability of an edible film to retain flavor compounds through drying and storage processes is critical. An important use of edible films involves encapsulation of flavorings (Reineccius 1994). Flavor encapsulation produces dry, free-flowing flavors from liquid flavors. This process stabilizes flavors by preventing interaction with other food ingredients that will produce negative effects during storage. Edible film efficacy for these purposes depends on film composition and the process used to encapsulate the flavoring. A majority of flavors are encapsulated using water-soluble carbohydrates including gum acacia, modified starches, maltodextrin and corn syrup solids. The suitability of an edible film as an encapsulant is determined by its ability to form and stabilize an emulsion, retain flavors during encapsulation, protect the flavor from oxygen during storage, and release the flavor to the final food product at the appropriate time or under desired conditions. To achieve this purpose, several factors play an important role:

- Edible film used as an encapsulant must provide good emulsification properties in beverage products.
- Film must retain flavor during the drying process to provide the whole range of desired notes during consumption.
- The film has to protect flavor compounds from oxygen during storage; the flavor release must occur very slowly with water contact or with heating.

11.6   Regulatory Status

Biopolymers, the main ingredients of coatings and films, are considered by law as additives. They may also carry smaller molecules such as plasticizers, adhesion agents, emulsifiers, active or functional additives, etc. However, their main function remains to act as a water barrier at the food-air interface. The variety of applications for edible films and coatings has been reported in literature since the seventeenth century. Patents on edible films developed for food use have been reported since 1942. Edible coatings have been used for extending product shelf life, maintaining product crispness, preventing dehydration, enhancing appearance and preventing the loss of gloss.

The Food and Drug Administration (FDA) defines a food additive as "any substance the intended use of which results or may reasonably be expected to result, directly or indirectly, in its becoming a component or otherwise affecting the characteristics of any food." The same institution defines an antimicrobial as "a compound or substance that kills microorganisms or prevents or inhibits their growth and reproduction and is included in a product at a concentration sufficient to prevent spoilage or prevent growth of inadvertently added microorganisms, but does not contribute to the claimed effects of the product in which it is included."

11.7 Techniques of Introducing Additives into Edible Films and Coatings

The formulation of films and coatings must generally include at least one component capable of forming a cohesive structural matrix. Edible films made from two or more matrix substances may be used to exploit specific functional properties of each individual component and to minimize their disadvantages. Most composite films studied are composed of a hydrophobic compound, often a lipid, and usually a hydrocolloid structural matrix.

Film-forming substances are able to form a continuous structure by interactions between matrix molecules, after addition of a chemical, or by physical treatment. Film and coating formation involves one of the following processes (Debeaufort et al. 1998):

- Melting and solidification of solid fats, waxes or resins.
- Simple coacervation, where a hydrocolloid dispersed in aqueous solution is precipitated or gellified by removal of solvent; by addition of a non-electrolyte solute, in which the polymer is not soluble; by addition of an electrolyte substance inducing a "salting out" effect; or by pH modification.
- Complex coacervation, where two hydrocolloid solutions with opposite charges are combined, inducing interactions and precipitation of the polymer mixture.
- Thermal gelation or coagulation by heating macromolecule solution which involves denaturation, gellification and precipitation, or by rapid cooling of a hydrocolloid solution that induces a sol-gel transition.

11.7.1 Multilayer Coatings

Lipids and biopolymers can be associated to form 'emulsified' or bilayer films and coatings. For emulsion-based edible packaging, water vapor barrier efficiency depends on the nature of incorporated lipids (i.e., chain length of the fatty acids) and structure of the dried emulsion constituting the film. Bilayer films have 10–1,000 times better barrier efficiency against water transfer than emulsified films.

Functional properties of lipid-based edible films are partially explained by a combination of the lipid nature (polarity) and structure (solid fat content, crystal type), though these parameters have always been studied independently. The most hydrophobic lipids have the best barrier efficiency. Martin-Polo and Voilley (1990) reported that permeability of lipid layers decreased substantially (100 times) when the solid fat content was higher than 20% for paraffin waxes and pure alkanes. Moreover, it appeared that the more stable the lipid crystal type (orthorhombic), the lower the moisture permeability (Debeaufort et al. 2000).

Multilayer coatings can be formulated to combine advantages of lipid and hydrocolloid components. When a barrier to water is desired, the lipid component can serve this function, while the hydrocolloid component provides necessary structure and durability (Greener Donhowe and Fennema 1994). Properties of multilayer films have been studied extensively (Greener and Fennema 1989; Martin-Polo and Voilley 1990). Concerning inclusion of food additives, multilayer films offer the advantage of both a lipophilic and a hydrophilic phase, allowing the coating to carry compounds of different polarity.

11.7.2 Micro- and Nanoemulsions

Though emulsified films have been less studied than other types of edible films, they are 100 times less efficient than bilayer films due to nonhomogeneous distribution of lipid substance. However, they have the advantage of needing only one drying step instead of the two needed for bilayer films. They can also be applied on food at room temperature. The presence of both hydrophilic and lipophilic components allows adhesion of emulsion-based films or coatings onto any support, whatever its polarity. Moreover, a lipid layer possesses really poor mechanical resistance, where a lipid-hydrocolloid emulsion-based film exhibits good mechanical resistance.

Two models have been proposed to describe transfer through emulsified films. The first is the 'microvoid model,' which suggests that mass transfer of gases and vapors occurs through microvoids, formed between microparticles of the hydrophobic material and the hydrocolloid matrix during emulsion drying. An alternative model, the 'micropathway model,' suggests that mass transfer occurs through the polymer matrix itself. This phenomenon can occur, because proteins and polysaccharides are often quite compatible with moisture and gases, and offer little resistance to their transmission. Emulsion characteristics significantly influence mechanical and barrier properties of emulsified films: the smaller the diameter and the more homogeneous the distribution of lipid globules, the better the barrier properties. Water vapor permeability decreases as hydrophobicity of components increases; waxes were found to be the most effective in lowering water permeability of edible films and coatings. The length of fatty acid chain influences permeability; palmitic and stearic acids showed the best efficiency when compared to capric or behenic acids. Creaming of the emulsion during film drying was responsible for the nonisotropic nature of emulsified films; one side of the film is shiny, due to increased protein content, whereas the other side is opaque, due to increased lipid content. It was also observed that films oriented with the lipid-enriched side facing the high relative humidity environment during water vapor transmission testing exhibited lower permeability. An increase in lipid content led to a loss of mechanical efficiency for emulsified wheat gluten-based films. For lipid concentrations higher than 35–40%, emulsified films became more brittle and permeable (Quezada-Gallo et al. 2000).

The emulsified type of coating provides a good moisture barrier, but they can impart unacceptable properties to the product. Lipids are dispersed in water as a macroemulsion (particle size range: 2×10^3–10^5 Å), as a microemulsion (particle size 1,000 – 2,000 Å), or as a nanoemulsion (particle size < 1,000 Å). Formulation of microemulsions requires selection of suitable emulsifiers that are compatible with both the dispersed and continuous phases. A surfactant that is partially soluble in the dispersed and continuous phases is normally used, whereas a co-surfactant is typically an alcohol (Hernandez 1994). Microemulsions are desired in edible coatings, because the small droplet size enables the lipid phase to coalesce into a uniform film.

Use of high pressure valve homogenizers or microfluidizers produces emulsions with droplet diameters of 100–500 nm. Such emulsions are often referred to as "nanoemulsions." Functional food components can be incorporated within the droplets, at the interfacial region or in the continuous phase (Weiss et al. 2006). Functional components could slow down chemical degradation processes, if the rate of permeation at interfaces is well-known and engineered.

Micelle-forming properties of chitosan derivatives have been studied by Zhang et al. (2006) to better understand their potential in controlled release and targeted delivery of hydrophobic bioactive food components.

11.7.3 Microcapsules and Liposomes

Microencapsulation is a process of entrapping particles, droplets or gases in a polymeric coating (Gennadios et al. 1994), which is applied to foods, pharmaceuticals, cosmetics and pesticides. Microencapsulated ingredients include acidulants, flavors, colors, leavening agents, salts, enzymes, artificial sweeteners, minerals and vitamins (Karel 1990). Gums, starch, and starch derivatives, cellulosic derivatives, lipids, proteins and inorganic materials are employed in the production of microcapsules.

Use of multiple emulsions can create delivery systems with novel encapsulation and delivery properties. For example, functional food components may be encapsulated within the inner water phase of a w/o/w emulsion, while other components that might adversely react with these emulsified components could be dissolved in the outer water phase. This method could be used to protect and release components at a specific site within the digestive system (Weiss et al. 2006).

Liposomes or lipid vesicles are formed from polar lipids that are available in abundance in nature, mainly phospholipids from soy and egg. They can incorporate functional components within their interiors, and can be used to encapsulate both water- and lipid-soluble compounds. Liposomes are spherical, polymolecular aggregates with a bilayer shell configuration (Chen et al. 2006). Several studies on liposomes report encapsulation of lactoferrin and nisin Z for bacteriostatic applications in dairy products; the antioxidant phosvitin and vitamin C have been encapsulated in liposomes to improve dairy and meat product quality (Taylor et al. 2005).

11.8 Irradiation

Addition of $CaCl_2$ combined with irradiation at doses ≥ 16 kGy resulted in homogeneous caseinate gels which increased mechanical properties of resulting films (Ressouany et al. 1998) due to an increase in cross-linking, an effect of the γ-irradiation dose. The biodegradability of calcium caseinate films, using a strain of *Pseudomonas aeruginosa*, showed that overall degradation modes were similar for films receiving 4 or 64 kGy of γ-irradiation. However, a difference was noted in their degradation time, which was delayed by 8 days for film having the highest extent of cross-linking (64 kGy) (Mezgheni et al. 1998).

Films obtained from irradiated solutions of calcium caseinate and whey protein isolate (to induce cross-linking) exhibit better mechanical properties than the non-irradiated films. Cross-linking has a direct impact on microbial resistance (Letendre et al. 2002).

Gamma irradiation at doses below 10 kGy is used to control fungal growth in food, but would undergo spoilage once packaging material is breached. However, radiation treatment accelerates lipid peroxidation; to overcome this challenge, Rao et al. (2005) proposed the use of an irradiated chitosan film as an antioxidant. This substantially retarded peroxidation and prevented microbial spoilage.

When cross-linked, by heating or by γ-irradiation, and entrapped in cellulose, whey proteins can generate insoluble biofilms with good mechanical properties and high resistance to attack by proteolytic enzymes. They can also increase puncture strength and decrease water vapor permeability of films. Cross-linking by γ-irradiation appears to modify conformation of proteins, which become more ordered and stable after treatment (Le Tien et al. 2000).

A 10 kGy dose of gamma irradiation increased tensile strength and decreased elongation (32% reduction) of wheat gluten films. Increasing the radiation dose to greater than 10 kGy reduced the observed effect, which was attributed to a decrease in insoluble glutenin polymers. The positive effect of γ-irradiation on mechanical properties of gluten films irradiated by 10 kGy was explained by formation of dityrosine cross-links. Cross-linking within polypeptide macromolecules has led to changes in film mechanical properties based on other proteins, such as caseinates (Micard et al. 2000).

11.9 Functional Properties of Edible Films and Coatings containing Food Additives

11.9.1 Barrier Properties

Incorporation of nisin into edible protein-based films can effectively inhibit *Listeria monocytogenes*. Edible films with higher hydrophobicities and added nisin in an acidic environment exerted a greater inhibitory effect. Water vapor permeability varied for each film tested, suggesting that it depends on interactions between the protein and antimicrobial agent.

Addition of oregano essential oil into a chitosan film decreased water vapor permeability and showed an antimicrobial effect in vitro when applied on bologna slices (Zivanovic et al. 2005). The same biopolymer offers an advantage in preventing *Aspergillus niger* surface growth, even at a very low concentration and minimized food dehydration (Sebti et al. 2005). Water vapor permeation of pectin or chitosan films, made with lactic acid to prevent fungal growth, was not different compared to that for films without lactic acid (Hoagland and Parris 1996).

Whey protein isolate films, incorporating the lactoperoxidase system (LPOS), inhibited *Salmonella enterica* and *Escherichia coli* completely on contact; in addition, the oxygen barrier property was improved (Min et al. 2005). Addition of ascorbyl palmitate and α-tocopherol changed diffusivity and solubility of oxygen through whey protein isolate films. It was also noted that the effect of additive incorporation on oxygen barrier properties of whey protein films was small (Han and Krochta 2007). Hydrolyzed whey protein isolate makes good films, with oxygen permeability values similar to whey protein isolate films with the same glycerol content, but it has more flexibility. Hydrolyzed whey protein isolate films and coatings can be used to protect foods from oxygen while reducing the allergenicity caused by the whey protein (Sothornvit and Krochta 2000).

Calcium salts and glucono-δ-lactone-treated soy protein isolate films have lower vapor permeability than soy protein isolate film (Park et al. 2001).

11.9.2 Mechanical Properties

Inclusion of an additive can modify mechanical properties of edible films and coatings, because of physical changes induced in the network structure resulting from physicochemical changes in polymer-polymer interactions. Induced changes depend on factors such as the molecular size, polarity, shape, and affinity of the additive to the polymer molecules.

Addition of zein hydrolyzate and use of transglutaminase improved flexibility of a whey protein film without sacrificing water vapor permeability (Oh et al. 2004). Transglutaminase is effective in introducing covalent bonds into films obtained from slightly deamidated gluten, inducing greater insolubility of treated films, increasing film integrity and the ability to stretch (Larre et al. 2000). When combining whey protein isolate and the lactoperoxydase system, tensile properties of whey protein films decreased, limiting its applications to food surfaces (Min et al. 2005). Whey protein isolate films, containing p-aminobenzoic acid as an antimicrobial agent and applied as casing on hot dogs, has similar mechanical properties as those without this additive (Cagri et al. 2003).

Salts and organic acids added to edible films might have a cross-linking effect in addition to their known antimicrobial effect. To prevent fungal growth, lactic acid was used to replace glycerol in a pectin film without significant change in dynamic mechanical properties (Hoagland and Parris 1996). Cross-linking of soy protein isolate with calcium salts resulted in formation of a rigid, three-dimensional film structure with higher mechanical properties than those found in soy protein films without cross-linking. Glucono-δ-lactone contributed to the formation of a homogeneous soy protein-based film structure, due to increased protein-solvent attraction (Park et al. 2001).

Chemical modification of polymers has a direct effect on mechanical properties. Free cysteine thiol groups of keratin, extracted from chicken feathers, were partially carboxymethylated with iodoacetic acid, after which stable modified keratin dispersions were used for preparation of films by solution casting. The degree of crystallinity in films increased as more cysteine residues were carboxymethylated, increasing tensile strength, E modulus, and elongation at break (Schrooyen et al. 2001).

11.9.3 Sensorial Properties

Chitosan and protein-based films showed no effect on the color of salmon and arrowtooth fillets (Sathivel 2005). Carrots treated with edible cellulose-based coatings had a reduced level of white surface discoloration and higher sensory scores for orange color intensity, fresh carrot aroma, fresh carrot flavor and overall acceptability than non-coated carrots (Howard and Dewi 1995).

When compared to hot dogs prepared with collagen or natural casings, sensory panelists gave hot dogs prepared with whey protein isolate–p-aminobenzoic acid casings superior scores for texture, juiciness, flavor and overall desirability (Cagri et al. 2003).

Fusion lamination and coating with drying oil were used to prepare zein films plasticized with oleic acid. This treatment resulted in clearer, tougher, more flexible and smoother finish films than those of untreated sheets. Moreover, lamination decreased O_2 and CO_2 permeability by filling voids and pinholes in the film structure. Coating increased tensile strength, and elongation percentage, and decreased water vapor permeability by forming a highly hydrophobic surface that prevented film wetting (Rakotonirainy and Padua 2001).

11.9.4 Delivery Properties

The movement of a food additive into food and/or between different components of the food could affect its quality. Release kinetics has been modeled for some active packaging systems using food applications. Buonocre et al. (2003) developed a model for lysozyme release kinetics from cross-linked polyvinylalcohol (PVOH).

Methylcellulose and hydroxypropyl methylcellulose mixed with lauric, palmitic, stearic and arachidic acid significantly lowered the potassium sorbate release rates as compared to films containing no fatty acids. This approach represents a good option for preserving antimicrobial concentration on food surfaces (Vojdani and Torres 1990). However, release of potassium sorbate increased with increasing water activity in the same edible film, while release of sorbic acid increased as pH increased (Rico-Peña and Torres 1991).

11.10 Applications in Foods

11.10.1 Fruits and Vegetables

Applications of antimicrobials and fungicides in emulsions and water suspensions have been studied on citrus fruit. Several authors have reported good results in applying imazalil onto Valencia oranges (Radina and Eckert 1988); sorbic acid, thiabendazole and benomyl onto grapefruit; 2, 6-dichloro-4-nitroaniline (DCNA) onto peaches and nectarines (Wells 1971); thiabendazole onto papaya (Couey and Farias 1979); iprodine and chitosan onto strawberries (Ghaouth et al. 1991); and captan and thiram onto tomatoes (Domenico et al. 1972).

Even though 1-methylcyclopropene showed better properties to reduce ethylene-induced effects on senescence in minimally-processed lettuce, alginate-based coating preserved crispness, creating the possibility of using both for combined benefits (Tay and Perera 2004). Zein coating has been shown to be an effective barrier by itself, revealed by a reduction in the *Listeria monocytogenes* population on coated sweet corn; addition of sorbic acid, as a surface antimicrobial in zein coating, did not provide any additional benefit (Carlin et al. 2001).

Formulations of whey protein-based coatings containing glycerol (plasticizer), lecithin (surfactant), methyl paraben (antimicrobial), or vitamin E (antioxidant) were used to coat peanuts. Analysis of hexanal indicated that all coated samples using the above formulations were oxidized significantly slower than the uncoated reference (Lee and Krochta 2002).

Inexpensive, natural, edible coatings could be a good option to increase shelf life of guava. When comparing permeability to water vapor and aroma (2-pentanone), tests showed that efficiency of edible packaging to preserve guava depends on retention of hydrophobic elements more than water vapor (Quezada-Gallo et al. 2004).

11.10.2 Bakery Products

Sweet taste is generally the flavor most desired by consumers for breakfast cereals. This is the reason why the majority of coating materials used for cereals have a sugar base. Additives to coating formulations improve flavor and adhesion of the coating to the breakfast cereal products. Acetic acid and sodium acetate improve flavor. Thickening agents, emulsifiers (e.g. lecithin), humectants and plasticizers (sorbitol) are used to change consistency of the coating.

Use of chitosan with acetic acid, calcium propionate or potassium sorbate as an edible coating on precooked pizza delayed microbial growth, preserving the dough. Unfortunately, Maillard reactions resulted in reduction of its antimicrobial capacity (Rodriguez et al. 2003).

Application of different polysaccharide-based edible coatings on bakery products, with sodium benzoate showed an increase in product shelf life of at least 100%, reducing the water loss rate during the first 100 h by 50%. Kinetics of water loss showed a three-phase mechanism (1) water loss on the surface for 48 h, (2) a stabilization corresponding to migration of water from inside, and (3) a continuation of moisture loss in the product. The combination of a biopolymer and an additive applied as a coating gave better results when additives where distributed in both the product and coating. Based on results of the comparative sensory evaluation, the coating did not affect consumer preference (Quezada-Gallo et al. 2005).

11.10.3 Heterogeneous Foods

Certain ingredients of food products may undergo deleterious reactions with other ingredients that can bring about changes in color and/or nutritional quality. Often, the only practical method for separating components of a composite food is to provide an edible film barrier between components (Guilbert 1986). Performances of edible coatings have been widely studied in a variety of heterogeneous products – such as frozen pizza, ice cream, cereal and raisins, dried fruits in cakes and cheese (Katz and Labuza 1981; Rico-Peña and Torres 1990; Greener and Fennema 1989; Baranowski 1990).

11.10.4 Miscellaneous Food Applications

Inhibition of Clostridium perfringens on cooked ground beef and turkey was achieved by mixing meat with chitosan at a concentration of 3% (Juneja et al. 2006). A coating made of a styrene-acrylate copolymer, containing triclosan, can be an effective antimicrobial layer (Chung et al. 2003). Zein film coating with nisin shows promise in the control of Listeria monocytogenes on the surface of ready-to-eat meat products (Janes et al. 2002). Several authors have conducted research on properties of edible coatings and films containing and delivering antimicrobial agents (Longinos-Martinez et al. 2005).

Chitosan films had an inhibitory effect on the same microorganism inoculated on Emmental cheese (Coma et al. 2002). Edible casings containing 1% p-aminobenzoic acid (PABA) were more inhibitory towards *Listeria monocyogenes* than other commercial casings, with *Listeria* growth suppressed on hot dogs during 42 days of refrigerated storage (Cagri et al. 2003). Coating fresh fillets of Atlantic cod with chitosan-based coatings showed a significant reduction in relative moisture losses. Lipid oxidation was also reduced, as shown by its peroxide value, conjugated dienes, 2-thiobarbituric acid reactive substances and headspace volatiles. In addition, there was a reduction of its total volatile basic nitrogen (trimethylamine and hypoxantine) and a reduction of the total plate count in the fish model system when compared to uncoated samples.

11.11 Possibilities of Future Applications

11.11.1 Probiotic Agents

Probiotics are defined as live microbial food ingredients that have a beneficial effect on human health. The concept of probiotics evolved from the hypothesis that the healthy life of Bulgarian peasants resulted from their consumption of fermented milk products (Sanders 1999). In the last century, probiotic bacteria most commonly studied include members of the genera *Lactobacillus* and *Bifidobacterium*. *Saccharomyces boulardii, Escherichia coli* and *Enterococcus* strains are used as probiotics in non-food formats. Some potential or established effects of probiotic bacteria have been reported (Sanders and Huis in't Veld 1999) such as an aid to lactose digestion, increased resistance to enteric pathogens, an immune system modulator, and as inhibition of anti-urogenital infections and hepatic encephalopathy. Increased worldwide interest in probiotics has set the stage for expanded marketing of these products. The benefit of probiotics as healthful ingredients could be enhanced if used in combination with other health-promoting dietary strategies (Sanders 1999).

Microencapsulation and edible films and coatings have tremendous potential, because they are ideal delivery systems for a number of viable probiotics, either inside food products or on food surfaces. These techniques offer the possibility of controlling viable microorganism counts on an edible film, while simultaneously offering protection to changes in temperature, pH, chemical and enzymatic processes during food production, storage, consumption and digestion.

11.11.2 Nutraceuticals

Along with increased market demands on nutritionally-fortified foods, edible coatings and films with high concentrations of nutraceuticals would provide alternative ways to fortify foods that otherwise cannot be fortified, such as fresh fruits, fresh vegetables and other minimally-processed foods. Food products could either be coated or wrapped with nutritionally-fortified coatings or films. A study developed by Mei and Zhao (2003) demonstrated that water vapor permeability and tensile properties of milk protein-based edible films, with calcium caseinate and whey protein isolate, may be compromised when high concentrations of calcium and vitamin E are incorporated. Vitamin E incorporation increased elongation percentage at break, and reduced tensile strength of films, probably because of its hydrophobic nature, plasticizing effect or due to the heterogeneous structure of the film.

Some nanoscale phenomena have been utilized in nutraceutical and functional food formulation and manufacturing processes. New concepts, based on nanotechnology, are being explored to improve product functionality and delivery efficiency (Chen et al. 2006). Characterization in the nanometer resolution of systems with large interfacial areas – such as emulsion, dispersion, and structured fluid – offers a better understanding of their functionality. This understanding could lead to a potentially novel type of active packaging for food products that enhance solubility, facilitate controlled release, improves bioavailability, and protect stability of micronutrients and bioactive compounds during processing, storage and distribution.

11.11.3   Volatile Compounds

The barrier efficiency against vapor transfer through an edible film depends mainly on both sorption and diffusion. The sorption mechanism consists of adsorption, absorption, and/or desorption of penetrant molecules to reveal the polymer-volatile compound affinity. Diffusion, on the other hand, is related to mobility of volatile compounds within the polymeric network of the material. The volatile compound and film polymer characteristics must be taken into account to explain the transfer process. However, other structural factors can affect sorption and diffusion; in other words, the type of polymerization, polymer three-dimensional structure, material cohesion maintained by weak energy bonds and the polymer glass transition temperature can all affect sorption and diffusion (Quezada-Gallo et al. 1999). In addition, physicochemical characteristics of volatile compounds influence the film's permeability. Aroma compound's shape and size affect its diffusivity; solubility is influenced by the compound's nature, polarity, and ability to condense.

Solute polarity is an important factor in the sorption process. Some authors have shown that aroma compounds are adsorbed more easily on the polymeric film if their polarities are similar. However, very few authors have studied the effect of this factor on transfer through hydrophilic edible packaging. Furthermore, aroma barrier properties of edible films have not been as thoroughly studied as water and oxygen transfer (Miller and Krochta 1997). Transfer of three methyl ketones through methylcellulose-based films is more complex than that for low-density polyethylene plastic (LDPE). Their transfer rate increases with the chain length of the volatile compound, while their diffusion coefficient decreases.

There is a relationship between permeability and sorption and between permeability and saturated vapor pressure. Mass transfer seems to depend mainly on the affinity between the volatile compound and the polymer. This relationship reveals the presence of physicochemical interactions that modify the structure of the polymer (plasticization).

Applied nanotechnology can lead to development of new flavor delivery systems to improve food quality and functionality. Controlled release may eventually lead to in-situ flavor and color modification of products (Chen et al. 2006).

11.11.4 Combined Delivery

Chitosan-based coatings demonstrated an antifungal function on inoculated fresh strawberries, which had an additive inhibition activity in vitro testing of the film with potassium sorbate (Park et al. 2005). A combination of chitosan and protein coatings diminished spoilage of frozen salmon and arrowtooth fillets (Sathivel 2005).

The combination of nisin and stearic acid in a hydroxypropyl methylcellulose film can result in a synergistic inhibitory effect on microorganisms with a pH < 3 (Sebti et al. 2002).

11.11.5 Modified and Controlled Atmospheres

In general, use of controlled or modified atmospheres (CA/MA) is beneficial for minimally-processed products. Typically, a concentration of 5% to 10% CO_2 and 2% to 5% O_2 is applied to extend shelf life of these products. A combination of edible coatings carrying additives and CA/MA has shown interesting results and promise for fruit and vegetable preservation (Amanatidou et al. 2000).

References

Amanatidou A, Slump RA, Gorris LGM, Smid EJ (2000) High oxygen and high carbon dioxide modified atmospheres for shelf-life extension of minimally processed carrots. J Food Sci 65(1): 61–66

Baranowski ES (1990) Miscellaneous food additives. In: Branen AL, Davidson PM, Salminen S (eds) Food Additives. Marcel Dekker, New York, NY, pp. 511–578

Buonocre GG, Del Nobile MA, Panizza A, Bove S, Battaglia G, Nicolais L (2003) Modeling the lysozyme release kinetics from antimicrobial films intended for food packaging applications. J Food Sci 68(4): 1365–1370

Cagri A, Ustunol Z, Osburn W, Ryser ET (2003) Inhibition of Listeria monocytogenes on hot dogs using antimicrobial whey protein-based edible casing. J Food Sci 68(1): 291–299

Carlin F, Gontard N, Reich M, Nguyen-The C (2001) Utilization of zein coating and sorbic acid to reduce Listeria monocytogenes growth on cooked sweet corn. J Food Sci 66(9): 1385–1389

Chen H, Weiss J, Shahidi F (2006) Nanotechnology in nutraceuticals and functional foods. Food Technol 03(06): 30–36

Chung D, Papadakis SE, Yam KL (2003) Evaluation of a polymer coating containing triclosan as the antimicrobial layer for packaging materials. Int J Food Sci Tech 38: 165–169

Coma V, Martial-Gros A, Garreau S, Copinet A, Salin F, Deschamps A (2002) Edible antimicrobial films based on chitosan matrix. J Food Sci 67(3): 1162–1168

Couey HM, Farias G (1979) Control of postharvest decay of papaya. HortSci 14(6): 719–721

Cuq B, Gontard N, Cuq JL, Guilbert S (1997) Selected functional properties of fish myofibrillar protein-based films as affected by hydrophilic plasticizers. J Agric Food Chem 45: 622–626

Debeufort F, Quezada-Gallo JA, Voilley A (1998) Edible films and coatings: Tomorrow's packaging: A review. Crit Rev Food Sci 38(4): 299–313

Debeaufort F, Quezada-Gallo JA, Delporte B, Voilley A (2000) Lipid hydrophobicity and physical state effects on the properties of bilayer edible films. J Membrane Sci 180: 47–55

Domenico JA, Rahman AR, Westcott DE (1972) Effects of fungicides in combination with hot water and wax on the shelf life of tomato fruit. J Food Sci 37: 957–960

Ecker JW, Kolbezen MJ (1977) Influence of formulation and application method on the effectiveness of benzimidazole fungicides for controlling postharvest diseases of citrus fruit. Neth J Plant Path 83: 343–352

Fairley P, Monahan FJ, German JB, Krochta JM (1996) Mechanical properties and water vapor permeability of edible films from whey protein isolate and sodium dodecyl sulfate. J agric Food Chem 44: 438–443

Gennadios A, McHugh TH, Weller CL, Krochta JM (1994) Edible coatings and films based on proteins. In: Krochta JM, Baldwin EA, Nisperos-Carriedo M (eds) Edible Coatings and Films to Improve Food Quality. CRC New York, NY, pp. 201–278

Ghaouth AE, Arul J, Ponnampalam R, Boulet M (1991) Chitosan coating effect on storability and quality fresh strawberries. J Food Sci 56(6): 1618–1631

Greener I, Fennema O (1989) Evaluation of edible bilayer films for use as moisture barriers for food. J Food Sci 54(6): 1400–1406

Greener Donhowe I, Fennema O (1994) Edible films and coatings: Characteristics, formation, definitions, and testing methods. In: Krochta JM, Baldwin EA, Nisperos-Carriedo M (eds) Edible Coatings and Films to Improve Food QUALITY. CRC New York, NY, pp. 1–24

Guilbert S (1986) Technology and application of edible protective films. In: Mathlouthi M (ed) Food Packaging and Preservation: Theory and Practice. Elsevier Applied Sciece Publishing Co. London, pp. 371–394

Guilbert S (1988) Use of superficial edible layer to protect intermediate moisture foods: Application to the protection of tropical fruit dehydrated by osmosis. In: Seow CC (ed) Food Preservation by Moisture Control. Elsevier Applied Science Publishers Ltd, London. pp. 119–219

Han JH, Krochta JM (2007) Physical properties of whey protein coating solution and films containing antioxidants. J Food Sci 00(0): E1–E7

Hernandez E (1994) Edible coatings from lipids and resins. In: Krochta JM, Baldwin EA, Nisperos-Carriedo M (eds) Edible Coatings and Films to Improve Food Quality. CRC, New York NY. pp. 279–304

Hoagland PD, Parris N (1996) Chitosan/pectin laminated films. J Agric Food Chem 44: 1915–1919

Howard LR, Dewi T (1995) Sensory, microbiological and chemical quality of mini-peeled carrots as affected by edible coating treatment. J Food Sci 60(1): 142–144

Janes ME, Kooshesh S, Johnson MG (2002) Control of Listeria monocytogenes on the surface of refrigerated, ready-to-eat chicken coated with edible zein film coatings containing nisin and/or calcium propionate. J Food Sci 67(7): 2754–2757

Jeon YJ, Kamil JYVA, Shahidi F (2002) Chitosan as an edible invisible film for quality preservation of herring and Atlantic cod. J Agric Food Chem 50: 5167–5178

Juneja VK, Thippareddi H, Bari L, Inatsu Y, Kawamoto S, Friedman M (2006) Chitosan protects cooked ground beef and turkey against Clostridium perfringens spores during chilling. J Food Sci 71(6): M236–M240

Karel M (1990) Encapsulation and controlled release of food additives. In: Schwartzberg HG, Rao MA (eds) Biotechnology and Food Process Engineering. Marcel Dekker, New York, NY. pp. 277–294

Katz EE, Labuza TP (1981) Effect of water activity on the sensory crispness and mechanical deformation of snack food products. J Food Sci 46: 403

Kim SJ, Ustunol Z (2001) Solubility and moisture sorption isotherms of whey-protein-based edible films as influenced by lipid and plasticizer incorporation. J Agric Food Chem 49: 4388–4391

Larre C, Desserme C, Barbot J, Gueguen J (2000) Properties of deamidated gluten films enzymatically cross-linked. J Agric Food Chem 48: 5444–5449

Le Tien C, Letendre M, Ispas-Szabo P, Mateescu MA, Delmas-Patterson G, Yu HL, Lacroix M (2000) Development of biodegradable films from whey proteins by cross-linking and entrapment in cellulose. J Agric Food Chem 48: 5566–5575

Lee SY, Krochta JM (2002) Accelerated shelf life testing of whey-protein-coated peanuts analyzed by static headspace gas chromatography. J. Agric. Food Chem. 50: 2022–2028

Letendre M, D'Aprano G, Lacroix M, Salmieri S, St-Gelais D (2002) Physicochemical properties and bacterial resistance of biodegradable milk protein films containing agar and pectin. J Agric Food Chem 50: 6017–6022

Longinos-Martinez S, Mendoza-Chapulin MR, Quezada-Gallo JA, Pedroza-Islas R (2005) Películas antimicrobianas para carne y productos cárnicos. Mundo lácteo y cárnico 05(06): 12–17

Martin-Polo M, Voilley A (1990) Comparative study of the water permeability of edible film composed of arabic gum and glycerol monostearate. Sci Aliments 10: 473–483

Mei Y, Zhao Y (2003) Barrie rand mechanical properties of milk protein-based films containing nutraceuticals. J Agric Fodd Chem 51: 1914–1918

Mezgheni E, Vachon C, Lacroix V (1998) Biodegradability behavior of cross-linked calcium caseinate films. Biotechnol Prog 14: 534–536

Micard V, Belamri R, Morel MH, Huilbert S (2000) Properties of chemically and physically treated wheat gluten. J Agric Food Chem 48: 2948–2953

Miller KS, Krochta JM (1997) Oxygen and aroma barrier properties of edible films: A review. Trends Food Sci. Tecnol 8: 228–237

Min S, Harris LJ, Krochta JM (2005) Antimicrobial effects of lactoferrin, lysozyme, and the lactoperoxydase system and edible whey protein films incorporating the lactoperoxidase system against *Salmonella enterica* and *Escherichia coli* O157:H7. J Food Sci 70(7): M332–M338

Nisperos-Carriedo MO, Shaw PE, Baldwin EA (1990) Changes in volatile flavor components of pineapple orange juice as influenced by the application of lipid and composite films. J Agric Food Chem 38: 1382–1387

Oh JH, Wang B, Field PD, Aglan HA (2004) Characteristics of edible films made from dairy proteins and zein hydrolysate cross-linked with transglutaminase. Int J Food Sci Tech 39: 287–294

Park SK, Rhee CO, Bae DH, Hettiarachchy NS (2001) Mechanical properties and water-vapor permeability of soy-protein films affected by calcium salts and glucono-δ-lactone. J Agric Food Chem 49: 2308–2312

Park SI, Stan SD, Daeschel MA, Zhao Y (2005) Antifungal coatings on fresh strawberries (Fragaria x ananassa) to control mold growth during cold storage. J Food Sci 70(4): M202–M207

Quezada-Gallo JA, Debeaufort F, Voilley A (1999) Interactions between aroma and edible films. 1. Permeability of methylcellulose and low-density polyethylene films to methyl ketones. J Agric Food Chem 47: 108–113

Quezada-Gallo JA, Debeaufort F, Callegarin F, Voilley F (2000) Lipid hydrophobicity, physical state and distribution effects on the properties of emulsion-based edible films. J Membrane Sci 180: 37–46

Quezada-Gallo JA, Diaz-Amaro R, Gramin A, Pattyn C, Debeaufort F,Voilley A (2004) Biopolymers used as edible coating to limit water transfert, colour degradation and aroma lost in Mexican fruits. Acta Horticulturae 682: 1709–1716

Quezada-Gallo JA, Bon Rosas F, Ramírez Gómez M, Díaz Amaro MR, Noah N, d Datzenko A (2005) Performances of edible coatings combined with bread additives to preserve fried donuts and mexican white bread "bolillo". Proceedings of the annual meeting of the institute of food technologists. 20–25 June, New Orleans, LA.

Radina PM, Eckert JW (1988) Evaluation of imazilil efficacy in relation to fungicide formulation and wax formulation. In: Cohen R, Mendel K (eds) Citriculture proceedings of the sixth international citrus congress. Balaban Publishers, Philadelphia PA. pp. 1427–1434

Rakotonirainy AM, Padua GW (2001) Effects of lamination and coating with drying oils on tensile and barrier properties of zein films. J Agric Food Chem 49: 2860–2863

Rao MS, Chander R, Sharma A (2005) Development of shelf-stable intermediate-moisture meat products using active edible chitosan coating and irradiation. J Food Sci 70(7): M325–M331

Reineccius GA (1994) Flavor encapsulation. In: Krochta JM, Baldwin EA, Nisperos-Carriedo M (eds) Edible coatings and films to improve food quality. CRC, New York NY. pp. 105–120

Ressouany M, Vachon C, Lacroix M (1998) Irradiation dose and calcium effect on the mechanical properties of cross-linked caseinate films. J Agric Food Chem 46: 1618–1623

Rico-Peña DC, Torres JA (1990) Edible methylcellulose-based films as moisture-impermeable barriers in sundae ice cream cones. J Food Sci 55: 1468–1469

Rico-Peña DC, Torres JA (1991) Sorbic acid and potassium sorbate permeability of an edible methylcellulose-palmitic acid film: Water activity and pH effects. J Food Sci 56(2): 497–499

Rodriguez MS, Ramos V, Agulló E (2003) Antimicrobial action of chitosan against spoilage organisms in precooked pizza. J food Sci 68(1): 271–274

Sanders ES (1999) Probiotics. Food Technology 53(11): 67–77

Sanders ES, Huis in't Veld J (1999) Bringing a probiotic containing functional food to the market: Microbiological, product, regulatory and labeling issues. Antonie van Leeuwenhoek 76: 293–315

Sathivel S (2005) Chitosan and protein coatings affect yield, moisture loss, and lipid oxidation of pink salmon (Onchorhynchus gorbuscha) fillets during frozen storage. J Food Sci 70(8): E455–E459

Schrooyen PMM, Dijkstra PJ, Oberthür RC, Bantjes A, Feijen J (2001) Partially carboxymethylated keratins. 2. Thermal and mechanical properties of films. J Agric Food Chem 49: 221–230

Sebti I, Ham-Pichavant F, Coma V (2002) Edible bioactive fatty acid-cellulosic derivative composites used in food applications. J Agric Food Chem 50: 4290–4294

Sebti I, Martial-Gros A, Carnet-Pantiez A, Grelier S, Coma V (2005) Chitosan polymer as bioactive coating and films against aspergillus niger contamination. J Food Sci 70(2): M100–M104

Sothornvit R, Krochta JM (2000) Oxygen permeability and mechanical properties of films from hydrolyzed whey protein. J Agric Food Chem 48: 3913–3916

Tay SL, Perera CO (2004) Effect of 1-methylcyclopropene treatment and edible coatings on the quality of minimally processed lettuce. J Food Sci 69(2): C131–C135

Taylor TM, Davidson PM, Bruce BD, Weiss J (2005) Liposomal nanocapsules in food science and agriculture. Crit Rev Food Sci Nutr 45: 1–19

Torres JA, Karel M (1985) Microbial stabilization of intermediate moisture food surfaces. III. Effect of surface preservative concentration and surface pH control on microbial stability of an intermediate moisture cheese analog. J Food Process Preserv 9: 107–119

Torres JA, Bouzas JO, Karel M (1985a) Microbial stabilization of intermediate moisture food surfaces. II. Control of surface pH. J Food Process Preserv 9: 93–106

Torres JA, Motoki M, Karel M (1985b) Microbial stabilization of intermediate moisture food surfaces. I. Control of surface preservative concentration. J Food Process Preserv 9: 75–92

Vodjani F, Torres JA (1990) Potassium sorbate permeability of methylcellulose and hydroxypropyl methylcellulose coatings: Effect of fatty acids. J Food Sci 55(3): 841–846

Weiss J, Takhistov P, McClements DJ (2006) Functional materials in food nanotechnology. J Food Sci 71(9): R107–R116

Wells JM (1971) Heated wax-emulsion with benomyl and 2,6-dichloro-4-nitroaniline for control of postharvest decay of peaches and nectarines. Phytopathology 62: 129–133

Zhang C, Ding Y, Ping Q, Yu L (2006) Novel chitosan-derived nanomaterials and their micelle-forming properties. J Agric Food Chem 54: 8409–8416

Zivanovic S, Chi S, Draughon AE (2005) Antimicrobial activity of chitosan films enriched with essential oils. J Food Chem 70(1): M45–M51

Chapter 12
Application of Infrared Analysis to Edible Films

Charles M. Zapf

12.1 Introduction

In the modern analytical laboratory today, one will find an assortment of analytical instrumentation such as gas chromatography for the analysis of volatile compounds and liquid chromatography for the analysis of non-volatile chemicals. Many other computer-assisted instruments are used for the analysis of materials such as foods or other consumer products, as well as industrial materials. For many of these instruments, talent and experience is required to prepare the sample prior to analysis and this can be challenging. These procedures require time and even more time is needed to develop them as useful analytical methods.

Among these instruments is the infrared spectrometer that has many attractive performance features: easy to use, and require little maintenance and repair. Sample preparation can be simple and analysis is time rapid. When an automated database is available either in the system or on-line through the Internet, data evaluation can be rapid too. In this chapter, application of infrared technology to analysis of edible films as found in published literature will be described. However, the reader may quickly appreciate that the instrument has utility for analysis of materials in both research and in rapid, routine, quality assessment.

12.2 Instrumentation

Many vendors supply infrared instrumentation and offer systems suitable for research or quality assurance with attractive pricing compared to other analytical instruments used for material characterization. These will not be thoroughly described here, except for a simple introduction.

C.M. Zapf
Technical Innovation Center, McCormick & Company, Inc., 204 Wight Avenue,
Hunt Valley, MD 21031-1501, USA
e-mail: mike_zapf@mccormick.com

M.E. Embuscado and K.C. Huber (eds.), *Edible Films and Coatings for Food Applications*, 335
DOI 10.1007/978-0-387-92824-1_12, © Springer Science+Business Media, LLC 2009

Modern infrared instruments are often simply referred to as FTIR where the two leading initials are shorthand for the technique of fourier transform and the later two initials for the type of instrument. Earlier instruments were a different design and hence this commonly used distinction.

The bench will include a light source, detector and sampler options depending on the type of analysis required for transmission or absorbance analysis of liquids, powders or films. A configuration for the instrument is usually decided upon before purchasing, based on the type of samples to be analyzed. However, upgrades might be done on site by a service technician. Various sampling systems are simply installed in the instrument compartment by the operator when needed. The instrument is quite adaptable.

An example of a system for simple analysis is a bench for mid-infrared analysis using a global source for sample irradiation and a deuterated triglycine sulfate (DGTS) detector for signal collection. A robust sampler, suitable for films, might be an attenuated total reflectance (ATR), or horizontal attenuated reflectance platform. In the case of the ATR sampler, a flat crystal made of ZnSe for example, is constructed such that the incoming IR beam is bounced one or more times within the crystal where it interacts with the sample at the surface of the crystal. Radiation interacts with the sample in intimate contact at the surface, where a small penetration is sufficient. There are a variety of crystals, each with specific benefits. A popular type is the diamond-ATR, a very robust sample accessory. There are many other types of sample accessories (Coleman 1993), but the common transmission-sampling accessory is included with the purchase. A simple diagram of either the transmission or ATR experiment is shown in Fig. 12.1.

The computer and software completes the infrared system, and while a high-speed computer is often not needed, most of this investment might be for analysis

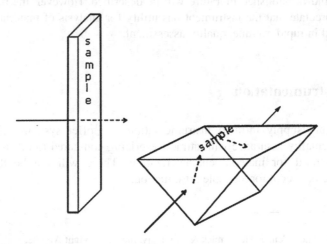

Fig. 12.1 Diagram of the transmission or ATR experiment

software and a possible materials database. The analysis software permits display of the sample spectrum and the computer analyzes the spectrum by human interpretation, by comparing it to a materials database search, or possibly through an automated analytical interpretation. Such software usually stores collected data, and permits system calibration for analytic quantitation when necessary.

As will be described later, sample preparation and presentation often involves a simple application of sample to the sampling apparatus. This important step dictates the success of the method. For the ATR crystal, a film sample can be applied horizontally on the face of the crystal and sufficient pressure is applied to ensure good contact. An instrument scanning procedure is decided upon, and the sample analysis is often complete in less than 5 min. It should be realized that the various vibrational spectroscopic techniques, including mid-IR near-IR, and Raman, operate in similar manner. Thus, for a reasonable investment cost and retention of the right talent, massive sample information can be acquired for either research or production sample types. Furthermore, once the procedure is optimized, technicians or process operators can collect chemical data to determine material quality either in the quality assurance laboratory or online.

12.3 Application of Infrared Analysis to Edible Films

One can find many references to successful application of IR technology to analysis of edible films, particularly in the area of research and applications. Application of IR analysis using simple preparation and quick analysis will be highlighted and used for some selected samples. Examples chosen are not exhaustive, but from the viewpoint of the author, they exemplify the application of infrared technology that will be useful to reduce of this book.

There are a number of edible films that can be created from natural biomaterials and after the addition of suitable plasticizers, one might wish to confirm the presence of the proper concentration of those chemicals and perhaps other ingredients incorporated in films such as flavors, antimicrobials, food additives, or pharmaceutical ingredients. Figure 12.2 shows a small collection of spectra recorded in the author's laboratory and Fig. 12.3 shows spectra of a commercial product containing flavor used as a breath freshener.

As shown in Fig. 12.2, matrices of edible films are made from carbohydrates, gums and proteins. Of interest to the food scientist or formulator may be the composition, stability and film thickness. Many of these have been researched, given the reliability of modern infrared instrumentation and ease of operation. These measurements are used in routine operation for product quality and consistency in quality control laboratories.

Composition of a film can be easily assessed by taking an infrared spectrum. The common ATR sample accessory makes this easy, and can be used for this purpose.

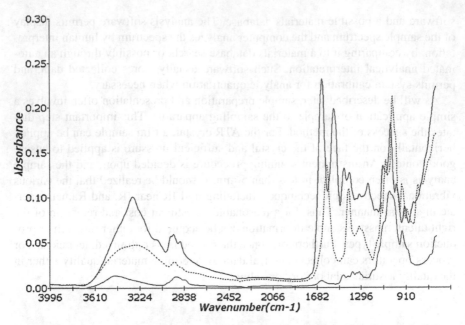

Fig. 12.2 Spectral comparison of calcium caseinate (*top*), sodium alginate (*dotted*), methylcellulose (*bottom*) (edible films courtesy of ERRC, USDA and TIC Gums)

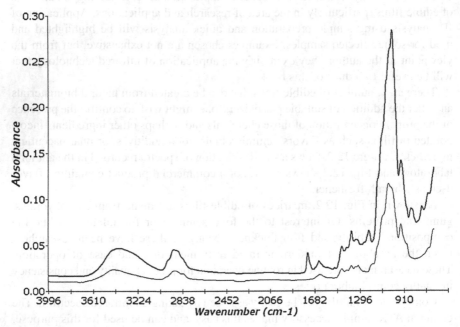

Fig. 12.3 Spectra of commercial cinnamon flavor film (*top*) and methylcellulose (*bottom*)

A film is placed onto the sampling crystal that is scanned, and using the spectrometer's default parameters, a useful spectrum can be obtained. Other infrared sampling techniques can be as simple.

An important component of edible films is the plasticizer. This is added to the basic film matrix (e.g., carbohydrates or proteins) to improve mechanical properties, stability and performance. Turhan et al. (2001) analyzed methylcellulose-based (MC) films plasticized with polyethylene glycol (PEG) using FTIR and the standard transmission sample holder to obtain spectra of MC films containing various levels of PEG's with different molecular weights. Several replicate samples of each formulation were collected, which permitted statistical analysis using analysis of variance (ANOVA). They studied the hydroxyl, O-H stretch region of the spectrum for the changes in band shape at 3,466 cm^{-1} to assess the effect of hydrogen bonding between the MC and PEG's. From the symmetry distortion of the band, they concluded that the MC hydrogen bonding was replaced by MC to PEG hydrogen bonding. The PEG's are reported to improve tensile strength of MC films.

In a similar experiment of methylcellulose films, Velazquez et al. (2003) used FTIR to monitor the O-H bending IR region of a series of spectra obtained after exposure of the film to increasing saturated salt solutions, thus increasing its water activity (a_w). Subsequent analysis using curve fitting of a composite O-H bending peak permitted assessment of bound and free water. A linear plot of moisture versus total peak area resulted from the analysis. This curve-fitting routine will be described later.

Wilhelm et al. (2003) studied the infrared spectra of Cara starch films containing mineral clay and plasticized with glycerol, using a simple scan of each film. As the ratio of clay increased, the change in composition of the film was detected in each spectrum. Physical measurements too were made of the films that could have been correlated with the spectral changes. Hence, the infrared spectra might estimate the film performance.

Evidence of blended ratios of food polymers was presented in the study by Li et al. (2006) for gelatin and Konjac glucomannan (KGM) where the proportion of KGM and gelatin was recorded as FTIR spectra of the powdered film, mixed in potassium bromide. In this infrared technique, a small amount of sample is well mixed with powdered KBr, and pressed into a clear pellet, yielding approximately a 1% composition. As the KGM proportion increased, infrared bands attributed to mannose in KGM were apparent at 807 and 892 cm^{-1}, and infrared bands attributed to gelatin, notably the amide structure, were detected. This is an example of a compositional analysis of the edible film using FTIR. Additional property analyses were measured, and although authors did not correlate properties such as tensile strength and elongation with the spectra collected, they disclosed that blend K4 (4:6 KGM to gelatin ratio) had the highest tensile strength. Hence correlation of film properties with the associated spectra may allow inference of film performance.

Indeed, Vicentini et al. (2005) used Chemometric Analysis of the IR spectra for cassava starch films to study the relationship between the spectrum of the films and their functional properties. Chemometrics is the mathematical procedure in which data matrices (i.e. multiple sample spectra and properties) are analyzed for systematic variances that might be successfully correlated (Malinowski 1991). This procedure is often presented in graphical summary plots where the analyst might detect trends between spectra and properties, thus facilitating the interpretation of multiple sample spectra and complementary data at once. One common procedure is Principle Component Analysis or PCA where new factors are mathematically derived from initial data to explain variance in complete spectral dataset, and are then correlated to physical or compositional data.

Authors used PCA to plot the 1, 2, 3, and 4% starch formulations into clusters formed on a two-way plot. Each cluster contained samples of the nominal concentration, sharing related variation in functional properties such as thickness, puncture-force, etc. However, the PCA method failed to separate a common cluster for the 3 and 4% level samples, giving only three clusters; 1, 2, and a 3 + 4 overlap.

The authors concluded that infrared spectra for a series is an indicator of the functional properties of films, and then applied a second chemometric procedure, Partial Least Squares (PLS), to create calibrations for properties such as puncture force and deformation using correlation. Each of these gave favorable statistical figures of merit for both the calibration model and a small sample set used for validation. The resulting model-regression vector revealed that bands at 994 cm^{-1} and 1,012 cm^{-1} were important to hydroxyl solvation and hydrogen bonding. Chemometric models require sample data having adequate variation in both composition and properties in order to be successful and useful in the final application.

12.3.1 Composition of Protein Films

Protein materials from many common sources can produce films with different properties, and have been derived from a number of sources as listed in Table 12.1.

Once a successful formulation is found, its composition is studied so that it can be defined by chemical and physical measurements. Specifications can then be used to ensure consistent formulations later. Here, an infrared analysis of the film is expedient, using the ATR sampler mentioned earlier for example, as a direct chemical measurement of the formulation, and as a means to estimate physical properties.

In the study of β-casein films, Mauer et al. (2000) used simple preparation of each film by casting film solution onto polyethylene IR cards (a thin film of PE sandwiched between cardboard) and drying it, thus forming a thin film upon the

Table 12.1 Protein sources

Source	Proteins
Milk	Casein, whey
Wheat	Gliadin
Sorghum	Kafirin

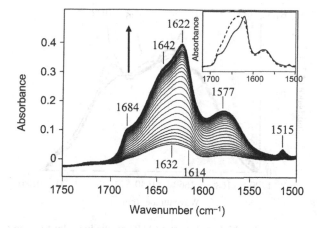

Fig. 12.4 ATR spectra in the amide I' region of a film-forming β-1g solution (10% w/v in D₂O, after heating 30 min at 80°C, β-1g:DEG weight ratio of 1:1) as a function of time during dehydration. The *arrow* indicates the direction of intensity changes as time increases. In *inset*, the broken and the full lines correspond to the first (film-forming solution) and last (final film) spectrum of the figure, respectively. They are normalized relative to the peak intensity maximum. (Reprinted with permission from Lefèvre (2005); copyright 2005 American Chemical Society)

polyethylene card. After some spectral data treatments and subsequent statistical correlation, comparisons of films made from frozen and room temperature caseins indicated that the two treatments did not result in differences in the amide region of spectra which are attributed to protein secondary structures as will be explained below.

Bovine milk β-lactoglobulin and diethylene glycol (DEG) used as a plasticizer in the ratio 1:1, were studied during dehydration (D₂O as solvent) as a function of time (Lefèvre et al. 2005) using the amide I′ infrared spectral region, employing an ATR sampler. In Fig. 12.4, the spectral trend with time can be followed as dehydration of the solution occurs. As noted by the authors, the maximum of the amide I′ band shifts from 1,632 to 1,622 cm⁻¹ revealing a simultaneous intensity change at 1,684 cm⁻¹ thus providing a dynamic means of monitoring formation of the film. An extensive description of the process is given.

These authors further extended infrared analysis to describe the resulting protein film secondary structure, by means of comparison to known band assignments after spectral band deconvolution and curve fitting. Curve fitting is a means of fitting characterized (i.e. Gaussian) isolated peaks to a composite profile in order to assess the possible contribution of simpler, underling bands, thus assisting interpretation of complex overlapping peaks.

In another example of the application of this technique to study film structure, Mangavel et al. (2001) studied the evolution of protein conformation in wheat gliadins, again during drying, using the ATR sampler controlled at 70°C. Here the focus was on resolution of the amide I (H₂O as solvent) band at a 1,626 cm⁻¹ – shoulder,

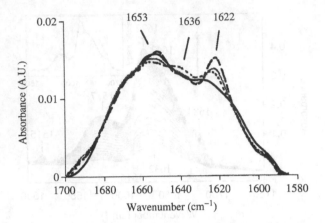

Fig. 12.5 Fourier deconvoluted infrared spectra (ATR) of gliadins film-forming solutions (16% w/w) at pH 11 during drying at 70°C: (*solid line*), initial time; (*dotted line*), after 30 min; (*shaded line*), after 45 min; (*dashed line*), after 60 min (film state). (Reprinted with permission from Mangavel et al. (2001); copyright 2001 American Chemical Society)

evolving to a better defined band at 1,622 cm^{-1}, which they attributed to appearance of a β-sheet having a high level of hydrogen bonding. The drying profile is shown in Fig. 12.5 during the first hour. Further analysis of the composite spectra was elaborated, as profiles of films dried at 25°C and 70°C are compared in Fig. 12.6. The spectral slices of two drying treatments were analyzed using a peak fitting routine to aid explanation of the resulting conformation for films produced at these two temperatures.

Biofilm formation of Kafirin from different extraction processes was studied using infrared with the ATR sampler. Kafirin is a prolamin plant protein similar to zein. Gao et al. (2005) measured the change of the ratio of the intensities at 1,650 cm^{-1}–1,620 cm^{-1} and suggested that protein aggregation occurred during the heat-drying process. This observation correlated to those films that were more difficult to disperse in the film-casting solution. Light microscopy images confirmed that protein aggregation had occurred.

An introduction to the assignment of protein secondary structures can be found in the book by Stuart (1997). Jackson and Mantsch (1995) discuss the pitfalls and successes of applying infrared spectroscopy for the determination of protein structure. A further caution is that films measured with ATR samplers may not have endured controlled moisture environments and, hence, effect on mechanical properties could be mistakenly interpreted.

12.3.2 Stability of Films

Among the variety of testing instrumentation used for measurement of edible films, the infrared apparatus can also be used to test stability of films. In this case, the film sample is degraded in a small heater or tubular oven. The oven is connected to the

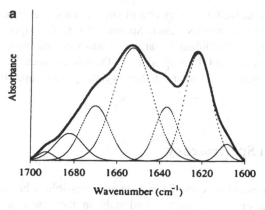

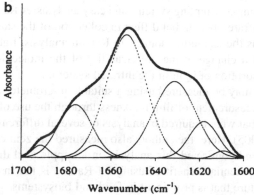

Fig. 12.6 Fourier deconvoluted amide I band of gliadin films fitted with sums of Gaussian bands: (**a**), film dried at 70°C, (reduced χ^2 = 6.74, r = 0.99750, standard error = 0.00003725; (**b**), film dried at 25°C, (reduced χ^2 = 30.4, r = 0.99627, standard error = 0.00004713. (Reprinted with permission from Mangavel et al. (2001); copyright 2001 American Chemical Society)

IR that analyzes gases evolved (i.e. NH_3, CO_2, H_2O) from decomposition, and is carried by a flowing gas such as nitrogen.

Barreto et al. (2003) utilized this technique to study stability of milk proteins (sodium caseinate) and gelatin using sorbitol as the plasticizer, and discovered that blends degraded at a lower temperature than native proteins, by employing thermogravimetric analysis (TGA). In a TG analyzer, the mass of the sample is monitored over time as heat is applied to assess the sample's thermal stability. The mass and time are recorded very accurately to ensure a quantitative measurement. For films, loss of low molecular weight gases is common.

Gases evolved from the oven were directed to the infrared instrument where identity of decomposition chemicals profiled the process. At 200°C, water was detected, and was explained as loss of absorbed moisture. Later, at 300°C, NH_3 and CO_2 were emitted, which resulted from degradation of protein. An IR analysis was also made of the residue from decomposition, and hence, through description of the gaseous products, as well as the spectra of the residues, a chemical interpretation supported the thermal analysis. Later, Soares et al. (2005) applied their infrared technique to analysis of xanthan/wheat starch and xanthan/maize starch blends in order to study their thermal decomposition in the tubular oven. The loss of CO and CO_2 were attributed to scission of carboxylate groups and the carbohydrate backbone.

12.4   Raman Spectroscopy

FT-Raman spectroscopy has also become more accessible, whereas in the past, the typical configuration was usually found only in the research laboratory. This method also uses simple sampling systems and easy analysis. A Raman spectrometer is similar to the infrared bench, but differs in collection of the sample spectrum. It also uses a laser as the excitation source. In Raman analysis, light is scattered by the molecule, and a change in the polarizability of the molecule is measured in contrast to the absorption of light in the infrared system.

Film thickness may be measured using a suitable micrometer device. However, Fig. 12.7 shows measurement of film thickness through the use of peak ratio using the same spectra that were acquired in analysis of several different compositions of zein (Hsu et al. 2005). Here, the authors also measured the zein coating applied to the intact apple, and confirmed this analysis on an apple peel dissected from the same apple. One potential benefit of using FT-Raman is the insensitivity of this technique to moisture that is prominent in food and biosystems.

12.5   Conclusion

The modern infrared spectrometer is a convenient laboratory apparatus because of its ease of use, availability of a variety of accessories for simple sample presentation, and application of the system for practical and research analyses. After some basic investigation into a particular problem, the analyst can quickly learn to set up and perform an analysis with care and confidence. Many spectral databases are available to aid in interpretation of the sample spectrum. References are readily available in print and on the Internet, so with some experience, the instrument can be applied to everyday analysis. Once a technically trained operator prepares the IR for use, one can obtain reliable analysis of everyday samples with ease and speed. As an example, after an expert spectroscopist has identified important structures of the various protein bands, these data can be cataloged in a spectral library for easy recall. These cataloged spectral data will permit accurate analysis by other support staff in application or quality laboratories. Library comparisons are standard practice often used for infrared analysis.

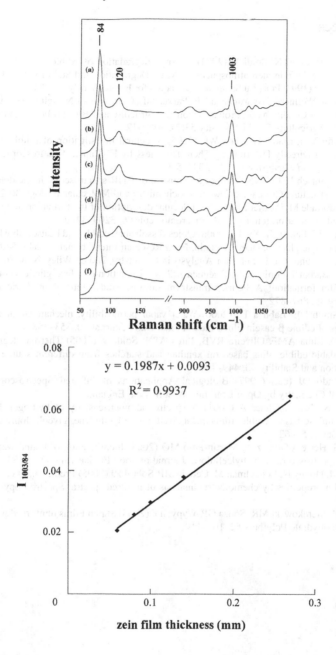

Fig. 12.7 Correlation of Raman intensity ratio of $I_{1003/84}$ vs. zein film thickness measured by micrometer (*Inset*: FT-Raman spectra normalized at the band of 84 cm^{-1} for different concentration of zein films: (**a**) 8%, (**b**) 10%, (**c**) 12%, (**d**) 14%, (**e**) 16%, and (**f**) 18%. Spectral intensity between 900 and 1,100 cm^{-1} was enlarged at the same proportion to easily observe the change of Raman intensity upon different zein thicknesses). (Hsu (2005); copyright 2005 American Chemical Society)

References

Barreto PLM, Pires ATN, Soldi V (2003) Thermal degradation of edible films based on milk proteins and gelatin in inert atmosphere. Polymer Degradation and Stability 79, 147–152

Coleman PB (ed.) (1993) Practical Sampling Techniques for Infrared Analysis. CRC, Boca Raton, FL

Gao C, Taylor J, Wellner N, Byaruhanga YB, Parker ML, Clare Mills EN, Belton PS (2005) Effect of Preparation Conditions on Protein Secondary Structure and Biofilm Formation of Kafirin. Journal of Agriculture Food Chemistry 53(2), 306–312

Hsu BL, Weng YM, Liao YH, Chen WL (2005) Structural Investigation of Edible Zein Films/ Coatings and Directly Determining Their Thickness by FT-Raman Spectroscopy. Journal of Agriculture *Food Chemistry* 53(13), 5089–5095

Jackson M, Mantsch HH (1995) The use and misuse of FTIR spectroscopy in the determination of protein structure. Critical Reviews in Biochemistry and Molecular Biology 30(2), 95–120

Lefèvre T, Subirade M, Pézolet M (2005) Molecular description of the formation and structure of plasticized globular protein films. Biomacromolecules 6, 3209–3219

Li B, Kennedy JF, Jiang QG, Xie BJ (2006) Quick dissolvable, edible and heatsealable blend films based on konjac glucomannan – Gelatin. Food Research International 39, 544–549

Malinowski, Edmund R. (1991) Factor Analysis in Chemistry, 2nd ed.Wiley, New York, NY.

Mangavel C, Barbot J, Popineau Y, Guéguen J (2001) Evolution of wheat gliadins conformation during film formation: A Fourier transform infrared study. Journal of Agriculture Food Chemistry 49, 867–872

Mauer LJ, Smith DE, Labuza TP (2000) Water vapor permeability, mechanical, and structural properties of edible β-casein films. International Dairy Journal 10, 353–358

Soares RMD, Lima AMF, Oliveira RVB, Pires ATN, Soldi V (2005) Thermal degradation of biodegradable edible films based on xanthan and starches from different sources. Polymer Degradation and Stability 90, 449–454

Stuart B, Ando DJ (ed.) (1997) Biological Applications of Infrared Spectroscopy. ACOL, Analytical Chemistry by Open Learning, West Sussex, England

Turhan KN, Sahbaz F, Güner A (2001) A spectrophotometric study of hydrogen bonding in methylcellulose-based edible films plasticized by polyethylene glycol. Journal of Food Science 66(1), 59–62

Velazquez G, Herrera-Gómez A, Martín-Polo MO (2003) Identification of bound water through infrared spectroscopy in methylcellulose. Journal of Food Engineering 59, 79–84

Vicentini NM, Dupuy N, Leitzelman M, Cereda MP, Sobral PJ (2005) Prediction of cassava starch edible film properties by chemometric analysis of infrared spectra. Spectroscopy Letters 38, 749–767

Wilhelm HM, Sierakowski MR, Souza GP, Wypych F (2003) Starch Films reinforced with mineral clay. Carbohydrate Polymers 52, 101–110

Chapter 13
Mechanical and Permeability Properties of Edible Films and Coatings for Food and Pharmaceutical Applications

Monique Lacroix

13.1 Introduction

Use of natural polymers, such as proteins and polysaccharides, as coating or film materials for protection of food has grown extensively in recent years. These natural polymers can prevent deterioration of food by extending shelf life of the product and maintaining sensory quality and safety of various types of foods (Robertson 1993). Generally, film and coating systems are designed to take advantage of barrier properties of polymers and other molecules to guard against physical/mechanical impacts, chemical reactions and microbiological invasion. In addition, the use of natural polymers presents added advantages due to their edible nature, availability, low cost and biodegradability. The latter particularly is of paramount interest due to demand for reducing the amount of non-biodegradable synthetic packaging. Furthermore, these polymers can be easily modified in order to improve their physicochemical properties for filming and coating applications.

Edible films are freestanding structures, formed and applied for specific packaging uses. Edible films can be used to maintain separation of various components within a single food product or applied directly to the external surface of a food to inhibit migration of moisture, oxygen, CO_2, aromas, lipids, etc. Use of edible films and coatings for separating different components in multi-component foods can improve the quality of a food product (Krochta and De Mulder-Johnston 1997). Edible films with adequate mechanical properties can conceivably also serve as edible packaging for select foods (Krochta and De Mulder-Johnston 1997). They are formed by casting and drying film-forming solution on a leveled surface; drying a film-forming solution on a drum drier; or using traditional plastic processing techniques, such as extrusion. However, it is also possible for an edible film to be formed by applying coating solution directly at the surface of a food.

M. Lacroix

INRS – Institut Armand-Frappier, Research Laboratory in Sciences Applied to Food, Canadian Irradiation Centre, 531 des Prairies, Laval, QC, Canada, H7V 1B7
e-mail: Monique.lacroix@iaf.inrs.ca

M.E. Embuscado and K.C. Huber (eds.), *Edible Films and Coatings for Food Applications*, 347
DOI 10.1007/978-0-387-92824-1_13, © Springer Science+Business Media, LLC 2009

Film requirements for foods are complex. Unlike inert packaged commodities, foods are often dynamic systems with limited shelf life and very specific packaging needs. In addition, since foods are consumed to sustain life, the guarantee of safety is a critical dimension of their packaging requirements. While issues of food quality and safety are first and foremost in the mind of the food scientist, a range of other issues surrounding development of any food package must be considered before a particular packaging system becomes a reality. Secondary packaging is often used to provide additional physical protection to the product. It may be a box surrounding a food packaged in a flexible plastic wrap. It could also be a corrugated box containing a number of primary packages for the purpose of easing handling during storage and distribution, improving stackability, or protecting primary packages from mechanical damage during storage and distribution.

Coatings are defined as thin layers of material applied to the surface of a food. Its purposes may be to provide a barrier against migration of moisture, oxygen (O_2), carbon dioxide (CO_2), aromas, lipids, etc., to carry functional food ingredients (e.g., antimicrobials, antioxidants, and flavor components), and/or to improve mechanical integrity or handling of the food product. In addition to increased barrier properties, edible films and coatings control adhesion, cohesion and durability to improve appearance of coated foods (Krochta 1997). Thus, coatings preserve attributes associated with food quality, as well as increase shelf life (Kester and Fennema 1986). Edible coatings also have potential for maintaining quality of a food after primary packaging is opened, by protecting against moisture change, oxygen uptake and aroma loss (Krochta 1997). Edible coatings form an integral part of the food product, and hence should not adversely impact on sensory characteristics of the food (Guilbert et al. 1996). Also, to be accepted commercially, components of an edible film should be generally recognized as safe (GRAS).

Edible coatings are applied and formed by addition of a liquid film-forming solution directly onto the food product. They may be applied with a paintbrush, by spraying, dipping or fluidizing (Cuq et al. 1995). Application of edible coating to meat and fish products may be produced by dipping, spraying, casting, rolling, brushing or foaming (Donhowe and Fennema 1994). Coatings used for meat products must have appropriate moisture barrier properties, water or lipid solubility, color, appearance, mechanical and rheological characteristics. Use of composite coatings may be necessary to improve gas exchange, adherence to coated products and moisture vapor permeability (Baldwin et al. 1995). Addition of vegetable oils can improve coating moisture barrier properties, while addition of glycerol, polyethylene glycol, and sorbitol can reduce film brittleness. Sealing meat with cross-linked sodium caseinate gel effectively preserved color of fresh meat, eliminated need for absorbent pads, and produced a juicer product by reducing drip loss (Ben and Kurth 1995). Thus, coated foods may not require high-barrier packaging materials, allowing the entire packaging structure to be simplified. Edible films and coatings can also incorporate active ingredients such as antioxidants, antimicrobial agents, colorants, flavors, fortified nutrients and/or spices (Floros et al. 1997). A composite film containing these active compounds can prevent rancidity and improve shelf life by controlling bacterial proliferation.

In summary, functional properties of an edible film or coating will be directly related to the composition and structure of its components. These parameters should be adapted appropriately so that functional properties of the film or coating are preserved through all processing, transport and storage conditions associated with a particular food. The challenge for successful use of biodegradable films and coatings is to stabilize their functional properties and adapt their mechanical properties in accordance with the intended application.

13.2 Film and Coating Composition, Structure and Functionality

Preparation of edible packaging generally involves addition of proteins, polysaccharides, plasticizers, lipids, and sometimes other components. Composite films may be heterogeneous in nature, and can be formed via mixtures of these materials. This approach allows for better exploitation of the unique functional properties inherent to each of the film's individual components.

According to Cuq et al. 1998, preparation of protein-based films requires formation of a continuous, low-moisture, and a more or less ordered macromolecular network, consisting of numerous and uniformly distributed interactions between polymer chains. The probability of forming intermolecular bonds mainly depends on shape of the protein (fibrous vs. globular) and on physicochemical conditions during processing. High molecular weight proteins (e.g., myosin) and fibrous proteins (e.g., myosin and F-actin) generally form films with good mechanical properties, while globular or pseudo-globular proteins (e.g., G-actin) need to be unfolded before film formation. Myofibrillar proteins are soluble in dilute salt solution or in low ionic strength solutions at neutral pH. Because of the hydrophilic nature of proteins, edible films made from them have an excellent oxygen barrier property, but they are susceptible to moisture (Gueguen et al. 1998; Stanley et al. 1994; Stefansson and Hultin 1994).

Proteins films are generally brittle and susceptible to cracking due to the strong cohesive energy density of the polymers (Lim et al. 2002). Addition of compatible plasticizers improves extensibility and viscoelasticity of films (Brault et al. 1997). According to Ressouany et al. (1998), protein based films containing sorbitol produce films with higher viscoelasticity. However, when polyethylene glycol and mannitol were added, a lower plasticizing effect was observed and more rigid films with lower elasticity properties were produced.

Edible coatings can be made from a variety of polysaccharides derived from plants (pectin and cellulosic derivatives), seaweed extracts to connective tissue extracts of crustaceans (alginate, chitosan, carrageenan) or bacteria (pullulan, levan, elsinan). Such coatings have been used to retard moisture loss of some foods during short term storage. However, polysaccharides, being hydrophilic in nature, do not function well as physical moisture barriers. The method by which they retard moisture loss is by acting as a sacrificial moisture barrier to the atmosphere, so that moisture content of the coated food can be maintained (Kester and Fennema 1986). In addition to preventing moisture loss, some types of polysaccharides films are less permeable to oxygen. Decreased oxygen permeability can help preserve certain foods.

Due to the hydrophilic nature of polysaccharide, polysaccharide-based films exhibit limited water vapor barrier ability (Gennadios et al. 1997). However, films based on polysaccharides like alginate, cellulose ethers, chitosan, carrageenan or pectins exhibit good gas barrier properties (Baldwin et al. 1995; Ben and Kurth 1995). The gas permeability properties of such films result in desirable modified atmospheres, thereby increasing shelf life of the product without creating anaerobic conditions (Baldwin et al. 1995). Polysaccharide films are used in Japan for meat products, ham and poultry packaging before smoking and steaming processing. The film is dissolved during the process and the coated meat exhibits improved yield, structure and texture and reduced moisture loss (Labell 1991; Stollman et al. 1994).

Multi-component edible films and coatings consisting of blends of various polymers, polysaccharides, proteins and/or lipids have been developed to have cooperative functionalities. Polysaccharides can produce structural cohesion to provide a supporting matrix. Proteins can also provide structural cohesion, while lipids contribute a hydro-repulsive character to a film (Wu et al. 2002). For example, under certain conditions, addition of a polysaccharide to a protein film formulation can improve moisture barrier resistance and mechanical properties of a film (Letendre et al. 2002a; Ressouany et al. 1998). These enhanced properties may be due in part to formation of cross-links between film components (Thakur et al. 1997) or formation of complexes between protein and polysaccharide constituents (Letendre et al. 2002a; Sabato et al. 2001; Thakur et al. 1997; Shih 1994; Imeson et al. 1977). Strong interaction between two components resulting with a higher bounded β-structure (β-sheet and β-strand content), a more ordered structure and a cross-linked protein–polysaccharide network was obtained after heating and irradiation treatment of caseinate-whey based films in the presence of carboxymethylcellulose (CMC). These structural modifications led to more rigid films with improved mechanical strength and barrier properties (Cieśla et al. 2006a, b; Letendre et al. 2002a, b; Le Tien et al. 2000).

13.3 Mechanical Properties

Many food proteins such as casein, wheat gluten, soy or whey isolate, and polysaccharides such as alginate, starch, cellulose, cassava starch or pectin, have been formulated into edible films or coatings (Henrique et al. 2007; Tapia et al. 2007; Letendre et al. 2002a; Le Tien et al. 2000; Vachon et al. 2000; Sabato et al. 2001). However, the highly hydrophobic nature of these proteins limits their ability to provide desired edible film functions.

Mechanical or physicochemical properties of films can be improved by application of extrusion or promotion of reaction between film components. Physical (heating, irradiation, pressure, ultrasound), chemical (chemical cross-linking, acids, alkali) or enzymatic treatments can improve functional properties of films (Conca 2002; Letendre et al. 2002a; Rhim et al. 1999; Yildirim and Hettiarachchy 1998; Stuchell and Krochta 1994). Cross-linking between proteins and polysaccharides, as well as other treatments used to improve interactions between these two types of macromolecules, may be used to improve functional properties and resistance of protein-based films (Letendre et al. 2002a; Le Tien et al. 2000; Ressouany et al. 2000; Mezgheni et al. 1998b). For example, pectin may form cross-links with proteins under certain conditions (Thakur et al. 1997). Autoclaving enhances protein–polysaccharide, pectin–protein and agar–protein interactions, resulting in a three-dimensional network with improved mechanical properties (Letendre et al. 2002a, b). According to Thakur et al. (1997), high temperatures generate a less ordered structure, making more functional groups available. According to Letendre et al. (2002b) when calcium, whey proteins and polysaccharide are mixed together, a fully disordered and dissociated polymer chains could interact more favorably together. The addition of glycerol in a film formulation reduces interactions between the protein and water (Letendre et al. 2002b). It is possible that hydrophilic structure of glycerol might be involved in associations with hydrophilic sites of proteins (H-bonding), thus limiting protein hydration i.e., protein–water interaction. The presence of plasticizers can enhance formation of protein cross-links. It was observed by Brault et al. (1997) and Mezgheni et al. (1998a) that the presence of glycerol, propylene glycol or triethylene glycol in caseinate based – film formulations enhanced formation of cross-links during irradiation treatment. Environmental and processing conditions affect composition and structure of polymers, which directly affect functionality of the resulting films. For example, use of homogenization and microfluidization for particle size reduction of whey protein based film formulations resulted in an increase in the interfacial area, providing a stronger film with reduced water vapor permeability (Perez-Gago and Krochta 2001).

13.4 Barrier Properties

Barrier properties of edible films include water vapor permeability, gas permeability (O_2 and CO_2), volatile permeability and solute permeability. Both O_2 and CO_2 permeability is important when respiration or oxidation reactions could affect quality of the food (e.g., fresh or pre-cut fruits and vegetables). Water vapor permeability is an important factor to consider when crispness of a food needs to be maintained during storage. The volatile and solute permeability is an important property to control when diffusion of a compound is to be limited. Barrier properties of edible films prepared from polar polymers (e.g., polysaccharides) are sensitive to humidity. However, protein-based films traditionally have the worst barrier properties (Cuq et al. 1995). Since lipid is an excellent moisture barrier, incorporation of lipid into a film can reduce water vapor permeability. The structure of the polymer chains and the lipid distribution within the matrix play a significant role in permeability of films. As described previously, a less ordered structure can permit better interactions between the polymers (Letendre et al. 2002b). The increase of cohesion between protein polypeptide chains and a uniform dispersion of lipid in the polymer matrice could also be effective for improvement of the barrier properties of films (Sabato et al. 2001; Perez-Gago and Krochta 2000). Molecular crystallinity also significantly affects permeability and the solubility of films. Lipids can be used in numerous crystalline forms. In general, higher the degree of crystallinity, the lower the permeability of a film (Perez-Gago and Krochta 2000; McHugh and Krochta 1994).

13.4.1 Gas Barrier Properties

Many foods require specific atmospheric conditions to sustain their freshness and overall quality during storage. Packaging a food product under a specific mixture of gases, known as modified atmosphere packaging (MAP), can help maintain quality and safety of such products. To ensure a constant gas composition inside the package, packaging material needs to exhibit certain gas barrier specificity. In most MAP applications, the gas mixture inside the package consists of carbon dioxide, oxygen or nitrogen, or some combination of these gases. Though scientific literature provides a vast amount of information on barrier properties of biomaterials, direct comparisons between different biomaterials are complicated and sometimes not possible, due to use of different types of equipment and dissimilar analytical measurements within various studies. A brief review of some of these literature reports is provided below.

Films based on polysaccharides, such as alginate, cellulose ethers, chitosan, carrageenan or pectin, generally exhibit good gas barrier properties (Baldwin et al. 1995; Ben and Kurth 1995). Addition of lipids or starch to a film formulation can improve both oxygen and oil barrier properties (Kroger and Igoe 1971; Morgan 1971; Sacharow 1972; Wu et al. 2002). The gas permeability properties of such films result in desirable modified atmospheres, thereby increasing the shelf life of the product without creating anaerobic conditions (Baldwin et al. 1995). Addition of a fatty acid through an emulsion with proteins can increase the O_2 and CO_2 permeability of the resulting film, while addition of acetylated monoglycerides actually provides the reverse effect (Wu et al. 2002).

The conventional approach to producing high-barrier edible films or coatings to maintain food in a protective atmosphere involves use of multilayers of different films to collectively obtain the required properties. An example of this multilayer approach might be a biobased laminate, consisting of an outer layer of plasticized chitosan film combined with another layer of polyhydroxyalkanoate (PHA) or alginate, to obtain an appropriate gas barrier.

Bilayer coatings, has been shown to reduce gas exchange, and result in higher internal carbon dioxide and lower oxygen concentrations in cut apple pieces (Wong et al. 1994). Unfortunately, little or no data exists showing the barrier property effects of bilayer coatings on whole fruits or vegetables.

13.4.2   Water Vapor Permeability

The predominantly hydrophophilic character of natural polymers, including many proteins and polysaccharides, result in poor water barrier characteristics (Kester and Fennema 1986; Peyron 1991; Gennadios et al. 1997; McHugh 2000). According to Henrique et al. (2007), water vapor permeability (WVP) can be directly related to the quantity of –OH group on the molecule. Also, environmental conditions can significantly affect the WVP. In general, a high relative humidity (90% RH) and a low (–30°C) storage temperature improve WVP. For example, increasing the relative humidity gradient at a constant temperature increased transfer of moisture through films based on hydroxypropyl methylcellulose, stearic and palmitic acids (Kamper and Fennema 1984), and polyethylene which are more permeable to water vapor at –30°C than at 35°C (Labuza and Contreras-Medellin 1981). Polysaccharide films are used in Japan for meat products, ham and poultry packaging prior to smoking and steam processing. The film is dissolved during this process, and the coated meat exhibits improved yield, structure and texture, and reduced moisture loss (Labell 1991; Stollman et al. 1994). Perez-Gago and Krochta (2000) have investigated the effect of drying temperature on WVP of whey protein isolate based-films, and observed that WVP decreased significantly as drying temperature increased.

Lipids have a low affinity for water, and can significantly reduce the water vapor permeability (WVP) of films. For example, it was found that WVP of pure casein-ate films could be reduced by over 70% through incorporation of lipid materials (beeswax) (Avena-Bustillos and Krochta 1993). However, the polarity of lipids used is an important factor to consider. Solid lipids, such as palm oil, stearic acid, beeswax or paraffin, yielded much smaller WVP values than cellulose ether based films containing liquid lipids, such as oleic acid (Wu et al. 2002). Fatty acids embedded within the cellulose ether matrix may function to enhance adhesion of the wax layer to the underlying film, thus, resulting in a bilayer film with a lower WVP (Wu et al. 2002). For films cast from aqueous, lipid emulsion solutions, the process is complex, and incorporation of lipids requires addition of emulsifying agents or surfactants to improve emulsion stability. Perez-Gago and Krochta (1999) showed that pH may also have an influence on film WVP. In protein-stabilized emulsions, as the net charge approaches zero at the protein isoelectric point (pI), electrostatic repulsion between emulsion droplets becomes weak, and probably a lowered lipid mobility due to protein–protein aggregation occurred and reduced interconnectivity among lipid droplets, resulting in a higher WVP. At pH values above and below the protein pI, droplets have a net charge and electrostatic repul-sive forces are present between droplets. The best water barrier properties were observed in protein films at neutral pH values (Perez-Gago and Krochta 1999; McHugh 2000). Decreased mean particle diameters of the emulsion or of the lipid particles resulted in linear decreases in WVP values (Perez-Gago and Krochta 2001; McHugh and Krochta 1994) although it is interesting to note that protein–lipid films are often difficult to obtain. For example, bilayer film formation requires the use of solvents and/or high temperatures, making production more costly. Furthermore, separation of the layers may occur with time.

Laminant films can be formed by applying another material (e.g., lipids) as a laminant over a polysaccharide-based film. Bilayer films, such as these, tend to delaminate over a period of time, developing pinholes or cracks, exhibiting poor strength and non-uniform surface and cohesion characteristics (Sherwin et al. 1998). Bilayers are often brittle, and not practical for use in many applications (Fairley et al. 1997). In addition, bilayers require multiple drying steps, whereas emulsion-based films require only one dehydration step (Debeaufort and Voilley 1995). Application of the lipid layer often requires use of solvents or high temperatures, making production more costly and less safe. However, bilayer films are desirable, in that they generally exhibit better barrier properties than emulsion films (Fairley et al. 1997; Sherwin et al. 1998). Chitosan/pectin laminated films can yield transparent films that exhibit increased resistance to water dissolution (Hoagland and Parris 1996). Yan et al. (2001) obtained similar results with chitosan/alginate laminated films. Additionally, these authors demonstrated that films prepared from low molecular weight chitosan were twice as thin and transparent, as well as 55% less permeable to water vapor, than those prepared with standard chitosan. The improved mechanical properties observed for these chitosan laminate films could be due to electrostatic interaction between carboxylate groups of pectin or alginate and the amino groups of chitosan (Dutkiewicz and Tuora 1992). Better WVP and mechanical properties were obtained after lamination of methylcellulose/corn zein–fatty acid films (Park et al. 1994). Laminated chitosan–cellulose and polycaprolactone films can be used in modified atmosphere packaging of fresh produce (Makino and Hirata 1997). Laminates with effective water barriers were also developed (Park et al. 1994) by coating a hot solution of corn zein with different types of fatty acids onto a methylcellulose-based film. Resulting films were water insoluble, and the WVP was reported to decrease as both length and concentration of the fatty acid increased. It is clear that homogeneous distribution of the hydrophobic material plays a key role in decreasing WVP in both emulsified and laminated film applications.

Increased cohesion between protein polypeptide chains is thought to improve moisture barrier properties of the protein-based films. For instance, cross-linking of proteins by means of chemical, enzymatic (transglutaminase) or physical treatments (heating, irradiation) was reported to improve water vapor barrier characteristics, as well as the mechanical properties and resistance to proteolysis, of protein films (Sabato et al. 2001; Ouattara et al. 2002; Ressouany et al. 1998, 2000; Brault et al. 1997). When cross-linked whey proteins are entrapped within a cellulose matrix, insoluble films with good mechanical properties, high resistance to proteolytic enzymes and decreased WVP are generated (Le Tien et al. 2000).

Food quality deterioration due to physicochemical changes or chemical reactions is often caused by mass transfer or between the food and its surrounding medium, or within the product itself. Compounds whose migrations need to be controlled are water, oxygen and flavor, with the main concern being water. Application of a coating containing hydrophobic substances (oils, fats, waxes, lacs, varnishes, resins, emulsifiers, essential oils, etc.) can be used to isolate or maintain separation of components differing in water activity within a composite food. Oils and fats are widely used in confectionery, cereal and dessert industries to prevent moisture absorption. Waxes are often used to coat fruits and vegetables in order to prevent desiccation during transport and storage. According to the literature, paraffin, candelilla, cellulose and beeswax are the most effective compounds for reduction of water vapor permeability ($0.02–0.06 \times 10^{-11}$ g m^{-1} s^{-1} Pa^{-1}), as compared to carnauba wax, glycerol, acetyl acylglycerols and myristic acid, for which a WVP of $1–148 \times 10^{-11}$ g m^{-1} s^{-1} Pa^{-1} was observed (Lovegren and Feuge 1954; Greener and Fennema 1994; Koelsch and Labuza 1992; Hugon 1998). Emulsifiers can also be added to improve adherence, as a barrier to gas and moisture, or as an emulsifier (e.g., to permit good homogenization of essential oils) (Morillon et al. 2002). According to Karbowiak et al. (2007), a thermal process may be used to favor formation of a bilayer structure in emulsion-based edible films, leading to better water barrier efficiency

Bilayer coatings are the edible coatings or films of the future. The use of two polymer types allows beneficial properties of the two materials to be combined to create a superior film. Composite films of the future may consist of hydrophobic particles distributed within a hydrophilic matrix. Such a configuration could yield a water-soluble coating with good water vapor barrier properties (Baldwin 1991). Bilayer coatings, which already have been used to a limited extent, combine the water barrier properties of lipid coatings with the greaseless feel and good gas permeability characteristics of polysaccharide coatings.

Improvement of water vapor barrier properties represents a major challenge for the manufacturer of film and coating materials intended for food applications. When comparing water vapor transmittance of various natural polymers, it becomes clear that it is difficult to produce an edible film or coating with water vapor permeability rates compared to those provided by conventional plastics. If a high water vapor barrier material is required, there are very few biomaterials that can be used. Consequently, research efforts are currently focused on this problem with the realization that future biomaterials must be able to mimic water vapor barrier characteristics of conventional materials (Butler et al. 1996).

13.5 Solubility

As described previously, most biopolymers in their native states are sensitive to humidity, and are soluble in water (Vachon et al. 2000). Lipid components of edible origin can be incorporated into films and coatings to alter their susceptibility to water

(Greener and Fennema 1989; Martin-Polo et al. 1992; Debeaufort and Voilley 1995; Lacroix et al. 2001). However, composite films of proteins and lipids can decrease their water-solubility. As for good water vapor properties, a bilayer film would permit to obtain film with better water resistance. However, the process needs to be improved to obtain better stability. Good homogenization will be able to avoid separation of the layers.

Formation of cross-links between polymers can also reduce significantly solubility of edible film constituents. Vachon et al. (2000) was able to reduce the overall film solubility by more than 75% by promoting cross-linking through irradiation treatment. Incorporation of polycaprolactone into alginate-based films following immersion in calcium salt chloride (2 or 20%) decreased water-solubility of the film, exhibiting a recovery yield of 86% after water immersion for 24 h (Salmieri and Lacroix 2006). It is also expected that polymer–polymer interactions via hydroxyl and carbonyl (H-bonds or covalent reaction) moieties can enhance film insolubility (Salmieri and Lacroix 2006; Wang et al. 2001). When applied on beef, bologna or ham, such films were resistant to high humidity conditions, and were able to allow a controlled release of the active compounds (e.g., essential oils (Tables 13.1–13.3) from the film to the product during storage. Results of these studies have shown that the release rate of compounds from the film is related to calcium treatment and the food and essential oil composition (Oussalah et al. 2006a, b).

Other modifications can also be done to reduce solubility of biopolymer films. A functionalization agent may be incorporated onto a film or coating polymer to increase hydrophobicity of the polymer, and increase the insolubility and moisture barrier properties of the film. The functionalization agent may also increase mechanical properties of the film or coating. A functionalization agent can be defined as a substance that is covalently linked to the polymer matrice, with or without the aid of a coupling agent. Examples of functionalization agents include glyceraldehyde, acyl chlorides, fatty acids and anhydrides. Covalent modification of polysaccharides and proteins is generally accomplished via ester and ether linkages (Mezgheni et al. 1998a; Ressouany et al. 2000; Lacroix et al. 2001; Mulbacher et al. 2001; Brode 1991; Brode et al. 1991; Rutenberg and Solarek 1984). Use of cellulose xanthate as a matrix for entrapment of cross-linked whey protein can produce insoluble films with good mechanical properties, good stability and increased resistance to enzymatic attack (Le Tien et al. 2000). The potential application of these films used as packaging and wrappings can be of interest for various materials, and is probably compatible with several types of foods.

Le Tien et al. (2003b) and Lacroix et al. (2001) demonstrated that viability of lactic acid bacteria and the functionality of active compounds can be preserved when encapsulated in functionalized alginate (succinylated or N-palmitoylaminoethylated) or chitosan (succinylated), due to the film's insolubility and resistance to gastric fluid (Fig. 13.1). N-palmitoylaminoethyl alginate and N-acylated chitosan were also used to immobilize bioactive compounds (e.g., enzymes) to protect their bioactivity and to facilitate a controlled release of drugs over period of 30–90 h, depending on degree of functionalization (Le Tien et al. 2003a, 2004).

Table 13.1 Availability of total phenolic compounds or aldehydes in films containing essential oils during storage of meat at 4°C

Films[b]	Percentage of active compounds availability (%)[a]				
	Day 1	Day 2	Day 3	Day 4	Day 5
O + 2% CaCl$_2$	100.00 a C	71.91 ± 2.13 b B	70.85 ± 1.82 c B	61.21 ± 4.64 c A	60.62 ± 4.65 c A
S + 2% CaCl$_2$	a C	53.75 ± 0.28 a B	53.34 ± 2.18 b B	39.75 ± 2.39 b A	39.13 ± 4.44 b A
C + 2% CaCl$_2$	a D	60.92 ± 6.74 a C	30.04 ± 1.19 a B	23.59 ± 3.37 a A	22.78 ± 0.28 a A
O + 20% CaCl$_2$	100.00 a C	50.16 ± 4.40 b B	47.07 ± 1.74 b B	37.39 ± 2.71 b A	34.86 ± 1.89 b A
S + 20% CaCl$_2$	a C	63.60 ± 1.20 c B	59.41 ± 2.48 c B	39.51 ± 2.38 b A	37.84 ± 1.93 c A
C + 20% CaCl$_2$	A E	42.74 ± 3.57 a D	24.11 ± 2.06 a C	16.86 ± 0.85 a B	9.94 ± 0.35 a A

[a]Means for each treatment and in the same column bearing the same *lower case letters* are not significantly different ($p > 0.05$). Means in the same row bearing the same *upper case letters* are not significantly different ($p > 0.05$)

[b]O films containing Spanish oregano essential oil; S films containing winter savory essential oil; C films containing Chinese cinnamon essential oil. Alginate based edible films were treated in 2 or 20% (w/v) of CaCl$_2$ solution

Table 13.2 Availability of main active compounds (total phenolic compounds or aldehydes) in films containing essential oils during storage of bologna at 4°C

Films[a]	Percentage of active compounds availability (%)[b]				
	Day 1	Day 2	Day 3	Day 4	Day 5
O + 2% CaCl$_2$	100.00 a D	66.75 ± 3.96 c C	49.21 ± 2.06 b B	20.24 ± 1.32 c A	18.84 ± 0.68 c A
S + 2% CaCl$_2$	a D	62.55 ± 1.23 b C	47.36 ± 3.27 b B	15.37 ± 1.08 b A	17.08 ± 0.22 b A
C + 2% CaCl$_2$	a E	26.63 ± 1.30 a D	14.06 ± 1.47 a C	9.00 ± 0.69 a B	7.11 ± 0.33 a A
O + 20% CaCl$_2$	100.00 a E	27.60 ± 0.94 b D	21.97 ± 1.86 c C	18.28 ± 0.38 c B	12.58 ± 0.97 c A
S + 20% CaCl$_2$	a E	46.72 ± 1.78 c D	24.53 ± 1.07 b C	14.57 ± 5.73 b B	9.59 ± 0.43 b A
C + 20% CaCl$_2$	a D	18.95 ± 0.36 a C	14.26 ± 1.50 a B	5.04 ± 0.33 a A	4.14 ± 0.45 a A

[a] Alginate based edible films were treated with 2 or 20% (w/v) CaCl$_2$ solution. O films containing Spanish oregano essential oil; S films containing winter savory essential oil; C films containing Chinese cinnamon essential oil

[b] Within each treatment and each column, means with the same *lowercase letters* are not significantly different ($p > 0.05$). Within each row, means with the same *upper case letters* are not significantly different ($p > 0.05$)

Table 13.3 Availability of main active compounds (total phenolic compounds or aldehydes) in films containing essential oils during storage of ham at 4°C

Films[a]	Percentage of active compounds availability (%)[b]				
	Day 1	Day 2	Day 3	Day 4	Day 5
O + 2% CaCl$_2$	100.00 a C	83.04 ± 8.62 b B	37.08 ± 3.19 c A	34.25 ± 3.01 c A	33.65 ± 0.54 c A
S + 2% CaCl$_2$	a E	80.10 ± 3.01 b D	27.61 ± 3.51 b C	21.70 ± 2.15 b B	11.09 ± 0.29 b A
C + 2% CaCl$_2$	a D	37.75 ± 3.92 a C	15.56 ± 0.66 a B	15.41 ± 2.08 a B	4.35 ± 0.28 a A
O + 20% CaCl$_2$	100.00 a D	49.77 ± 6.02 c C	47.79 ± 5.47 c C	32.46 ± 3.38 c B	21.65 ± 1.59 c A
S + 20% CaCl$_2$	a D	32.51 ± 3.00 b C	32.03 ± 3.51 b C	22.79 ± 0.90 b B	14.10 ± 1.61 b A
C + 20% CaCl$_2$	a D	20.47 ± 1.27 a C	19.14 ± 2.44 a C	12.04 ± 0.86 a B	3.64 ± 0.06 a A

[a] Alginate based edible films were treated with 2 or 20% (w/v) CaCl$_2$ solution. O films containing Spanish oregano essential oil; S films containing winter savory essential oil; C films containing Chinese cinnamon essential oil

[b] Within each treatment and each column, means with the same *lowercase letters* are not significantly different ($p > 0.05$). Within each row, means with the same *upper case letters* are not significantly different ($p > 0.05$)

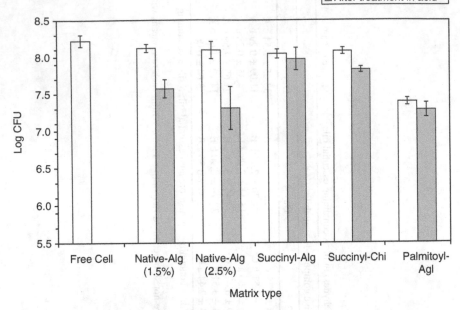

Fig. 13.1 Viability (%) of *L. rhamnosus* free and immobilized on different functionalized matrices and in simulated gastric fluid (pH 1.5) for 30 min (*FC* free cell; *NA* native alginate; *SA* succinylated alginate; *SC* succinylated chitosan and *PA* N-palmitoylaminoethyl alginate)

13.6 Thermal Properties

In the packaging industry, heat sealing is widely used to join polymer films (Mueller et al. 1998). In the heat sealing process, two films are presented together between heated plates or dies. As heat is applied, the surface of the crystalline polymer melts. Application of pressure results in interfacial interactions across joint surfaces to seal adjoining pieces of film. During cooling, a heat-sealed joint forms due to recrystallization of the polymer (Kim and Ustunol 2001). Seal strength is an indicator of seal quality. According to Lee 1994, modification of functional groups such as hydroxyl (OH), aldehyde (CHO), and carboxylic acid (COOH) moieties present in polyimide-based films, are responsible for the adhesion strength differences of various films. An increase in interactions between nitrogen- and oxygen-containing functional groups in polyethylene films is responsible for enhancement of seal strength (Possart and Deckhoff 1999). According to Kim and Ustunol (2001), hydrogen and covalent bonds involving C–O–H and N–C, may be the main forces responsible for sealed joint formation in films.

13.7 Potential Perspectives of Microemulsions and Bionanocomposites

Microemulsions have found numerous applications over a wide range of areas, including food systems (Flanagan and Singh 2006). However, application of microemulsions in foods systems is limited by the types of available surfactants, which are used to facilitate microemulsion formation. Microemulsions have attracted particular interest due to their ability to increase efficiency of antioxidants. Water-in-oil (w/o) microemulsions, formed with soybean oil and monoglycerides as the surfactant, were stable against oxidation for 70 days when 5% (w/w) ascorbic was incorporated into the water phase, When microemulsions were prepared without ascorbic acid, oxidative stability was reduced to less than 20 days (Moberger et al. 1987). A synergistic effect between ascorbyl palmitate and lecithin solubilized in the water phase, and a-tocopherol solubilized in the oil phase of the w/o microemulsion, were strongly synergistic in delaying peroxidation of lipid in fish oil (Hamilton et al. 1998).

Application of nanocomposites also promises to extend the uses of edible and biodegradable films (Sinha and Bousmina 2005). This new generation of composite films exhibits significant improvements in modulus, dimensional stability and solvent or gas resistance with respect to the pristine polymer. Nanocomposites also offer extra benefits like low density, transparency, good flow, better surface properties and recyclability. It is important to note that these improvements are obtained at very low filler contents (generally lower than 5% w/v). This enhancement of properties resides in fundamental length scales dominating the morphology of these materials, in which inorganic particles have at least one dimension in the nanometer (from 1 to 100 nm) range. It means that a uniform dispersion of these particles can lead to an ultra-large interfacial area between constituents. The very large organic/inorganic interface alters molecular mobility, relaxation behavior and consequent thermal and mechanical properties of the resulting nanocomposite material. Various inorganic nanoparticles have been recognized as possible additives to enhance polymer performance. Some examples of these particles are synthetic polymer nanofibers, cellulose nanowhiskers and carbon nanotubes. However, until now, only layered inorganic solids like clay have attracted some attention (Sorrentino et al. 2007). Only a few studies have suggested the possibility of incorporating nanoparticles to improve physical properties of food packaging. Mangiacapra et al. (2005) demonstrated the possibility of lowering diffusion of oxygen by adding clay montmorillonite to pectin to create nanocomposite materials. Nanocomposite films prepared from gelatin and montmorillonite exhibit significantly improved mechanical properties (Zheng et al. 2002). Coverage of nanocomposite films based on chitosan has demonstrated the increase in their stability during storage (Darder et al. 2003). These nanoparticles also can be used to stabilize food additives and efficiently control their diffusion into food. The controlled release of these compounds is important for long term storage of foods or for imparting specific desirable characteristics, such as flavor, to a food system (Sorrentino et al. 2007). However, more studies should be done to assure the innocuity of nanomaterials with regard to human health. Development of this technology will likely further facilitate development of stable, functional and intelligent packaging systems. Also, use of nanomaterials will most likely enhance ability to produce more efficient active packaging systems.

13.8 Conclusions

Edible films and coatings have great potential for use in a wide variety of applications. They can be used to extend shelf life of fruits, vegetables, seafood, meat and confectionery products by preventing dehydration, oxidative rancidity, surface browning, oil diffusion and for their ability to modify internal atmosphere in fruits and vegetables. Moreover, when applied on fruits and vegetables, polysaccharide-based coatings can improve physicochemical, nutritional and sensorial properties of food products. Several studies have demonstrated that incorporation of active compounds, such as antimicrobials, antioxidants or bacteriocins, within edible films and coatings can improve microbial safety, shelf life and quality of meat, fish, fruit and vegetable products during storage. Development of new technologies (functionalization, cross-linking, nanomaterials, microemulsions, etc.) to further improve film properties (control release, higher bioactivities, bioactivily protection, resistance to water etc.) of active packaging, coatings or encapsuled materials represent a need for the future. The same may be needed as well for foods, nutraceutical, pharmaceutical and other applications.

References

Avena-Bustillos RJ, Krochta, JM (1993). Water vapor permeability of caseinate-based edible films as affected by pH, calcium crosslink and lipid content. J. Food Sci. 58, 904–907

Baldwin EA (1991). Edible coatings for fresh fruits and vegetables: past, present, and future. In: JM Krochta, EA Baldwin, M Nisperos-Carriedo (Eds.), Edible coatings and films to improve food quality. Technomic Publishing Company, Lancaster, PA. pp. 25–64

Baldwin EA, Nisperos MO, Baker RA (1995). Use of edible coating to preserve quality of lightly and slightly processed products. Crit. Rev. Food Sci. Nutr. 35, 509–524

Ben A, Kurth LB (1995). Edible film coating for meat cuts and primal. Meat 95, The Australian Meat Industry Research Conference, CSIRO, September 10–12

Brault D, D'Aprano G, Lacroix M (1997). Formation of free-standing sterilized edible films from irradiated caseinates. J. Agric. Food Chem. 45, 2964–2969

Brode GL (1991). Polysaccharides: natural for cosmetic and pharmaceuticals. In: GG Gebelein, TC Cheng, VC Yang (Eds.), Cosmetic and pharmaceutical applications of polymers. Plenum, New York, NY. pp. 105–115

Brode GL, Goddard ED, Harris WC, Sale GA (1991). Cationic polysaccharides for cosmetics and therapeutics. In: CG Gebelein, TC Cheng, VC Yang (Eds.), Cosmetic and pharmaceutical applications of polymers. Plenum, New York, NY. pp. 117–128

Butler BL, Vergano PJ, Testin RF, Bunn JM, Wiles JL (1996). Mechanical and barrier properties of edible chitosan films as affected by composition and storage. J. Food Sci. 61, 953–956

Cieśla K, Salmieri S, Lacroix M (2006a). γ-Irradiation influence on the structure and properties of calcium caseinate-whey protein isolate based films. Part 1. Radiation effect on the structure of proteins gels and films. J. Agric. Food Chem. 54, 6374–6384

Cieśla K, Salmieri S, Lacroix M (2006b). γ-Irradiation influence on the structure and properties of calcium caseinate-whey protein isolate based-films. Part 2. Influence of polysaccharide addition and radiation treatment on the structure and functional properties of the films. J. Agric. Food Chem. 54, 8899–8908

Conca KR (2002). Protein-based films and coating for military packaging applications. In: A Gennadios (Ed.), Protein-based films and coatings. CRC, Boca Raton, FL. pp. 551–577

Cuq B, Gontard N, Guilbert S (1995). Edible films and coatings as active layers. In: ML Rooney (Ed.), Active food packaging. Blackie Academic and Professional, Glasgow. pp. 111–142

Cuq B, Gontard N, Guilbert S (1998). Proteins as agricultural polymers for packaging production. Cereal Chem. 75, 1–9

Darder M, Colilla M, Ruiz-Hitzky E (2003). Biopolymer clay nanocomposites based on chitosan intercalated in montmorillonite. Chem. Mater. 15, 3774–3780

Debeaufort F, Voilley A (1995). Effect of surfactants and drying rate on barrier properties of emulsified edible films. Int. J. Food Sci. Technol. 30, 183–190

Donhowe IG, Fennema O (1994). Edible films and coatings: characteristics, formation, definitions and testing methods. In: JM Krochta, EA Baldwin, MO Nisperos-Carriedo (Eds.), Edible coatings and films to improve food quality. Technomic Publishing Company, Lancaster, PA. pp. 1–24

Dutkiewicz J, Tuora M (1992). New forms of chitosans polyelectrolyte complexes. In: CJ Brine, PA Standford, JP Zikakis (Eds.), Advances in Chitin and Chitosan. Elsevier, London. pp. 496–505

Fairley P, Krochta JM, German JB (1997). Interfacial interactions in edible emulsion films from whey protein isolate. Food Hydrocolloids, 11, 245–252

Flanagan J, Singh H (2006). Microemulsions: a potential delivery system for bioactives in food. Crit. Rev. Food Sci. Nutr. 46, 221–237

Floros JD, Dock LL, Han JH (1997). Active packaging technologies and applications. Food Cosmet. Drug Packaging. 20, 10–16

Gennadios A, Hanna MA, Kurth B (1997). Application of edible coatings on meats, poultry and seafoods: a review. Lebensm. Wiss. Technol. 30, 337–350

Greener IK, Fennema, O. (1989). Evaluation of edible, bilayer films for use as moisture barriers for food. J. Food Sci. 54, 1400–1406

Greener IK, Fennema O (1994). Edible films and coatings: characteristics, formation, definitions and testing methods. In: JM Krochta, EA Baldwin, M Nisperos-Carriedo (Eds.), Edible coatings and films to improve food quality. Technomic Publishing Company, Lancaster, PA. pp. 1–24

Gueguen J, Viroben G, Noireaux P, Subirade M (1998). Influence of plasticizers on the properties of films from pea proteins. Ind. Crops Prod. 7, 149–157

Guilbert S, Gontard N, Gorris LGM (1996). Prolongation of the shelf-life of perishable food products using biodegradable films and coatings. Lebensm. Wiss. Technol. 29, 10–17

Hamilton RJ, Kalu C, McNeill GP, Padley FB, Pierce JH (1998) Effects of tocopherols, ascorbyl palmitate, and lecithin on autoxidation of fish oil. J. Am. Oil Chem. Soc. 75(7), 813–822

Henrique CM, Teófilo RF, Sabino L, Ferreira MMC, Cereda MP (2007). Classification of cassava starch films by physicochemical properties and water vapor permeability quantification by FT-IR and PLS. J. Food Sci. 72(4), 184– 189

Hoagland PD, Parris N (1996). Chitosan/pectin laminated films. J. Agric. Food Chem. 44, 1915–1919

Hugon F (1998). Étude et maîtrise des transferts d'eau dans des céréales enrobées. D.R.T. ENSBANA, Université de Bourgogne, Dijon, France

Imeson A, Ledward DA, Mitchell JR (1977). On the nature of the interaction between some anionic polysaccharide acid proteins. J. Sci. Food Agric. 28, 661

Kamper SL, Fennema, O (1984). Water vapor permeability of an edible, fatty acid, bilayer film. J. Food Sci. 49, 1482–1484

Karbowiak T, Debeaufort F, Voilley A (2007) Influence of thermal process on structure and functional properties of emulsion based edible films. Food Hydrocolloids 21(2), 879–888

Kester JJ, Fennema OR (1986). Edible films and coatings: a review. Food Technol. 12, 47–59.

Kim SJ, Ustunol Z (2001). Thermal properties, heat sealability and seal attributes of whey protein isolate/lipid emulsion edible films. J. Food Sci. 66(7), 985–990

Koelsch CM, Labuza TP (1992). Functional, physical and morphological properties of methyl cellulose and fatty acid-based edible barriers. Lebensm. Wiss. Technol. 25, 404–411

Krochta JM (1997). Edible protein films and coatings. In: S Damodaran, A Paraf (Eds.), Food proteins and their applications in foods. Marcel Dekker, New York, NY. pp. 529–549

Krochta JM, De Mulder-Johnston C (1997). Edible and bi-odegradable polymer films. Challenges and opportunities. Food Technol. 51, 61–74

Kroger M, Igoe RS (1971). Edible containers. Food Prod. Dev. 5, 74, 76, 78–79, 82

Labell F (1991). Edible packaging. Food Process. Eng. 52, 24

Labuza TP, Contreras-Medellin R (1981). Prediction of moisture protection requirements for foods. Cereal Foods World 26(7) 335–343

Lacroix M, Mateescu MA, Le Tien C, Patterson G (2001). Biocompatible composition as carriers or excipients for pharmaceutical formulations and for food protection. PCT/CA00/01386

Lee KW (1994). Modification of polyimide surface morphology: relationship between modification depth and adhesion strength. J. Adhes. Sci. Technol. 8(10), 1077–1092

Letendre M, D'Aprano G, Lacroix M, Salmieri S, St-Gelais D (2002a). Physicochemical properties and bacterial resistance of biodegradable milk protein films containing agar and pectin. J. Agric. Food Chem. 50, 6017–6022

Letendre M, D'Aprano G, Delmas-Patterson G, Lacroix M (2002b). Isothermal calorimetry study of calcium caseinate and whey protein isolate edible films cross-linked by heating and γ- irradiation. J. Agric. Food Chem. 50, 6053–6057

Le Tien C, Letendre M, Ispas-Szabo P, Mateescu MA, Delmas-Patterson G, Yu HL, Lacroix M (2000). Development of biodegradable films from whey proteins by cross-linking and entrapment in cellulose. J. Agric. Food Chem. 48, 5566–5575

Le Tien C, Lacroix M, Ispas-Szabo P, Mateescu MA (2003a). N-acylated chitosan: hydrophobic matrices for controlled drug release. J. Control. Release 93, 1–13

Le Tien C, Lacroix M, Ispas-Szabo P, Mateescu MA (2003b). Modified alginate and chitosan for lactic acid bacteria immobilization. J. Control. Release 93, 1–13

Le Tien C, Millette M, Lacroix M, Mateescu MA (2004). Modified alginate matrice for the immobilization of bioactive agents. Biotechnol. Appl. Biochem. 39, 189–198

Lim LT, Mine Y, Britt IJ, Tung MA (2002). Formation and properties of egg white protein films and coatings. In: A Gennadios (Ed.), Proteins-based films and coatings. CRC, Boca Raton, FL. pp. 233–252

Lovegren NV, Feuge RO (1954). Permeability of acetostearin products to water vapor. J. Agric. Food Chem. 2, 558–563

Makino Y, Hirata T (1997). Modified atmosphere packaging of fresh produce with a biodegradable laminate of chitosan-cellulose and polycaprolactone. Postharvest Biol. Technol. 10, 247–254

Mangiacapra P, Gorrasi G, Sorrentino A,Vittoria V (2005). Biodegradable nanocomposites obtained by ball milling of pectin and montmorillonites. Carbohydr. Polym. 64(4), 516–523

Martin-Polo M, Mauguin C, Voilley A (1992). Hydrophobic films and their efficiency against moisture transfer. 1-Influence of the film preparation technique. J. Agric. Food Chem. 40, 407–412

McHugh TH (2000). Protein-lipid interactions in edible films and coatings. Nahrung 44, 148–151

McHugh TH, Krochta J (1994). Permeability properties of edible films. In: J Krochta, EA Baldwin, M Nisperos-Carriedo (Eds.), Edible coatings and films to improve food quality. Technomic Publishing Company, Lancaster, PA. pp. 139–188

Mezgheni E, D'Aprano G, Lacroix M (1998a). Formation of sterilized edible films based on caseinates: effects of calcium and plasticizers. J. Agric. Food Chem. 46, 318–324

Mezgheni E, Vachon C, Lacroix M (1998b). Biodegradability behaviour of cross-linked calcium caseinates films. Biotechnol. Prog. 14, 534–536

Moberger L, Larsson K, Buchheim W, Timmen H (1987). A study of fat oxidation in a microemulsion system. J. Dispers. Sci. Technol. 8, 207–215

Morgan BH (1971). Edible packaging update. Food Prod. Dev. 5, 75–77, 108

Morillon V, Debeaufort F, Blond G, Capelle M,Voiley A (2002). Factors affecting the moisture permeability of lipid-based edible films: a review. Crit. Rev. Food Sci. Nutr. 42(1), 67–89

Mueller C, Cappacio G, Hiltner A, Baer E (1998). Heat sealing of LLDPE: relationships to melting and interdiffusion. J. Appl. Polym. Sci. 70(11), 2021–2030

Mulbacher J, Ispas-Szabo P, Lenaerts V, Mateescu MA (2001). Cross-linked high amylose starch derivatives as matrices for controlled release of high drug loadings. J. Control. Release 76, 51–58

Ouattara B, Giroux M, Smoragiewicz W, Saucier L, Lacroix M (2002). Combined effect of gamma irradiation, ascorbic acid and edible film on the improvement of microbial and biochemical characteristics of ground beef. J. Food Prot. 6, 981–987

Oussalah M, Caillet S, Salmieri S, Saucier L, Lacroix M (2006a). Antimicrobial effects of alginate-based film containing essential oils for the preservation of whole beef muscle. J. Food Prot. 69(10), 2364–2369

Oussalah M, Caillet S, Salmieri S, Saucier L, Lacroix M (2006b). Antimicrobial effects of alginate-based films containing essential oils on Listeria monocytogenes and Salmonella Typhimurium present in bologna and ham. J. Food Prot. 70(4), 901–908

Park JW, Testin RF, Park HJ, Vergano PJ, Weller CL (1994). Fatty acid concentration effect on tensile strength, elongation, and water vapor permeability of laminated edible film. J. Food Sci. 59, 916–919

Perez-Gago MB, Krochta JM (1999). Water vapor permeability of whey protein emulsion films as affected by pH. J. Food Sci. 64(4), 695–698

Perez-Gago M, Krochta J (2000). Drying temperature effect on water vapor permeability and mechanical properties of whey protein-lipid emulsion films. J. Agric. Food Chem. 48, 2687–2692

Perez-Gago M, Krochta J (2001). Lipid particle size effect on water vapor permeability and mechanical properties of whey protein/beeswax emulsion films. J. Agric. Food Chem. 49, 996–1002

Peyron A (1991). L'enrobage et les produits filmogènes: un nouveau mode d'emballage. Viandes Prod. Carnes 12, 41–46

Possart W, Deckhoff S (1999). Adhesion mechanism in a cyanurate prepolymer on silicon and on aluminium. Int. J. Adhes. Adhes. 19, 425–434

Ressouany M, Vachon C, Lacroix M (1998). Irradiation dose and calcium effect on the mechanical properties of cross-linked caseinate films. J. Agric. Food Chem. 46, 1618–1623

Ressouany M, Vachon C, Lacroix M (2000). Microbial resistance of caseinate films crosslinked by gamma irradiation. J. Dairy Res. 67, 119–124

Rhim JW, Gennadios A, Fu D, Weller CL, Hanna MA (1999). Properties of ultraviolet irradiated protein films. Lebensm. Wiss. Technol. 32, 129–133

Robertson GL (1993). Food packaging. Principles and practice. Marcel Dekker, New York, NY. 686 p

Rutenberg MW, Solarek D (1984). Starch derivatives: production and uses. In: X Chap, RL Whistler, JN Be Miller, EF Paschall (Eds.), Starch: chemistry and technology, 2nd Edition. Academic, Orlando, FL. pp. 311–387

Sabato SF, Ouattara B, Yu H, D'Aprano G, Lacroix M (2001). Mechanical and barrier properties of cross-linked soy and whey protein based films. J. Agric. Food Chem. 49, 1397–1403

Sacharow S (1972). Edible films. Packaging, 43, 6, 9

Salmieri S, Lacroix M (2006). Physicochemical properties of alginate/polycaprolactone-based films containing essential oils. J. Agric. Food Chem. 54, 10205–10214

Sherwin CP, Smith DE, Fulcher RG (1998). Effect of fatty acid type on dispersed phase particle size distributions in emulsion edible films. J. Agric. Food Chem. 46, 534–4538

Shih FF (1994). Interaction of soy isolate with polysaccharide and its effect on film properties. J. Am. Oil Chem. Soc. 71, 1281–1285

Sinha RS, Bousmina M (2005). Biodegradable polymers and their layered silicate nanocomposites: in greening the 21st century materials world. Prog. Mater. Sci. 50, 962–1079

Sorrentino A, Gorrasi G, Vittoria V (2007). Potential perspectives of bio-nanocomposites for food packaging applications. Trends Food Sci. Technol. 18, 84–95

Stanley DW, Stone AP, Hultin HO (1994). Solubility of beef and chicken myofibrillar proteins in low ionic strength media. J. Agric. Food Chem. 42, 863–867

Stefansson G, Hultin HO (1994). On the solubility of cod muscle proteins in water. J. Agric. Food Chem. 42, 2656–2664

Stollman U, Hohansson F, Leufven A (1994). Packaging and food quality. In: CMD Man, AA Jones (Eds.), Shelf life evaluation of foods. Blackie Academic and Professional, New York, NY. pp. 52–71

Stuchell YM, Krochta JM (1994). Enzymatic treatments and thermal effects on edible soy protein films. J. Food Sci. 59, 1332–1337

Tapia MS, Roias-Graü EJ, Rodriguez J, Ramirez J, Carmona A, Martin-Belloso O (2007). Alginate and gellan based edible films for probiotic coatings on fresh cut fruits. J. Food Sci. 72(4), 190–196

Thakur BR, Singh RK, Handa AK (1997). Chemistry and uses of pectin- a review. Crit. Rev. Food Sci. Nutr. 37, 47–73

Vachon C, Yu HL, Yefsah R, Alain R, St-Gelais D, Lacroix M (2000). Mechanical and structural properties of milk protein edible films cross-linked by heating and gamma irradiation. J. Agric. Food Chem. 48, 3202–3209

Wang L, Khor E, Lim LYS (2001). Chitosan-alginate-CaCl$_2$ system for membrane coat application. J. Pharm. Sci. 90, 1134–1142

Wong DWS, Tillin SJ, Hudson JS, Pavlath AE (1994). Gas exchange in cut apples with bilayer coatings. J. Agric. Food Chem. 42, 2278–2285

Wu Y, Weller CL, Hamouz F, Cuppett SL, Schnepf M (2002). Development and applications of multicomponent edible coatings and films: a review. Adv. Food Nutr. Res. 44, 347–394

Yan XL, Khor E, Lim LY (2001). Chitosan-alginate films prepared with chitosans of different molecular weights. J. Biomed. Mater. Res. 58(4), 358–365

Yildirim M, Hettiarachchy NS (1998). Properties of films produced by cross-linking whey proteins and 11S globulin using transglutaminase. J. Food Sci. 63, 248–252

Zheng JP, Li P, Ma YL, Yao KD (2002). Gelatine/montmorillonite hybrid nanocomposite. I. Preparation and properties. J. Appl. Polym. Sci. 86, 1189–1194

Chapter 14
Commercial Manufacture of Edible Films

James M. Rossman

14.1 Introduction

Edible films are defined by two essential elements. First, edible implies that the film, including all of its components, must be safe to eat or that they are "generally recognized as safe" (GRAS), as defined by the U.S. Food and Drug Administration (FDA). Second, edible films must be composed of a film-forming material, typically a water-soluble hydrocolloid or polymer. Further, these films must be manufactured on equipment and in a facility suitable for processing food products.

Edible coatings applied to fruit, vegetables, meat and other food products for improved protection or appearance are often referred to as edible films. For the purposes of this chapter in the book, these applications are not included in this chapter. Edible films can be consumed directly (e.g., a breath freshener film) or can be used to wrap food products, form a pouch or become a component layer in other food products. In most cases, these films are water-soluble and dissolve rapidly in water or in the mouth. However, some components of edible films (e.g., shellac and soy protein) can be insoluble in water, but are digestible when consumed.

Discussions in this chapter relate to the selection of ingredient mixtures or formulations that can be developed into solutions suitable for casting as thin films. The equipment and processes needed to manufacture films on a commercial scale will also be described.

14.2 History

Products defined loosely as edible films have been in use for centuries. Sausage casings, seaweed (nori) for wrapping sushi and rice paper for wrapping sticky candies, are just a few examples. Commercial manufacture of edible, polymer-based

J.M. Rossman
Rossman Consulting, LLC, 1706 W. Morrison Avenue, Tampa, FL 33606, USA
e-mail: rossmanj@prodigy.net

M.E. Embuscado and K.C. Huber (eds.), *Edible Films and Coatings for Food Applications*, 367
DOI 10.1007/978-0-387-92824-1_14, © Springer Science+Business Media, LLC 2009

films using modern casting and drying methods began about 1960. Hydroxypropyl methylcellulose, a food grade polymer manufactured by the Dow Chemical Company, was used to produce an edible film. This film was made into pouches containing pre-measured weights of vitamin and mineral enrichment blends for the baking industry. This pioneering application has remained in continuous use up to present. Many other films have been developed over the years, employing almost every known edible polymer. Cellulose ethers, starches, pectins, alginates, gelatin, pullulan and blends of these materials are commonly found in commercial use. Many other natural and synthetic polymers, as well as protein extracts (e.g., soy and whey), are used commercially in smaller volumes.

A small number of companies in the United States and around the world have developed and built equipment and facilities for the production of edible films. This equipment is installed in plant locations suitable for commercial processing of food, personal care, and over-the-counter drug products. All of these facilities conform to good manufacturing practices (GMP) and established standards.

14.3 Applications

Commercial applications for edible films continue to emerge at such a rapid pace that no comprehensive list can be current and complete. Improvements in film formulation and methods of manufacture contribute to the expanded usefulness of these films. Most applications fall into a few broad categories, as shown in Table 14.1.

14.4 Product Development

It is useful to examine the methodology employed in developing an edible film from conception of the product through formulation, laboratory evaluation and, finally, from pilot plant operations to full production. Careful progress through each of these developmental stages is crucial to efficient manufacture of high quality films. Development of a product begins with a clearly stated idea or concept. The challenge is to then select the ingredients, solution preparation method and drying process that will produce a film with the desired performance characteristics at an acceptable cost. Typically, this development follows a well-defined sequence of activities:

Table 14.1 Edible film applications

Categories	Application examples
Packaging	Vitamins, enzymes, food colors, food additives, beverage mixes, soup
Freestanding films	Breath freshener, toothpaste inclusions, confections, labels, nutraceuticals, over the counter drugs (OTC), contraceptives
Wraps	Vitamins, meat curing, sushi, enrobing, meat glazes, spice blends

Laboratory

- Define product objectives
- Select ingredients
- Simulate manufacturing processes
- Prepare and test film formulations
- Proceed to pilot scale evaluation

14.4.1 Laboratory

For any new product to move from a concept to a commercially-useful film, it must undergo a rigorous examination in the laboratory. A well-equipped laboratory will have a broad selection of film-making ingredients in its inventory. Water-soluble polymers of many types and grades, a full selection of food grade plasticizers, surfactants, fibers, colors, flavors, sweeteners and other standard additives will be available. Solution preparation equipment will provide for processing materials over a wide range of viscosities and shear.

The laboratory will be capable of preparing uniform wet films using suitable equipment, such as Meyer rods or adjustable draw-down blades. The solution is poured onto a substrate, usually a glass plate, and is drawn down over the plate to a uniform thickness. The coated plate is placed in a convection oven to dry the film to a point at or near its equilibrium moisture content (e.g., 6–8% by weight), after which the dry film is stripped from the plate for testing. A well-designed laboratory will have film testing equipment to measure physical properties of the films, equilibrium moisture content, dissolution rates in water and other application-specific properties. Selection of ingredients, preparation of aqueous solutions and order of addition of the ingredient, ingredient compatibility, solution solids, viscosity and solution stability must all be determined with the commercial manufacturing process in mind. Film thickness, both wet and dry, and maximum allowable drying temperatures are process elements that can be established in the laboratory to guide both pilot scale and full scale manufacturing.

14.4.2 Formulation

All ingredients in edible films must be GRAS or be otherwise approved for use in the intended application. This requirement significantly limits both the choice of ingredients and their levels of usage in films. Formulation of edible films begins with the choice of the polymer or mixture of polymers that will form the base film. Most commercial polymers are available in many grades, varying in viscosity and in their chemical functionality. It is necessary to select polymers that will be compatible with the key ingredients, both in the manufacture and storage of the films, and will provide properties essential for the end use. The polymers most commonly used for the commercial production of edible films are listed in Table 14.2. BeMiller and Whistler (1992), Nussinovitch (2003) and Steinbuchel and Rhee (2006) discussed different polysaccharides and polymers, their structure–function relationships and applications.

Plasticizers are added to these polymer solutions to enhance softness, flexibility, elongation, clarity and heat seal characteristics. It is not unusual for multiple plasticizers to be used in a single formulation, since each plasticizer will respond differently to the polymers being used. Solubility parameters will vary with each combination of plasticizer/polymer, and the plasticizing efficiency of each plasticizer decreases sharply when addition levels exceed about 10% based on weight of the polymer. Total plasticizer content of a film will range from 5% to as much as 50%, based on weight of the polymer and the desired physical performance of the film. It would be highly unusual for a single plasticizer to be stable, miscible and functional at an addition level as high as 50%. The very limited selection of commonly used, food grade plasticizers is found in Table 14.3.

Surfactants perform many critical functions in preparation of solutions and in manufacture of films, including:

Table 14.2 Hydrocolloids commonly used in commercial edible films

Polysaccharides/polymers	Proteins
Carrageenan	Casein
Carboxymethyl cellulose	Gelatin
Gum acacia (Arabic)	Soy protein
Hydroxypropyl methylcellulose	Whey protein
Hydroxypropyl cellulose	
Modified starches	
Pectin	
Pullulan	
Sodium alginate	
Xanthan gum	
Polyvinylpyrrolidone	
Polyethylene oxide	

Table 14.3 Food grade plasticizers

Name	Molecular formula	Form	Other names
Glycerol	$C_3H_5(OH)_3$	Liquid	Glycerin, glycerine, propane-1,2,3-triol
Propylene glycol	$C_3H_8O_2$	Liquid	Propane-1,2-diol
Polyethylene glycol	$C_{2n}H_{4n+1}O_{n+1}$	Solid	
Acetylated monoglycerides			
1,6 butanediol			
Triacetin	$C_9H_{14}O_6$	Liquid	1,3-diacetyloxypropan-2-yl acetate
Sorbitol	$C_6H_{14}O_6$	Solid	
Hydrogenated starch hydrolysates		Solid	

- Emulsion stabilization for formulations containing oil- or wax-based ingredients
- Antifoaming during solution preparation (surfactant is typically the first ingredient added to batch water)
- Wetting agent to ensure uniform wetting of the substrate (steel belt, release liner, polyester film); to reduce surface tension of the solution below the surface tension of the substrate
- Leveling agent to minimize surface defects in the wet film (bubbles, pinholes, craters)
- Defoaming to assist in deaeration of the solution before solution casting

A broad range of food grade surfactants is available for use, usually at addition levels of 1% or less, based on weight of the polymer. Surfactants commonly used are sorbitan monooleate, sorbitan monostearate sodium lauryl sulfate, glycerol monooleate and lecithin.

Both soluble and insoluble fibers are found in edible film formulations to provide film strength, dimensional stability, matte finish (appearance), a desired dissolution characteristic or as agents to bond with and restrict the mobility of active ingredients (e.g., an oil-based vitamin). In some cases, fiber is present to provide a nutritional or digestive benefit. Addition levels are determined by the desired function, and will range from 5 to 25% based on polymer solids. Some fiber sources are shown in Table 14.4.

Active or functional ingredients are additives that determine the usefulness and end-use applications for edible films. While some films, such as those used for soluble packaging, have no active ingredients, most edible films are designed to be high performance delivery systems for ingredients with a functional purpose. In their simplest form, these actives may be colors, flavors and sweeteners that enhance appearance and provide a pleasant experience when placed in the mouth. Applications include confections (candy) and breath fresheners. In higher performance uses, choice of active ingredients covers an amazing range. A few examples of actives and their end uses are shown in Table 14.5.

Addition levels for actives vary depending on the end use. For nutritional snacks, the intent is to maximize the active ingredient. Addition levels can be 100% or more based on polymer solids. For ethical drug delivery, the active level must be equal to a precise, specified dosage for each piece of film. Addition levels would typically be more than 5% to perhaps 20%, based on the polymer solids. An example of a film formulation incorporating a vitamin blend is shown in Table 14.6. It will help to summarize and clarify this description of ingredient selection.

Table 14.4 Fibers used in edible films

Soluble fiber	Insoluble fiber
Fibersol II	Purified wood cellulose
Inulin	Microcrystalline cellulose
Psyllium	Oat bran fiber

Table 14.5 Actives for edible films

Actives	End use
Fruit and vegetable puree	Nutritional snacks
Nutraceuticals	Health supplement
Vitamins	Vitamin delivery
Cough and cold medicine	OTC (Over the counter) drug delivery
Hydrogen peroxide	Teeth whitening
Pharmaceuticals	Drug delivery

Table 14.6 An edible film formulation containing vitamin supplements

Ingredient	Function	Weight (g)	% solids
HPMC (low molecular weight)	Polymer	38.95	38.95
Glycerol	Plasticizer	10.00	10.00
Propylene glycol	Plasticizer	6.00	6.00
Polyethylene glycol	Plasticizer	6.00	6.00
Fibersol II	Soluble fiber	5.00	5.00
Avicel	Insoluble fiber	5.00	5.00
PGMS	Surfactant	0.50	0.50
Vitamin blend	Actives	20.00	20.00
Water	Solvent	800.00	–
Menthol	Flavor	2.00	2.00
Flavor oil	Flavor	5.00	5.00
Polysorbate 80	Surfactant	0.50	0.50
Xanthan gum	Stabilizer	0.50	0.50
FD&C Red	Colorant	0.05	0.05
Sucralose	Sweetener	0.50	0.50
Water	Solvent	100.0	–
		1,000	100.0%

The procedure is as follows: Add HPMC (hydroxypropyl methylcellulose) and PGMS (propylene glycol monostearate) to the bulk of the water in the main solution tank. Mix with moderate shear to fully hydrate the polymer. Add all other ingredients in the formula. It is typical, that active ingredients would be added last to the main portion of the batch. The flavor is mixed with very high shear in a separate tank to create a stable emulsion. This emulsion is then added to the main solution tank. The finished formulation will have 10% total solids.

Most formulations will require one or more processing aids to assist in the manufacturing process. Releasing agents, typically stearate-based materials, such as PGMS, are used to ensure easy release of the dry film from the steel belt or other casting material. Preservatives against microbial growth can be used both to protect the solution as it awaits processing into dry film and to protect the finished film during storage. In most formulations, these preservatives are limited to methylparabens, propylparabens and sodium benzoate as antimicrobials. Butylated hydroxytoluene (BHT), butylated hydroxyanisole (BHA) and ascorbic acid are typically used as antioxidants.

14.4.3 Solution Preparation

Once the formulation has been defined, it is necessary to dissolve or disperse ingredients into a uniform, stable solution suitable for processing into a finished film. This effort always begins in the laboratory, where many combinations of materials and many process conditions can be studied quickly on a small scale. The objective of the preliminary laboratory screening is to create a formulation and preparation process of the solution that can be converted to a film with desired performance properties and that can be easily reproduced on a larger, commercial scale. The significant differences between laboratory and full scale production equipment mean that this scale-up process is not an exact science. It is normal for formulations that produce excellent films on a laboratory scale to produce unacceptable films on commercial equipment due to dynamics of solution handling, casting, and continuous (and often aggressive) drying conditions. For this reason, most manufacturers have pilot scale equipment and processes to provide transition from lab scale to production with ability to return to the laboratory for refinement as needed. This progression from laboratory to full production is illustrated in Fig. 14.1.

Establishing a solution protocol for a new formulation follows a routine that can be varied as needed:

- Identify the essential end product properties

    - Film strength
    - Solubility characteristics
    - Level of actives
    - Appearance
    - Flavor
    - Texture in the mouth

- Select a polymer or blend of polymers based on desired attributes

    - Film-forming attributes
    - Compatibility with actives
    - Solubility/dissolution characteristics
    - Film stability in storage

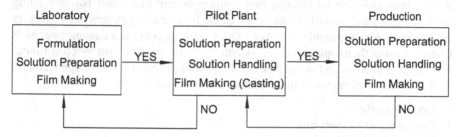

Fig. 14.1 Film development progression

- Determine the order of addition of ingredients

  - Water
  - Surfactant (antifoam)
  - Polymer (needs maximum water and shear to fully dissolve)
  - Fiber
  - Active ingredients
  - Flavor/color (prepare in separate emulsion if needed)
  - Sweetener
  - Processing aids (belt release, preservatives)
  - Water (adjust final viscosity and solids levels as needed)

- Select type of mixing equipment

  - Moderate shear (ensure dissolution of polymers)
  - Uniform dispersion of ingredients
  - High shear to create stable emulsions
  - Vacuum deareation of solution
  - Low shear for continuous mixing

Quality and handling considerations influence the level of solids and the viscosity of solutions. There must be sufficient water present to fully hydrate the polymers for optimal film strength. Viscosity must be high enough to maintain stability of the solution, keep insoluble ingredients in suspension and preserve any emulsions. Viscosity must be also low enough to permit uniform mixing, to pump solutions through filters, to transfer lines to the film casting line, and to spread solution smoothly into a uniform, thin film during casting. For most edible film processes, solution solids will be roughly 20–35% with viscosities of 1,000–10,000 centipoise. The temperature of the solution will also impact viscosity of these fluids, with increasing temperatures normally resulting in decreasing viscosity.

14.4.4   Film Conversion

Converting aqueous solutions into edible films is essentially a coating operation, and is carried out on equipment designed for this purpose. While there are many diverse technologies available for creating thin coatings or thin films, only two converting methods are typically used to manufacture edible films: casting on steel belt conveyors and casting on a disposable substrate (e.g., release paper) on a coating line. Melt extrusion using thermoplastic, edible polymers is also possible, but inherent limitations in the materials and in the process render it impractical for most applications.

The essential elements of these coating processes are similar:

- Coating methods
- Conveying of the wet film
- Ovens or drying chambers
- Secondary treatments

- Release from the drying operation
- Wind-up into master rolls of film
- Subsequent conversion

14.4.4.1   Steel Belt Conveyors

Solutions are cast or spread uniformly on a continuous steel belt that passes through a drying chamber to remove water. The dry film is stripped from the steel belt and is wound into mill rolls for later conversion.

Fully-equipped steel belt conveyor lines are designed, built and supported by many companies around the world. In the United States, Berndorf Belt Systems (Elgin, IL) and Sandvik Process Systems, Inc. (Totowa, NJ) are the primary sources of this equipment (Table 14.7). A typical steel belt line is shown in the schematic in Fig. 14.2 and in the photograph in Fig. 14.3.

The steel belt rotates around large drums situated at either end of the line. Solution is applied uniformly at one end of the line using conventional coating equipment and this is carried through a drying chamber to remove excess water. As the then dried film exits the drying chamber, it may receive a thin, secondary coating; be dusted with a powder (e.g., starch) to prevent sticking or blocking; be

Table 14.7 Equipment manufacturers and suppliers

	Website	Location
Steel belt conveying lines		
Sandvik Process Systems, LLC	http://www.sandvik.com	Totowa, NJ
Berndorf Belt Systems, Inc.	http://www.berndorf-usa.com	Elgin, IL
Web coating lines		
Faustel, Inc.	http://www.faustel.com	Germantown, WI
Radiant Energy Systems, Inc.	http://www.radiantenergy.com	Hawthorne, NJ
Werner Mathis USA, Inc.	http://www.mathisag.com	Concord, NC
Drying chambers		
Radiant Energy Systems, Inc.	http://www.radiantenergy.com	Hawthorne, NJ
Glenro	http://www.glenro.com	Patterson, NJ
C. A. Litzler Co., Inc.	http://www.calitzler.com	Cleveland, OH
Heat and Control, Inc.	http://www.heatandcontrol.com	Hayward, CA
Extrusion coating dies		
Extrusion Dies, Inc.	http://www.extrusiondies.com	Chippewa Falls, WI
TSE Troller AG	http://www.tse-coatings.ch	Murgenthal, Switzerland
Cloeren, Inc.	http://www.cloeren.com	Orange, TX
Other coating equipment		
New Era Converting Machinery	http://www.newera.com	Hawthorne, NJ
Solution preparation equipment		
A&B Process Systems, Corp.	http://www.abprocess.com	Stratford, WI
Charles Ross & Son Company	http://www.mixers.com	Hauppauge, NY
Pfaudler, Inc.	http://www.pfaudler.com	Rochester, NY
Admix, Inc.	http://www.admix.com	Manchester, NH

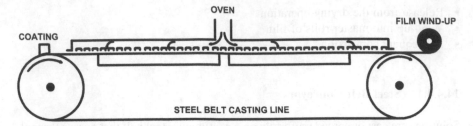

Fig. 14.2 Steel belt casting line

Fig. 14.3 Steel belt casting line

printed with edible inks for decoration or identification; or receive a variety of other treatments before being stripped from the steel belt and wound into large master rolls. For edible films, these conveyor lines are typically 50–100 ft in length measured from the midpoint of the two drums. Steel belt widths range from 20 to 60 in.

One of the highly desirable features of steel belt conveyors is the ability to casting aqueous solutions directly on the belt surface. This optimizes uniformity, heat transfer, and drying efficiency, while eliminating expense of a separate carrier web such as polyester film or coated paper. However, many coating formulations will adhere strongly to the steel belt. This requires that release coatings, as previously discussed, be applied to the steel belt or be incorporated into the coating formulation to ensure uniform release of the film before the wind up station. Some edible films, especially formulations with high loadings of active ingredients, will be too weak or fragile to strip from the steel belt. These formulations can be coated on a carrier substrate (e.g., polyester film) that is lightly adhered to the steel belt. The coated substrate is then stripped from the belt and wound into a master roll.

14.4.4.2   Web Coating Lines

Solutions are cast or spread uniformly onto a carrier web, usually a polyester film or coated release paper. The coated substrate is passed through a drying chamber to remove the water. The dry film, still adhering/stuck to the substrate, is wound into rolls. The edible film is normally separated from the carrier web or substrate in a secondary operation.

Web coating lines are designed to apply aqueous coatings to any flexible substrate – plastic films, paper, textiles, nonwovens – and to transport the coated substrate through a drying chamber. Support rolls are provided to maintain web stability through the dryer. In a typical non-food application, a pressure-sensitive adhesive solution would be coated onto a label paper and dried in the oven. A release paper would be laminated to the dry, sticky adhesive layer, and the finished label stock would be wound into a roll for later processing.

For edible films, a carrier substrate would be selected with the following features:

- Surface tension that permits uniform coating of the aqueous formulation
- Excellent dimensional stability of the carrier to ensure that no process stress is transferred to the film
- Adhesion of both the wet and dry films that will hold the film in place until separation of the film and carrier web is desired
- Minimum of static build-up anywhere in the process

A laboratory or pilot scale coating line, depicted in Fig. 14.4, shows the key elements of the process moving from left to right:

- The carrier substrate is unwound and fed to the coating operation
- The edible film solution is applied to the carrier substrate
- The coated substrate rides on support rollers as it enters and passes through two separate, 10-ft long oven drying zones
- The coated substrate is wound into a finished mill roll; the illustrated lamination feature is rarely used with edible films

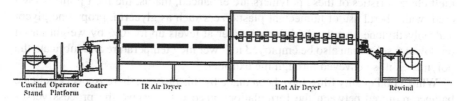

Fig. 14.4 Pilot scale web coating line (Courtesy of Radiant Energy Systems, Hawthorne, NJ)

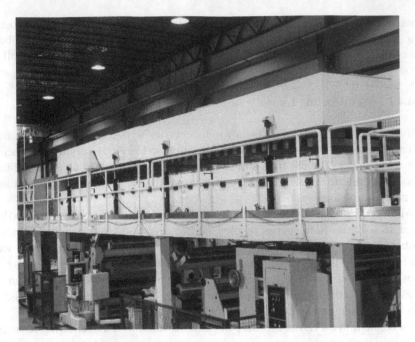

Fig. 14.5 Commercial web coating line (Courtesy of Faustel, Inc., Germantown, WI)

The dryer on this web coating line can incorporate many different, well-known drying technologies utilizing diverse energy sources. Solution coatings can be applied by a variety of coating technologies, typically, knife over roll and slot die extrusion methods provide the best results for thin films made on this equipment. Drying and coating method technologies will be discussed in more detail later in this chapter.

A full-scale commercial coating line is shown in Fig. 14.5.

14.4.4.3 Melt Extrusion

Commercial extrusion of edible polymers has been demonstrated with hydroxypropyl cellulose and with several highly modified corn starches. The thermoplastic melt characteristics of these polymers are enhanced, that is, the melt point is lowered, with liberal use of humectant plasticizers such as glycerin, propylene glycol and polyethylene glycols. Addition of water at levels up to 20% by weight based on polymer solids can also be employed to lower melt temperature and lubricate the polymer as it is conveyed through the extruder.

While good quality films have been made by this process, there is a very delicate balance required between the formulations needed to survive the process versus achieving the desired end product performance in the film. An excess of plasticizer will migrate or "bloom" to the surface of the film over a period of time. An excess of water in the formulation will either cause foaming in the film or boiling off during the process, creating steam in the process area. The high operating temperature, typically 300–400°F, is very near the degradation point of these polymers and limits ingredients to those that will remain stable under these process conditions.

14.4.5 Coating Methods

Laboratory coating methods are used to simulate as closely as possible methods used for commercial film manufacturing. For practical reasons, either wire-wound rods or adjustable doctor knife coaters are used in the laboratory. Wet films are formed on glass plates at desired thickness and are dried in air-circulating convection ovens.

Rod coating. A wire-wound rod is used to doctor or uniformly spread a pool of coating solution onto a glass plate. The rod is drawn smoothly over the plate with the rod rotating in the direction of the draw. The wire-wound rod creates many narrow ridges running down the casting surface that correspond to the wires on the rod. Smooth, uniform coatings are achieved as these ridges flow and level out in the dryer/oven. Wire rods are available from most of the sources supplied by the laboratory.

Adjustable knife. For laboratory use, a rigid metal frame is used to support a coating knife. At both ends of the knife, micrometer adjustments are provided to set the knife at a precise distance from the surface to be coated. A pool of coating solution is poured behind the knife and the knife assembly is drawn smoothly over the surface to form a uniform coating. The coated plate is placed in the oven to dry. Adjustable knife assemblies are readily available from laboratory supply sources.

Commercial coating methods are very diverse. The method selected is highly dependent on the desired end product, solution rheology, web speed, and drying technology in use. Examples of coating methods include spraying, dip coating, gravure coating, curtain coating and reverse roll coating. For production of edible films, other technologies are more commonly used, including knife over roll and slot die extrusion.

Knife coating. A stationary, rigid knife doctors a precise layer of solution onto a web moving under the knife. Micrometer adjustments are provided to adjust height of the knife above the web and to increase/decrease the wet film thickness. Since a roll or drum is normally positioned beneath the knife, this process is known as knife-over-roll coating. Excess solution is held in a pool behind the knife. Knife coating fills in any depression/variation in the surface being coated, producing a smooth, uniform film. These coaters are uncomplicated devices that are simple to operate and need little maintenance. They are also easy to clean and to set up for small production runs.

14.4.5.1 Slot Extrusion Coating

Highly specialized extrusion dies have been designed for applying aqueous solutions to a moving substrate. The extrusion die is split into top and bottom sections that are bolted together. Enclosed between these sections is a flow channel

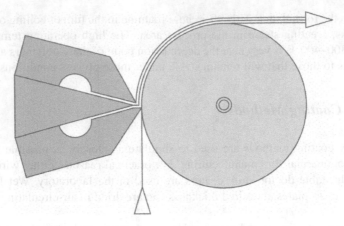

Fig. 14.6 Coating from a low pressure, dual-layer slot die (Courtesy of TSE Troller AG, Murgenthal, Switzerland)

machined into one of the sections. This flow channel delivers coating solution to the die lips (or exit slot) in a uniform, streamlined flow. The die lips are positioned very close to the moving web that pulls film from the exit slot. A vacuum source is often built into the die to stabilize the wet film and minimize coating defects/variations (e.g., trapped air). Film thickness is achieved by adjusting the gap between the die lips or by increasing the speed of the substrate that causes the wet film to be drawn to a thinner layer. Due to the more elaborate set-up and cleaning of extrusion dies, this process lends itself to long production runs of a single product or to products demanding very tight film thickness tolerances. Extrusion dies are designed to operate across a wide range of temperatures and pressures. Dies can be designed to simultaneously lay down multiple layers of wet film. The low pressure, low temperature slot die shown in Fig. 14.6 illustrates two layers of wet film being deposited on a support substrate; it is more typical for this die to be built for single layer coatings.

Figure 14.7 illustrates a high-pressure extrusion die positioned to extrude a single layer of wet solution onto a substrate that will pass over the integral roller in this casting station.

14.4.5.2 Basic Coating Considerations

The first step in setting up the coating operation requires a definition of the coating solution, the coating equipment available and the desired properties of the film. Key variables to consider are

- Coating formulation
- Number of layers

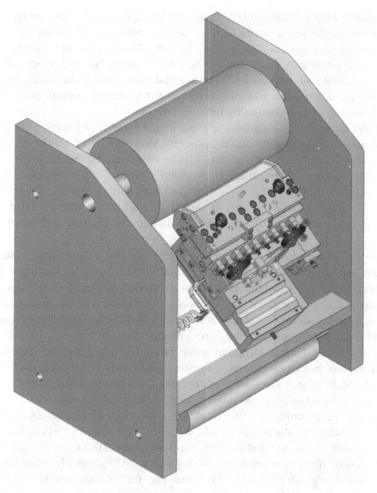

Fig. 14.7 Extrusion coating station (Courtesy of EDI, Inc., Cincinnati, OH)

- Wet film thickness
- Viscosity
- Solution solids
- Solution temperature
- Coating accuracy required
- Coating substrate

Coating formulation. Understanding characteristics of the coating solution is critical to designing an efficient coating process. If the formulation is highly shear-sensitive, temperature-sensitive, surface-active or contains volatile ingredients, adjustments must be made to ensure that physical handling of the solution does not create degradation or defects in the finished film.

Number of layers. Most edible films are produced in a single, homogenous layer. To have several layers, it would be necessary to make several passes through the coating and drying process. This exposes the early layers to multiple heat histories, which could potentially cause damage to flavors, colors and other heat-sensitive ingredients. However, both the knife coating and the extrusion die coating methods can be modified to apply two or more layers of wet solution in a single pass. Unique edible films can be produced using these techniques.

Wet film thickness. Coating thickness is a major variable in selecting the coating method and in designing the coating process. Very thin films will be difficult to coat uniformly and will tend to easily over dry, creating weak, fragile films. Depending on viscosity and solution, the solid's, ideal wet film thickness will be in the range of 0.010–0.030 in. (254–762 μm), and will yield a dry film thickness of 0.0015–0.0030 in. (38–76 μm). It should be understood that a final dry film thickness below 0.001 in. (25 μm) will be difficult to produce and handle after drying.

Viscosity. The viscosity of all polymer solutions depends on shear rate, temperature, and solution solids. While solution viscosity can be identified in the laboratory under static conditions, it is far more difficult to monitor and control viscosity under dynamic shear conditions observed in most production systems. It is important to be aware of the potential for process variation when solution temperature and shear conditions fluctuate during production.

Solution solids. Once a formulation has been developed and the viscosity for processing has been determined, solution solids will be defined. Solids will remain fixed as long as viscosity is acceptable for solution coating. Any addition of water to reduce viscosity will cause a decrease in solution solids. Solids content is a factor in determining wet film thickness, since wet film thickness multiplied by the percentage of the solids equals dry film thickness. For example, a 10% solids solution if cast at 0.020 in. (508 μm) would equal 0.002 in. (51 μm) in dry film thickness. If solution solids are too high (e.g., 50%), wet film thickness may be too low for good process control and the formulation would need to be adjusted to reduce solids.

Solution temperature. Maintaining a constant temperature during processing is critical to accurate film production. As temperature varies, solution viscosity, fluid flow characteristics, drying profile will also change, and will have a profound impact on the process and quality of the finished film product.

Coating accuracy. The surface to be coated is critical to quality of the film produced. Whether the surface is a steel belt, polyester film, coated paper, or other substrate, coating solution must wet the substrate and level quickly to create a smooth, uniform wet film. If surface activity of the coating solution is lower than surface tension of the substrate, the solution will flow out smoothly. If the opposite is true, solution will tend to bead up on the substrate and yield films of very poor quality. Surface active agents (surfactants) can be added to solutions to modify wetting and leveling properties of the solution. The substrate can also be coated to improve its wetting properties, or the surface can be treated with a flame, corona discharge or other processes to increase surface tension and improve coatability.

14.4.5.3 Drying

All edible films begin as aqueous solutions and are applied to substrates in a fluid form, but the films are used in the dry state. Drying is the critical processing step that makes this transformation possible. Drying can either enhance or damage properties of the coating. Coating and drying processes interact dynamically; consequently, limitations in drying restrict optimization of the coating process and vice versa.

Drying of coatings removes solvent used to suspend, dissolve, or disperse ingredients that make up the film, so that only the desired film solids remain on the substrate. For most purposes, water is the solvent or vehicle of choice. Very small amounts of volatiles contained in flavors and plasticizers are also typically lost in the drying process. Drying requires application of heat to vaporize water in a controlled process. Care must be taken to remove the water without damaging the formulation or interfering with development of the desired physical properties of the film.

A discussion of film defects that occur as a result of poor quality of the solution, inappropriate casting methods and dryer limitations will be provided later. Prior to describing the types of drying equipment normally used in the production of edible films, it is important to note that most coating processes operate with preferred conditions of wet film thickness, line speed, solution viscosity and so on. However, it is essential that dryer capacity and efficiency can handle the drying load and wet film residence time for these optimal conditions. For example, if the drying load (i.e., the quantity of water that must be removed) exceeds capacity of the oven, the film will be wound up wet, unless line speed and/or wet film thickness are reduced.

Drying is one of the oldest chemical engineering operations. Calculations for efficient evaporation of water at varying temperatures and air movements can be found in basic chemical engineering texts and are readily available from oven/dryer equipment manufacturers. Gutoff and Cohen (1992) wrote an excellent review of this subject.

Drying fundamentals. There are many dryer designs and many energy sources used in equipment provided for removal of water from thin films. All of these involve input of energy to raise the temperature of the wet film and evaporate water. How this energy is applied and the rate at which the wet film is heated has everything to do with both drying rate/efficiency and dry film quality. Primary dryer designs used in the manufacture of edible films include:

- Hot air convection
- Hot air impingement
- Steam
- Infrared
- Hot air flotation
- Zoned drying

In all of these processes, the wet solution is cast onto the substrate (steel belt or carrier web) at a predetermined temperature – usually ambient plant temperatures or slightly higher. As the wet film enters the drying chamber, it begins the constant drying rate phase, where the rate of water evaporation remains relatively constant and is limited primarily by external mass transfer resistance. All of the heat input is available to evaporate the water, which can move freely to the surface boundary layer where it can be carried away. The constant rate phase is characterized by:

- Transition
  - Wet film has low temperature
  - Wet film has high water content
  - Rate of evaporation is low
  - Rate of heating is high
  - Wet film temperature increases

- Constant Rate
  - Film equilibrates at a level that balances the rate of heating with evaporative cooling
  - Vapor pressure, drying rate, and rate of evaporative cooling all rise to the equilibrium point
  - Drying rate is largely independent of water content

- Falling Rate
  - Internal mass transfer resistance increases, resulting in an increase in film temperature and a decrease in the drying rate

As illustrated/shown in the ideal illustration in Fig. 14.8, moisture content of a typical solution decreases from 70% water content to a final equilibrium point of 8% in the finished dry film. Drying takes place primarily in the constant rate phase,

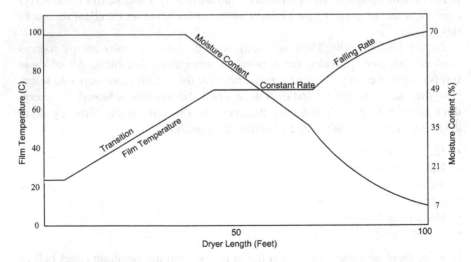

Fig. 14.8 Drying rate graph

and care must be taken to adjust line speed, dryer/ temperatures and film thickness to maintain the targeted equilibrium moisture in the dry film. Over-drying the film usually results in irreparable damage.

In reality, most drying processes are more complex than the simple process depicted in Fig. 14.8. Predryers (e.g., infrared) are used to rapidly raise the temperature of the wet film to enter more quickly into the constant rate phase. Most dryers use multiple drying zones to apply heat at different temperatures and rates. Such arrangements allow the drying profile to be adapted to film ingredient characteristics and temperature stability. Careful zone drying is also critical to the production of high quality films that are free from process defects (Fig. 14.9).

Hot air convection. With simple convection dryers, hot dry air is blown into the drying chamber, and flows counter to the movement of the substrate. Hot air can be introduced above the web, below the web, or on both sides simultaneously. As hot, dry air flows over the wet film, it "scrubs away" moisture-laden air at the boundary layer, increases in moisture content, and decreases in its drying efficiency as air approaches the web entry point and is removed by an exhaust fan. It is possible to design this dryer with two or more zones for improved control and efficiency (Fig. 14.10).

Hot air impingement. Impingement dryers are designed with a series of slots or nozzles that extend the width of the dryer and are positioned every few inches along the length of the dryer. The slots are fed from a plenum that controls the temperature, volume, and velocity of the hot air that enters the oven. Impingement ovens usually have multiple zones with independent controls that permit excellent adjustment

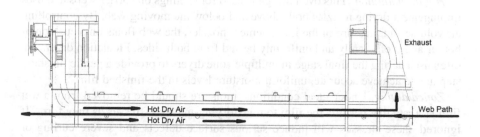

Fig. 14.9 Hot air convection oven

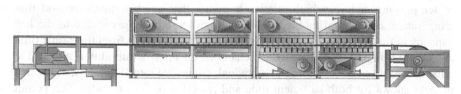

Fig. 14.10 Hot air impingement line design (Courtesy of Sandvik Process Systems, Totowa, NJ)

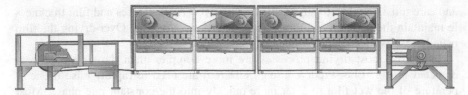

Fig. 14.11 Hot air and steam impingement line design (Courtesy of Sandvik Process Systems, Totowa, NJ)

of the drying profile. Impingement nozzles can be used both above and below the coated substrate.

Impingement nozzles direct the dryer air at the wet film surface causing turbulence that efficiently scrubs away the layer of high moisture content air at the wet boundary layer and enhances dryer efficiency. Early in the drying process while the wet film is still highly mobile, high air velocities can cause movement of the wet film, creating serious defects in the dry film. Care must be taken to balance drying efficiency with optimal film quality (Fig. 14.11).

Steam drying. Some drying processes that use steel belt conveyors combine the use of hot air impingement applied to the wet film with live steam applied to the bottom surface of the belt. The excellent heat transfer efficiency of steam to the steel develops a highly desirable "bottom up" drying of the wet film, while minimizing tendency of the wet film to develop a dry skin. The hot air impingement on the wet film surface promotes rapid removal of moisture-laden air at the wet boundary surface. Both steam and impingement drying can be provided in multiple zones.

Hot air floatation. This oven design is used for coatings on carrier webs and uses impingement drying nozzles both above and below the moving web. By controlling air volume and pressure of the impingement nozzles, the web floats on a cushion of hot air and is efficiently and uniformly heated from both sides. Floatation dryers are often used during the final stage in multiple zone dryers to provide a gentle finishing step and to achieve accurate, uniform moisture levels in the finished film.

Zoned drying. Under ideal conditions, moisture should be removed from a wet film at a rate that permits the film to relax drying-induced stresses as they form. If ignored, these stresses will induce serious surface defects and severe curling or other distortions of the dry film once it is released from the substrate. The causes of these defects will be discussed in the next section. One effective method for minimizing process defects is use of zoned drying. In this process, the oven is designed with multiple drying chambers. The first chamber provides an atmosphere of temperature and very high relative humidity that quickly raises the wet film temperature, but minimizes the rate of water evaporation. This is due to the low capacity of the humid air to carry additional moisture coming from the edible film. As the wet film passes into the second chamber, the relative humidity is adjusted to a lower level to continue drying at a gradual, controlled rate. This slow, controlled process allows for both sufficient time and retention of film mobility that permit relaxation of stresses in the film as they develop. This process is repeated in as many chambers as necessary to produce a stress-free dry film. Process lines of this design can be 200 ft or more in length.

14.4.5.4   Dryer Controls

Equipment manufacturers have developed sophisticated controls to monitor temperature, humidity and velocity of air streams in their dryers. By using noncontact, infrared thermometers to measure film temperatures at various points in the drying chamber, it is possible to vary inlet air conditions, line speed, and wet film thickness to adjust drying rate and produce quality films. However, many types of defects can be created in the wet film by motion of the air flow in the dryer, by aggressive drying conditions and by formulation errors that do not show up in laboratory and pilot plant conditions. Gutoff and Cohen (2007) discussed these coating and drying defects.

Mottle. Gross, nonuniform disturbances in the wet film are defined as mottle. This defect is caused by the motion of air in the very early stages of the dryer when the wet solution is still highly mobile and can be easily pushed around by these forces. Mottle can be eliminated by reducing air velocity in the early drying zones or by increasing viscosity of the coating solution.

Fat edges. Typically, the wet film is thinner at its edges due to wetting and leveling actions. In the dryer, evaporation takes place at the same rate across the entire wet film. Solids will concentrate faster at the edges and, under some conditions, surface tension gradients will draw solution toward these edges, resulting in fat edges. This defect can be reduced or eliminated by coating thinner layers or by using higher solution viscosity to increase resistance to flow. Addition of surfactants will also resolve this defect by minimizing surface tension gradients.

Curling. As the uniform wet film dries, it shrinks in volume. Since the film is adhered to the substrate and cannot shrink across the web, all shrinkage takes place in the film's thickness. As the wet film dries and solidifies, stresses from shrinkage can no longer be relieved by flow and will be displayed across the web. In thin, flexible films, the film will curl inward to relieve the stress. Stress buildup can be reduced with addition of plasticizers to soften films and make them more flexible and easier to stretch. Rate of drying also affects stress buildup with slower rates allowing more stress relief before film structure is permanently set.

Holes and craters. Holes in the film can usually be traced to dirt, oil or airborne particulates that have been deposited on the coating substrate, or have been blown onto the wet film by the air flow of the dryer. Ensuring that the substrate is clean and isolated from sources of contamination is critical. Using only clean, filtered air for drying will also help eliminate these defects. Craters are almost always surface-tension related defects, caused by airborne contamination (silicones and other volatile oils are particularly troublesome) or by phase separation in the coating solution. Protecting the wet film from contamination before it enters the oven and in the early drying stages is essential. If the problem persists, the coating formulation must be modified to eliminate breakdown in the quality of the solution.

Blisters. In the early stages of the drying process, bubbles will form when coating is heated to a temperature above the boiling point of the solvent (e.g., with water/plasticizer blends). When bubbles burst, they are displayed as blisters. Reducing air temperatures in the dryer zones where blistering occurs will eliminate this defect. Air bubbles present in the coating solution can also lead to blisters. During drying, small air bubbles can expand to form a blister if they are unable to migrate easily to the coating surface. Careful deaeration of the solutions before casting is recommended.

Adhesion failure. In commercial manufacture of edible films, it is important that films adhere uniformly to the substrate until drying is complete. Premature or partial release of the film will result in a distorted, curled web as stresses are relieved in a nonuniform manner. Surfactants and release agents added to the solution or coated on the substrate are often the cause of this failure. Adjusting levels of these additives will alleviate the problem. Too rapid drying of the film can also cause adhesion failure. Efficient manufacturing requires that the film will have reached its equilibrium or target moisture level at the point of windup into a mill roll.

14.5   Future Trends

Most commercially available edible films are single layer, homogeneous compositions. To make them more user-friendly and convenient, they are cut to size and packaged for specific applications. In a few end uses, the films are printed with brand logos or decorative patterns, using edible inks and either ink jet printers or gravure printing equipment. These product forms are based on well-known processes and equipment, and represent the most basic applications for edible films. Work is already well-advanced onto the next level of development for edible films, as evidenced by intense patent activity and investments by film manufacturers. Future developments for edible films will introduce added value by combining performance characteristics of different films into new multifunctional structures. Key technologies will include:

- Barrier functions
- Foam structures
- Multiple layers
- Controlled solubility (release rates)
- Precision dosage of actives

14.5.1   Barrier Functions

Edible polymers provide many barrier functions in food product applications by enhancing resistance to moisture, fats and oils, oxygen and other gases. These barrier functions can be reproduced in edible films and used in unique packaging, food wraps and barrier layers to separate wet and dry components in prepared foods. A cellulose ether polymer-based film has been developed as a replacement for gelatin in enrobing tablets and capsules.

14.5.2 Foam Structures

Highly aerated or foamed films can be produced on existing film casting equipment. Film pieces are thicker, heavier and easier to handle without sacrificing solubility in the mouth. Foamed films are opaque. They maintain their capacity for binding and delivering active ingredients.

14.5.3 Multiple Layers

In the simplest form, multiple layers of the same formulation can be made with a drying step between each layer. The resulting film will be free of most dryer-induced defects and will be more stable. A more valuable form would use different formulations for each layer, with each providing a different function:

- Outer layer for protection (humidity, oxidation, UV) or to prevent loss of actives from inner layers.
- A second layer with rapid solubility and release of intense flavor
- A center layer containing a high loading of actives
- Additional layers as needed

A promising variation on multiple layers is to form tiny, bite sized "packages" by capturing an active ingredient mixture (powder, paste, gel) between two layers of colored, flavored film and die cutting the "lamination" into small pouches. This could also be done in a simple thermoforming process.

14.5.4 Controlled Release

Most edible films are designed to have the fastest possible solubility rate to ensure that flavor and actives are released before the user senses a negative gumminess in the mouth. The physical strength of the film is sacrificed and thickness is minimized to achieve rapid solubility. An alternative form of product can be created by:

- Building a thicker film from multiple layers. Each layer would contain one or more active ingredients and dissolve at a unique rate. The purpose is to create a long-lasting, "thick film" or "thin lozenge" that provides a pleasant experience, and delivers impact from several unique formulations.
- Selecting edible polymer mixtures that combine with active ingredients to dissolve slowly and uniformly over time. A desirable feature of these products

is to form a surface gel when the film becomes wet in the mouth. As this surface gel hydrates and dissolves away, new surfaces are exposed to renew the gel, thus, delivering controlled release.

14.5.5   Precision Dosage

For ethical pharmaceuticals and for many over the counter drug products, dosage per unit of use is critical and must be held to tolerances that are difficult to achieve with film manufacturing processes. A number of methods are being developed that precisely meter drug dosages onto a standard film:

- Ink jet printing utilizes existing edible inks, blended with pharmaceuticals, to "print" precise drug dosages onto film pieces.
- Precise quantities of drug powders, pastes, or gels are metered onto the film, and are then covered with a second layer of film to form a laminate or sandwich dosage form.

14.6   Support Sources

References

BeMiller J, Whistler RL (1992) Industrial gums. Academic, New York
Gutoff EB, Cohen ED (1992) Modern coating and drying technology. Wiley, New York
Gutoff EB, Cohen ED (2007) Coating and drying defects: troubleshooting operating problems. Wiley, New York
Nussinovitch A (2003) Water-soluble polymer applications in foods. Blackwell, Oxford
Steinbuchel A, Rhee SK (2006) Polysaccharides and polyamides in the food industry. Wiley, New York

Index